THE LAND BENEATH THE ICE

THE LAND BENEATH THE ICE

The Pioneering Years of Radar Exploration in Antarctica

DAVID J. DREWRY

PRINCETON UNIVERSITY PRESS
PRINCETON AND OXFORD

Published by Princeton University Press
41 William Street, Princeton, New Jersey 08540
99 Banbury Road, Oxford OX2 6JX

press.princeton.edu

ISBN 978-0-691-23791-6
ISBN (e-book) 978-0-691-23792-3

British Library Cataloging-in-Publication Data is available

Editorial: Ingrid Gnerlich and Whitney Rauenhorst
Production Editorial: Karen Carter
Text Design: Carmina Alvarez
Jacket/Cover Design: Daniel Benneworth-Gray
Production: Jacqueline Poirier
Publicity: Kate Farquhar-Thomson and Sara Henning-Stout
Copyeditor: Barbara Liguori

Jacket/Cover Credit: Lower Beardmore Glacier by David J. Drewry

This book has been composed in Minion Pro with Helvetica Neue

Printed on acid-free paper. ∞

Printed in the United States of America

1 3 5 7 9 10 8 6 4 2

This book is dedicated to the memory of two of
the principal pioneers of Radio-Echo Sounding

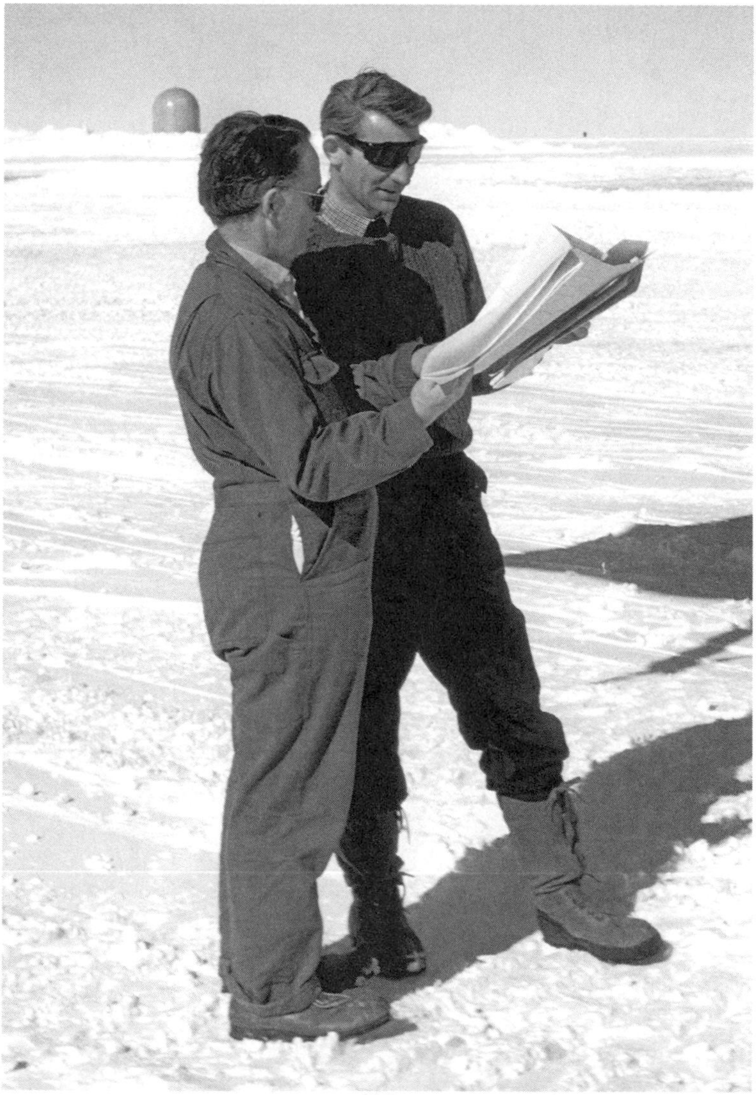

Gordon de Quetteville Robin (left) and Stanley Evans, McMurdo Station, 1969.

Contents

Illustrations

Color plates of the following figures follow page 158: 0.1, 2.1, 3.3, 4.12, 5.7, 5.12, 5.17, 5.20, 5.23, 7.1, 7.4, 7.8, 7.9, 8.4, 8.8, 8.9, 8.12, 9.5, 9.11, 11.3, 11.7, 11.8, 11.14, 12.2, 12.7, 12.16, 12.20, 14.3, 14.5, 14.7, 14.8, 14.9, 14.13, 16.4, 17.2, 17.3, 17.4, and 17.5.

Preface

This book has emerged from several years of rumination and reflection on how best to capture the excitement and pioneering contribution to our knowledge of the Antarctic continent of the radio-echo sounding (RES) programme conducted by the Scott Polar Research Institute at Cambridge University (henceforth referred to as SPRI). Various invitations over the last decade to present lectures on RES have helped focus my ideas on a more substantial work. I commenced some writing in 2015 but put the material to one side as other activities took precedence. I could observe the expanding scope of the book and the demands on my time as I recorded the development of the science and the burgeoning research emanating from our Antarctic campaigns. With the worldwide disaster of the COVID-19 pandemic, which began in 2020, and the enforced lockdown of many citizens, I spied the possibility of reviving the manuscript and completing the task. The audio interviews of some of the SPRI personnel involved in the RES programme by Dr Bryan Lintott in 2019 proved especially helpful. Access to other archival materials has proved difficult, as have discussions with former colleagues. Sadly, several have died; my indolence has deprived me of their wisdom and assistance. Nevertheless, I hope this volume will pay tribute in some manner to their contributions during a groundbreaking time in Antarctic geophysical exploration.

The style of this volume presented a problem. I have aimed to provide as thorough an account as possible of the scientific background and challenges that this programme generated; the financial, human, and political trials and encounters that were experienced; as well as a fulsome review of the results and achievements of the many individuals who participated. At the same time, I considered it paramount that the text should be accessible and engaging to more general readers who would be introduced to the nature of more contemporary exploration and to the fascinating, indeed intoxicating, environment of Antarctica. Whilst I have attempted to cover these many aspects as dispassionately as possible, a significant quantum of the writing reflects my own experiences and engagement. I make no excuse for this; a

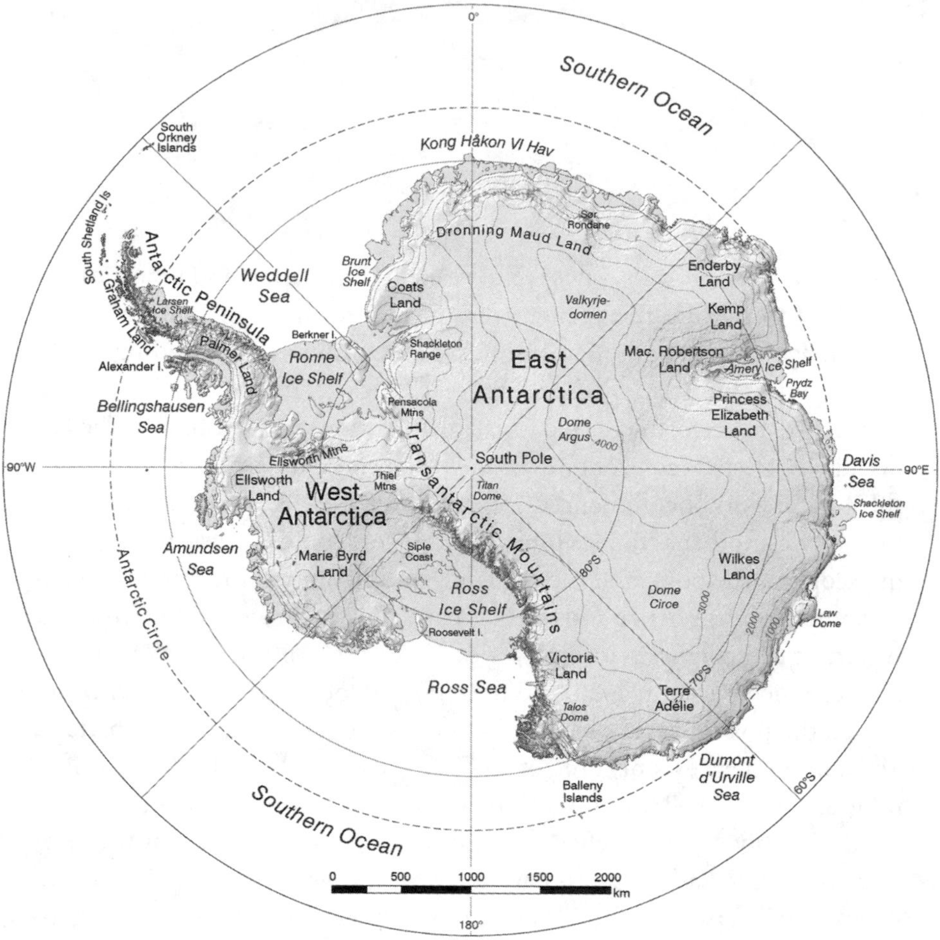

Figure P.1. Antarctica showing some of the principal features and place names referred to in the text. (Map produced by Mapping and Geographic Information Centre, British Antarctic Survey).

dry and impersonal rendering of these years of activity might suffice for an audit of polar science, but the intense experience of Antarctic research is also a very personal story. Consequently, the text is a compromise, and I hope there will be some measure of interest for specialists, historians of science, Antarctic enthusiasts, and general readers alike.

The book structure is largely chronological, charting the steady development of the RES technique in Cambridge and the many experiments in the field in the Arctic and the Antarctic. There is a focus on the latter, where

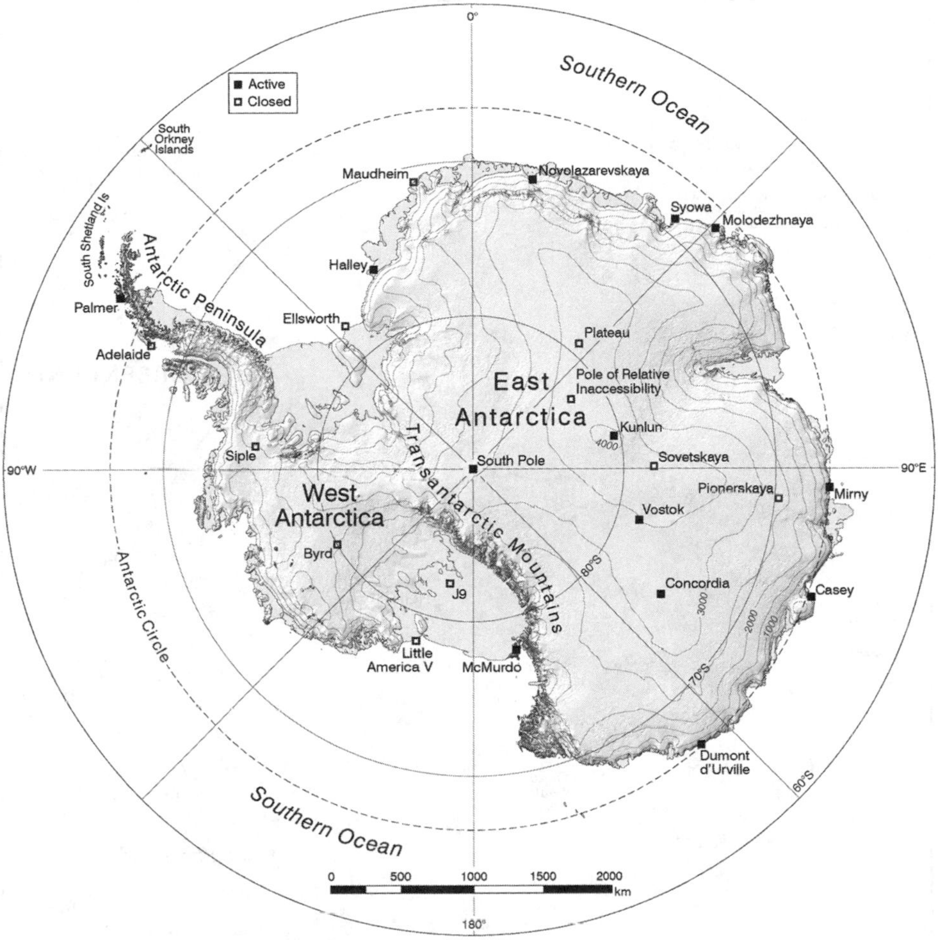

Figure P.2. Antarctica showing the research stations referred to in the text. (Map produced by Mapping and Geographic Information Centre, British Antarctic Survey).

a long-term programme to map the ice thickness of this vast continent—almost twice the size of the conterminous United States—engaged us from 1967 to 1983. The story follows the twists and turns of operating in this remote and hostile region, where planning and preparations must be detailed and exact only to be disrupted by bad weather, mechanical and electronic breakdowns, aircraft crashes, and human frailty. The narrative also reports on the high levels of international cooperation in Antarctica at a time of tension between East and West—the Cold War—demonstrating the veracity of the 1961 Antarctic Treaty, as well as the recognition that only by working

Figure P.3. Arctic islands of Canada and northwest Greenland.

together can problems of global significance be effectively tackled. Occasionally, the project revealed the dark underbelly of science, in which changing attitudes, moody politics, and institutional rivalries can distort otherwise exemplary relationships and convergent goals.

The Antarctic campaigns took place as interest and concern regarding global climate change was emerging. To those of us involved it was evident the environment of the planet was indeed changing and that these great ice

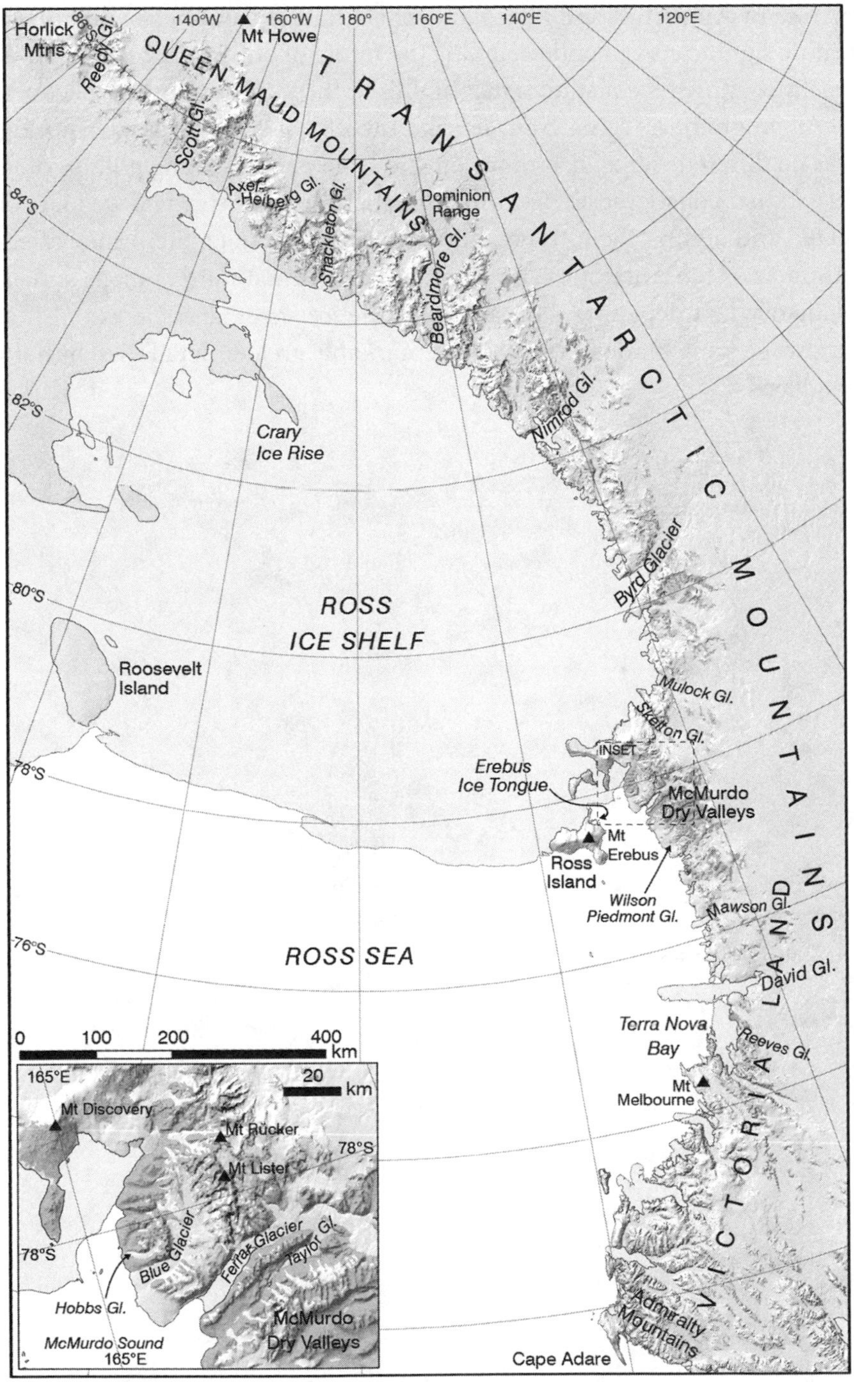

Figure P.4. Transantarctic Mountains and environs. (Map produced by Mapping and Geographic Information Centre, British Antarctic Survey).

masses in Antarctica and Greenland, their stability and behaviour, would feature significantly in understanding the future of our world. I have referred briefly to more recent studies that highlight these important perspectives.

All programmes come to an end, and the RES of Antarctica at Cambridge was no different, albeit driven by logistic reality and pivoting politics of international collaboration. Drawing together much of the data we had collected and adding them to previous work led to one of the principal end products of the Antarctic campaigns: a large-format folio of maps and explanatory text depicting the features of the ice sheet, and the astonishing landscape lying hidden beneath, in remarkably greater detail than hitherto produced.

Acknowledgements

Many people contributed to the scientific work that I describe: research students and staff at the University in Cambridge, administrators in various national Antarctic and Arctic programmes, air support personnel including pilots, ground crew, meteorologists and field party operatives, colleagues in many academic and government institutions, cartographers, photographers, electronics specialists, communications engineers, database managers, and computer specialists. Those involved in these many aspects have been acknowledged in detail in the Antarctic Glaciological and Geophysical Folio published by the SPRI in 1983.[1] Nevertheless, I deem it necessary to thank the internationally spirited support that the Cambridge team received from the US National Science Foundation (NSF) and its Antarctic Program. The provision of immensely costly aircraft logistics, airframe modifications, data recording facilities, accommodation in Antarctica, and myriad small considerations that enabled this extensive research programme to unfold successfully was of a magnitude rarely encountered. During the decade of working with the NSF the SPRI participants enjoyed the company of NSF staff and formed long-lasting relationships. In like fashion the professionalism of the US Navy VXE-6 air operation was of the highest order; many of the aircrew made contributions of incalculable value to the success of the RES programme. Several remained good friends for many years thereafter, despite the seeming cultural gulf between the military and academia.

Colleagues from the Technical University of Denmark at Lyngby proved outstanding partners with their first-class instrumental developments and assistance during several field operations. Their skills, hospitality, and good humour are acknowledged with gratitude.

The US Geological Survey supplied copious quantities of maps and charts of Antarctica on which much of our early navigation depended. In the UK the Natural Environment Research Council (NERC) were generous in their

[1] Drewry, David J (ed.) (1983) *Antarctica: Glaciological and Geophysical Folio*, Cambridge, SPRI.

financial support over many years for the RES programme. Their flexibility, patience, and understanding of the exigencies of operating in Antarctica with delays and changes of plan were greatly appreciated. In Svalbard the cooperation of the Norsk Polarinstitutt (NP) is readily acknowledged, as was the aircraft support by the British Antarctic Survey for the later airborne campaigns.

The SPRI as an establishment provided the stimulating intellectual environment in which this research activity flourished. It was the base for planning, preparation of instrumentation, reduction of returned data, and the writing of the research outcomes. It also embraced its staff and students with the personal warmth that arguably only a 'polar' establishment can offer. The Institute, and this RES programme in particular, proved a magnificent training ground for a band of active, inquisitive, and adventurous young scientists, many of whom have gone on to luminous and productive careers in glaciology, remote sensing, and industry. I pay tribute to the founding fathers of the radio-echo sounding programme at the SPRI, Drs Gordon Robin and Stan Evans. Both are sadly deceased, but their spirit, inspiration, and achievements live on. I trust I have revealed fully their decisive contributions in this book.

Others whom I would like to thank are David Petrie, Professor David T Meldrum, Drs Chris Neal, Julian Paren, Gordon Oswald, Beverley Ewen-Smith, and David Millar, all of whom agreed to read sections of the book and add their own reflections. Mr Michael Gorman worked extensively on the RES equipment and participated in field seasons in both the Arctic and Antarctic. His audio and video recordings that explain aspects of the SPRI RES equipment have been most helpful, as have his comments on several chapters. He also generously supplied photos of the Devon Island operations. Mr Paul Cooper and Mrs Susan Jordan assisted ably in reading texts and adding their informed views regarding databases and the SPRI folio. The folio required skilled, professional cartographic work. Mr David Fryer and Mrs Mary Spence MBE were very helpful in recalling details of this work and for photos. Andy Smith at the British Antarctic Survey kindly read and commented on a draft chapter. I thank Professor Julian Dowdeswell, the previous SPRI director, for his comradeship in the field, scientific assistance, and endurance in reading and commenting on an early draft of this book.

Illustrations: Several individuals have assisted with illustrations, and I thank them for their generous permission to share their material in the text.

I am very grateful to the Thomas H Manning Polar Archives at the SPRI. Naomi Boneham, Laura Ibbett, and Lucy Martin answered many questions and supplied photographs. These are acknowledged as 'Scott Polar Research Institute, University of Cambridge, with permission', which I have abbreviated to "courtesy SPRI". Reproduction of short sections of radio-echo sounding records and related flight-line maps and diagrams prepared by me, some of which have been used in other publications, have been given the citation 'courtesy SPRI'. I thank Dr Neil Arnold, acting director at the SPRI, for permission to access these materials. All other sources are acknowledged with gratitude. Adrian Fox and Laura Gerrish of The British Antarctic Survey Mapping and Geographic Information Centre were immensely helpful in draughting the bespoke maps of the continent and the Transantarctic Mountains.

Finally, I could not have completed this work over the past year without the untiring support and assistance of my wife, Gillian. For her enthusiasm, astute questions, patience, and companionship I thank her unreservedly.

THE LAND BENEATH THE ICE

1

The Antarctic Ice Sheet Puzzle

1.1 Prelude

Today Planet Earth is in trouble. Several decades of scientific observations and studies have revealed the progressive and rapid deterioration in the health of our world's natural environment. Increasingly, alarms are being sounded regarding the dependence of humanity on the use of fossil fuels. The emissions from their combustion have dramatically increased the quantity of greenhouse gases in the atmosphere. The result is a world that continues to lurch towards disastrous warming, and despite warning calls to governments, there has been insufficient application of mitigation measures or preparedness to adapt to a new order. The fate of millions of individuals—their lives, livelihoods, and heritage—hangs on a thread as sea levels rise inexorably, storms and extreme weather events become more prevalent, and heat waves and wildfires increasingly threaten cities and the countryside.

The great ice sheets in Greenland and Antarctica, the latter the size of Europe, play a key role in the climate story and occupy a central stage in the long-term well-being of our world. On the one hand they reveal, through physical and chemical analysis of their layers of accumulated snow and ice, a remarkable and detailed record of changes in climate extending back 800,000 years. On the other hand, the destiny of the ice locked away in the ice sheets is crucial to future sea levels. The reduction in size of the ice sheets is already contributing to a steady rise in sea level—20 mm in the last two decades.[2] This shrinking is not purely a matter of melting around the periphery in response to atmospheric and ocean warming. These external forcings

[2] IPCC: Climate Change 2021: The Physical Science Basis. Contribution of Working Group I to the Sixth Assessment Report of the Intergovernmental Panel on Climate Change, Masson-Delmotte, V; et al. (eds.), Cambridge: CUP.

are having complex effects on the flow and stability of immense ice drainage basins that have the potential to discharge substantial additional volumes of fresh water into the ocean. Furthermore, these current changes are committing our world to sea-level rise for many centuries to come.

To understand the response of the ice sheets to climate, sophisticated models have been developed and are continuously being refined. All require data on the glaciological characteristics of the ice sheets. Of fundamental importance are the shape, thickness, bed topography, basal conditions such as melting or freezing, the net gain or loss of mass in the form of snow and ice, and other internal indicators of past flow or changes of state. The technique of radio-echo sounding (RES) and the many surveys that were undertaken principally in Antarctica in the late 1960s and '70s that form the substance of this book yielded the first comprehensive database for many of these parameters. They are affording, in several instances, the baseline from which we can assess the changes in ice volume and behaviour that will continue to challenge the environmental conditions of our planet. Before embarking on the story of these explorations, however, it is salutary to look back to the early questions about Antarctica and the search for methods to probe its icy carapace.

1.2 Some History

As soon as humans spied and later set foot on the remote Antarctic continent in the second decade of the nineteenth century and became aware of its ice cover they quickly desired to know more of its extent, shape, thickness, and behaviour. Exploratory ventures of the early part of that century—for example, the expeditions of James Clark Ross, Charles Wilkes, and Dumont D'Urville—brought back tantalising reports to Europe and North America of this enormous frozen region (Figure 1.1[3]). Their findings and records fed the fecund minds of natural scientists and learned societies and gained prominence in contemporary texts about the natural world.

James Croll, Robert Ball, James Geikie, and others—seized by these accounts—speculated on their significance and interpreted their wider implications. Sir Robert Ball, Lowndean Professor of Astronomy and Geometry at the University of Cambridge, ventured his thoughts on the matter

[3] Captain Sir James Clark Ross (1847) *A Voyage of Discovery and Research in the Southern and Antarctic Regions, during the Years 1839–43*, in two vols, London: John Murray.

Figure 1.1. The Great Ice Barrier (now known as the Ross Ice Shelf). (From a drawing by Sir James Clark Ross; see footnote 3).

in a little book, *The Cause of an Ice Age*, published in 1892: 'It seems, however that in its [Antarctica's] vicinity lies an extensive tract which is crushed under an ice-sheet far transcending, both in area and thickness, the pall which lies over Greenland. From the dimensions of the Antarctic icebergs, it becomes possible to estimate the thickness of the layer of ice, from the fringe of which those icebergs have broken away. It is now generally believed that the layer of ice which submerges the Antarctic continent must have a thickness amounting to some miles'.[4] A few years earlier, James Croll had made calculations on the possible depth of the ice sheet. He based his estimate on rudimentary notions of the flow of an ice mass which gave a depth in the centre of the continent of 39 km![5] Croll did consider this value excessive and revised his numbers downwards, also referring to the known thickness of icebergs, and gave as his best guess a thickness of 4 mi (6 km). Both Ball and Croll were remarkably close to what we know today as the thickest ice, which is just a shade under 5 km deep.[6]

[4] Ball, R (1892) *The Cause of an Ice Age*, 2nd ed., London: Kegan Paul, Trench, Trübner, 180pp.

[5] Croll, J (1875) *Climate and Time in Their Geological Relation*, London: Stanford, 577pp.

[6] Robin, G de Q, quotes an 1879 article by Croll that gives the thickness as 3 mi (4.8 km). 'The thickest ice measured by the SPRI RES programme was 4776 m in the subglacial Astrolabe Basin of Wilkes Land (footnote 1). A new maximum from the same area was reported in 2013 of 4897 m by the

It was not until the advent of the twentieth century that the prospect materialised of being able to gain some more exact measure of these large ice masses. Notions of drilling through the ice were entertained but soon abandoned after the deepest holes extended only a few tens of metres. Eric von Drygalski during his 'Gauss' expedition (1901–1903) attempted to bore a hole and reached about 30 m; the technology was incapable of penetrating to any great depth.[7] However, by the 1920s, geophysicists had devised methods of sounding through rock strata using sound waves generated by near-surface explosions. Such seismic sounding was initially applied to the exploration for oil—by identifying suitable rock structures for later drilling—and it was not long before the technique's potential was appreciated for the depth sounding of glaciers. Initial early exploration in the European Alps confirmed that the albeit rudimentary technique held promise for the great ice sheets. The first to grasp both the significance and the opportunity was the legendary German meteorologist Alfred Wegener. Although noted for his exposition on continental drift, Wegener became fascinated in his later career by the polar regions and organised expeditions to explore the geophysical conditions in Greenland. Pioneering the seismic method with an early apparatus (Figure 1.2), Wegener's team was able to undertake the first dependable measurements of Greenland ice during his 1929–31 expedition, when the team achieved a reliable determination of 2000 m.[8]

Such experimental forays in Greenland were not pursued further until after World War II, when the French Expéditions Polaires Françaises commenced activities under the charismatic leadership of Paul Emile Victor.[9] Using more modern electronic equipment, Alain Joset and Jean-Jacques Holtzscherer made more than 400 spot soundings in the central regions of the ice cap and revealed depths of over 3000 m. These measurements demonstrated that within the interior of this large island the bedrock was below sea level.[10] Such expeditions provided clear expectations that similar

---

Bedmap consortium (see Fretwell et al. (2013) (footnote 368). The average ice thickness of the whole continent by the SPRI analysis was 2160 m, and by the Bedmap consortium, 2126 m.

[7] Fogg, G E (1992) *A History of Antarctic Science*, Cambridge: CUP, 483pp.

[8] Sorge, E (1933) The scientific results of the Wegener expeditions to Greenland, *Geographical Journal* 81 (4): 333–44; Brockamp, B; Sorge, E; and Wölken, K (1933) Bd. II: *Seismik, Wissenschaftliche Ergebnisse der Deutschen Grönland-Expedition Alfred Wegener 1929 und 1930–31*, Leipzig: F A Brockhaus.

[9] I had the privilege of a most convivial meeting and lunch with Victor many years later when he was a tax exile from mainland France, living on a motu in the lagoon of Bora Bora in French Polynesia.

[10] Joset, M A; and Holtzscherer, J-J (1954) Détermination des épaisseurs de l'inlandsis de Groenland, *Annales de Géophysique* 10:351–81.

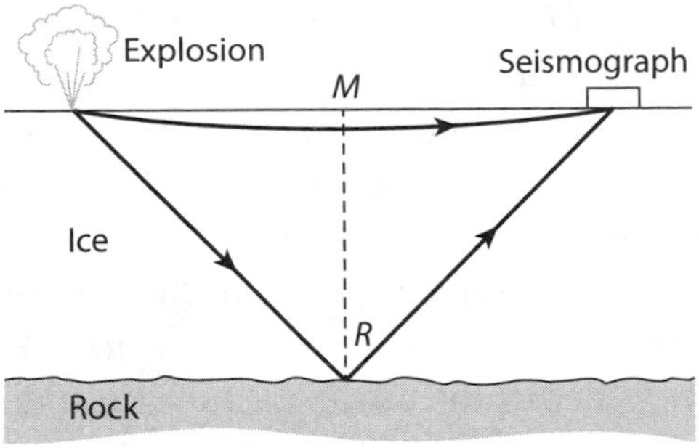

Figure 1.2. The seismic technique as used on the Wegener Expedition (R is the point of reflection at the bed, and M is the point vertically above it on the surface). (From Sorge (1933); see footnote 8).

thicknesses were to be encountered in Antarctica. But transferring the technology south was a much greater logistical and costly enterprise, so much so that Richard Foster Flint, writing in the first edition of his seminal textbook, *Glacial and Pleistocene Geology*, published in 1957, stated: 'The thickness of the ice sheet is virtually unknown except along a single seismic traverse, 600 km long, near the margin, where the maximum thickness is 2,400 m.' (p. 42). We shall return to these early seismic forays and the more extensive programmes of sounding conducted during and after the International Geophysical Year (IGY) (1957–58) in the next chapter, but we need to investigate further the 'single seismic traverse line' that Flint reported.

The Norwegian-British-Swedish Expedition (NBSE), which operated between 1949 and 1952, was a post–World War II collaborative operation that set the standards and logistic template for much of what was later undertaken in the IGY; it was also the first and one of the most successful examples of international scientific cooperation in Antarctica. The expedition was the brainchild of Hans W:son Ahlmann, professor of physical geography at the University of Stockholm, and one of an early and influential group of scientists with a keen interest in the study of the polar regions and glaciology. The expedition developed many of the techniques which would be adopted by all major scientific expeditions thereafter. Dr Albert 'Bert' Crary (chief scientist of the United States Antarctic Research Program in the 1960s) set the expedition in context some years later: 'The era of extensive exploration can be said to have had its beginning in the Norwegian-British-Swedish

Expedition'.[11] The story of the expedition was told in the official account by the leader, John Gaeiver, not long after its return[12] and latterly by Charles Swithinbank[13] in a very readable account from the perspective of one of the young scientists in the party. Several scientific reports were produced which are still of considerable value today.

A major objective of the expedition was to conduct seismic sounding of the ice thickness. This work was to be undertaken by tracked vehicles during an oversnow traverse across the floating ice shelf by the coast and thence onto the grounded ice sheet of the high polar plateau as far inland as the fuel supplies and terrain would allow. The person in charge of this programme was Gordon Robin, an Australian physicist who had previously worked as a meteorologist with the Falkland Islands Dependencies Survey (FIDS—the precursor to the British Antarctic Survey) on Signy Island in the South Orkney Islands.[14] Robin's painstaking and tireless efforts to achieve consistent and dependable seismic results were probably the crowning glory of the NBSE, and his pioneering techniques and experience were the model for later sounding campaigns. Robin's interest in probing the ice sheet and investigating its physical properties and behaviour did not diminish upon his return to Britain in 1953.

In 1955 Robin took the directorship of the SPRI at Cambridge University and continued to pursue his glaciological interests (Figure 1.3). With the appointment of Dr Stanley Evans to the Institute in 1959 Robin found another scientist with complementary experience in remote sounding (Figure 1.4). Evans had spent time at the British base of Halley Bay during the IGY, studying the ionosphere. It was his expertise in radio frequency research combined with Robin's glaciological background that spawned the development of a new and highly productive technique that revolutionised the study of glaciers and ice sheets—radio-echo sounding (RES). The RES method and its application engaged the author of this book as a young graduate student in the late 1960s and consequently dominated a significant part of his career. To tell the story fully of how this new technology evolved and became the standard for penetrating ice sheets and glaciers we

[11] Crary, A P (1962) The Antarctic, *Scientific American*, 207 (3): 60–73.

[12] Gaeiver, J (1954) *The White Desert*, London: Chatto and Windus, 304pp.

[13] Swithinbank, CWM (1999) *Foothold on Antarctica*, London: Longman, 260pp.

[14] Drewry, D J (2003) Children of the 'Golden Age': Gordon de Quetteville Robin, *Polar Record* 39 (208):61–78.

Figure 1.3. Dr Gordon de Quetteville Robin during his time as director of the SPRI at the University of Cambridge between 1955 and 1983. (Courtesy SPRI).

Figure 1.4. Dr Stanley Evans, in New Zealand, 1969.

must first travel back to the early days of seismic sounding and the work by many countries, but notably that of the United States and the then Soviet Union. Their efforts provided us with the first glimpse of the true dimension of the vast ice sheet of Antarctica and what lies beneath its icy shell, and that stimulated the development of alternative techniques.

2

Sounding through the Ice

Seismic methods that had been adapted from the petroleum exploration industry were used to make early measurement of the ice thickness of the great ice sheets. This was the primary technique employed through the period of the IGY and into the mid- to late-1960s, after which the use of electromagnetic systems revolutionised scientific investigations. A handy and approximate method for ice-thickness measurements is gravity surveying, which was often used to 'fill in' the ice thicknesses between control stations determined by seismic shooting.

Today seismic exploration is undertaken to investigate certain properties of the ice and the bedrock beneath that cannot be imaged by radar,[15] and typically in areas of limited geographical extent.[16] This includes determination of the depth of water beneath floating ice shelves and of subglacial lakes, the thickness of semi-consolidated moraine or till at the base of the ice sheet, and the deeper geological layers and structures of the Antarctic continent.

2.1 Seismic Measurements of Ice Thickness

The seismic technique, simply put, uses the time of the returned sound waves from the bottom of the ice sheet resulting from the detonation of an explosive charge at the surface. Knowing the speed at which sound waves travel through ice allows the derivation of the ice thickness. This 'reflection' method has been described extensively and documented particularly by Charles Bentley of the University of Wisconsin.[17] Bentley contributed more,

[15] Radar—radio detection and ranging.

[16] For example: Smith, A M; and Doake, CSM (1994) Sea-bed depths at the mouth of Rutford Ice Stream, Antarctica, *Annals of Glaciology* 20:353–56.

[17] Bentley, C R (1964) 'The Structure of Antarctica and Its Ice Cover', in Odishaw, H (ed.) *Research in Geophysics*, vol. 2: *Solid Earth and Interface Phenomena*, Cambridge, MA: MIT Press, 335–89.

Figure 2.1. Charles Bentley (left) and the author at a conference of the Scientific Committee on Antarctic Research (SCAR) in São Paulo, Brazil, in July 1990.

over his many decades of active research, to Antarctic geophysical exploration than any other individual, and his work alongside that of his numerous research students and associates enabled seismic investigations to be developed into a highly sensitive and sophisticated exploration tool for ice research (Figure 2.1).

Bentley and other geophysicists have described the important factors to be addressed in acquiring reliable seismic ice-thickness measurements, two of the most critical of which were interference from surface 'noise' that prevented the identification of echoes from the base of the ice, and uncertainties in the propagation velocity of elastic waves in ice. The first was a worrisome problem for the pioneers of the seismic technique. Gordon Robin on the NBSE encountered the masking effect of noise that originated as reverberations in the near-surface layers.[18] He experimented with several techniques to minimise them, such as detonating the charge above the surface in the air and setting off charges of different sizes at various depths in the snow. He noticed that quite often the explosion would result in the collapse of the upper layers of snow subject to *depth hoar* (a weak, porous

[18] Robin, G de Q (1958) *Glaciology III: Seismic Shooting and Related Investigations*, vol. 5 of *Norwegian-British-Swedish Expedition 1949–52: Scientific Results*, Oslo: Norsk Polarinstitutt, 134pp; Bentley, C R (1964) (footnote 17).

layer composed of very large ice crystals that develops owing to sublimation processes beneath it). Running a tracked vehicle back and forth to compact these layers achieved an improved coupling of the sound waves with the main body of firn[19] and newer ice layers. Combining this technique with a deep hole for the dynamite charge, he obtained good results. Bentley noted the only method of generally overcoming such ground reverberations was to use shot holes of 30 or more metres in depth.[20]

As for to the second complication, the speed of compressional or 'P' waves in ice, several studies found this to be dependent primarily upon the density of the ice, its temperature, and the structure and orientation of the ice crystals. Again, Robin and Bentley investigated these issues, along with the Swiss glaciologist Hans Röthlisberger.[21] Bentley undertook a project to determine velocities down the deep borehole (2164 m) at Byrd Station in West Antarctica and concluded that although crystal orientation has a measurable effect on the speed of propagation, it was difficult to extrapolate the results across the ice sheet of the whole continent. Overall, Bentley considered an error of ±2%–3% would be appropriate for depth determinations by seismic shooting in Antarctica.

The seismic method has proved reliable and effective and has yielded information on the internal properties of ice, as well as characteristics of the bedrock or soft sediments lying immediately beneath the ice (using a technique called 'refraction' sounding, which we do not explore further here). Nevertheless, it is slow and at times cumbersome. Most determinations were made during traverses across the ice sheet, although some were made by scientists airlifted to a few remote locations. Travel by Sno-Cat and tractor train can be ponderous and dependent upon safe ice conditions (Figure 2.2[22]). This means that regions of intense crevassing, steep slopes, and very rough surfaces are usually avoided and therefore go unexplored (Figure 2.3).

John Behrendt has described admirably the working conditions of typical oversnow scientific traverses and air-lifted activity in West Antarctica during and after the IGY.[23] At each sounding location along a traverse, say

[19] This is semiconsolidated snow with a density of between 400 and 800 kg m^{-3}.

[20] Bentley, C R (1965) 'The Land beneath the Ice', in Hatherton, T, *Antarctica*, London: Methuen, 59–277.

[21] Röthlisberger, H (1972) *Seismic Exploration in Cold Regions*, Cold Regions Research and Engineering Laboratory Monograph 11-A2a, 139pp.

[22] From Landis, C (2010) 'Investigating West Antarctica, Then and Now', https://beyondpenguins.ehe.osu.edu/issue/science-at-the-poles/investigating-west-antarctica-then-and-now.

[23] Behrendt, J C (1998) *Innocents on the Ice*, Niwot, CO: University Press of Colorado, 428pp; Behrendt, J C (2005), *The Ninth Circle*, Albuquerque: University of New Mexico Press, 240pp.

Figure 2.2. A typical oversnow traverse using tracked Sno-Cats pulling sledges. (Courtesy C R Bentley, Byrd Polar and Climate Research Center, The Ohio State University).

Figure 2.3. A Sno-Cat caught in a crevasse during an oversnow traverse. (Courtesy J C Behrendt).

at 50 km intervals, the vehicles would stop. A hole would be drilled, sometimes to a depth of 25 m or more, in which to place the explosive charge (typically 1 to 2 lbs of ammonium nitrate). This could take several hours of hard work. Lines of geophones would be extended over the ice surface to a distance of several hundred metres. The electronic apparatus with its various filters and the recording equipment would be in the Sno-Cat. The charge would be set off and the returning sound waves recorded on fast-moving photographic paper, which then had to be developed chemically and dried, a challenging task in sub-zero temperatures (Figure 2.4[24]).

[24] Bentley, C R; and Ostenso, N A (1961) Glacial and subglacial topography of West Antarctica, *Journal of Glaciology* 3 (29): 882–911.

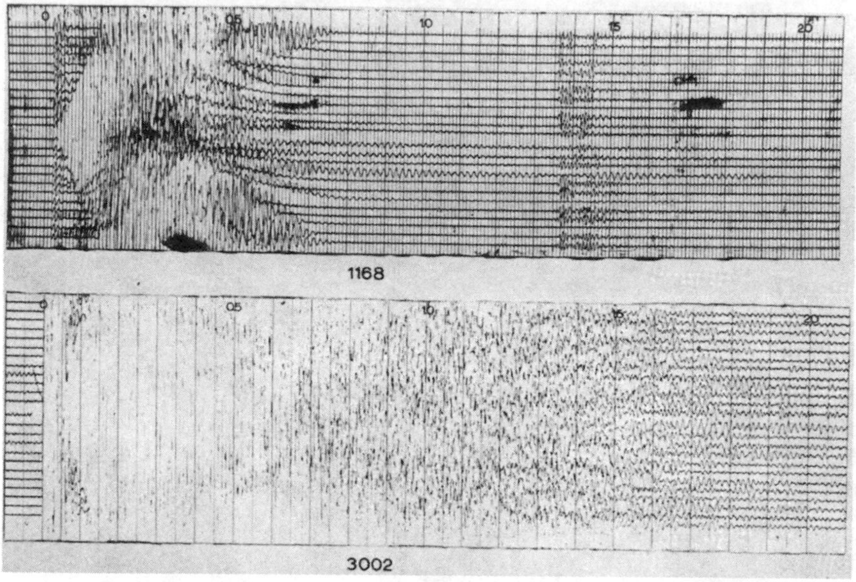

Figure 2.4. Representative seismic reflection records from early traverses in West Antarctica. Both were undertaken with similar charges and shot depths (0.45 kg and 4 m) and identical gain and filter settings. The upper record shows the rapid decay of the surface waves and a clear return from the ice/bedrock interface (at ~1.36 sec). The lower record displays prolonged surface noise masking the reflection from the bed. (From C R Bentley and Ostenso (Courtesy International Glaciological Society).

2.2 Gravity Measurements

Observations of gravity have been used extensively in Antarctica to obtain rapid estimates of ice thickness (Figure 2.5). The acceleration of gravity varies with the density of material beneath the point of observation. With the large contrast in density between ice and rock[25] it is possible to relate changes in the measured value of gravity to bedrock elevation and thence ice thickness if the surface height is also known with high accuracy.

A number of uncertainties are associated with this technique and arise mainly from instrumental errors often due to the extreme temperature environment in Antarctica, uncertainty in elevation, and assumptions about the density and inclination of the sub-ice bedrock. In early gravity surveys altitudes were derived by means of sets of aneroid barometers using a vari-

[25] The average density of crustal rocks is 2700–2800 kg m^{-3}. The average for ice is much more consistent and better known, at 910 kg m^{-3}.

Figure 2.5. Reading a gravity meter at Skelton Glacier, 1957. (With permission ©Antarctica New Zealand Pictorial Collection 1957–59 (TAE0507)).

ety of schemes to cross-check instruments, distribute closure errors, and take account of meteorological conditions. The results, however, could not be guaranteed to better than 50 m, with the consequent introduction of an uncertainty in the calculated ice thickness and bedrock heights. Of course, with the rocks covered by several kilometres of ice it was, and is still not, possible to assign a precise value for their density, so intuitive estimates were used. Bentley drew attention to the additional problem arising from the presence of dispersed layers of low-density moraine at the base of the ice sheet. Robin considered that the errors overall from such density estimates should be less than 30% and more typically 15%. In general, gravity measurements have been shown to be useful in providing some detail of ice thickness in the absence of other data and can function as a 'bridge' between seismic points. The assumptions and issues described can lead to errors of between 5% and 20% overall in areas of irregular relief, poorly determined surface elevations, and rapid changes in bedrock materials.

2.3 Early Tests

Thomas C. Poulter[26] was the first scientist to undertake seismic measurements in Antarctica during the second expedition of Rear Admiral Richard E Byrd in 1933–35. Byrd established his base at the location of his earlier

[26] Poulter, T C (1950) *Geophysical Studies in the Antarctic*, Palo Alto, CA: Stanford Research Institute.

expedition, "Little America" at the Bay of Whales, close to where Roald Amundsen had built 'Framheim'—the encampment for his successful attempt on the South Pole in 1911 (Figure P.2). Poulter tested his equipment on the adjacent floating ice shelf and also on some grounded ice farther inland—Roosevelt Island. These measurements proved the efficacy of the seismic method, although the photographic plates for recording proved somewhat unwieldy and their development time-consuming. Poulter was able to show that the ice shelf was several hundred metres thick and that the grounded area that constituted Roosevelt Island was of the order of 1600 m. Although only a handful of measurements were made, they hinted at the likely depths of the inland ice sheet.

After the confrontations of World War II, the idea of an international expedition to Antarctica was embraced readily in Europe. The Norwegian-British-Swedish Expedition (NBSE) has already been described in section 1.2 as one the most successful to Antarctica in the twentieth century. Robin's seismic sounding provided, for the first time, a reliable profile of the ice sheet across many hundred kilometres. Robin had undertaken comprehensive testing on a glacier near Finse in Norway in 1949, together with the NBSE senior glaciologist, Valter Schytt. The trials enabled him to determine the type of energy source, to select the variety of recorders then available, and to plan an ice thickness sounding campaign. Robin's team reached its furthest point from base Maudheim (Figure P.2), some 620 km distant (74.3°S) at an altitude of 2710 m, achieving a remarkably detailed profile of the ice and subglacial terrain (Figure 2.6). Charles Bentley always referred to Robin's report as 'the bible'.[27]

Whilst Robin was busy in Dronning (Queen) Maud Land, on the opposite side of the continent in that part of Antarctica lying south of Tasmania, Bertrand Imbert a member of the Expéditions Polaires Françaises (EPF) was embarking on a number of seismic experiments.[28] Paul Emile Victor had encouraged an expedition to Terre Adélie—the slender slice of 'Antarctic brie' claimed by France. Imbert's trials were not so comprehensive in scope or scale as those of the NBSE but were nevertheless an important contribution.

[27] C R Bentley archives, Byrd Polar and Climate Research Center, The Ohio State University, Columbus.

[28] Imbert, B (1953) Sondage séismiques en Terre Adélie, *Annales de Géophysique* 9 (1) : 85–92.

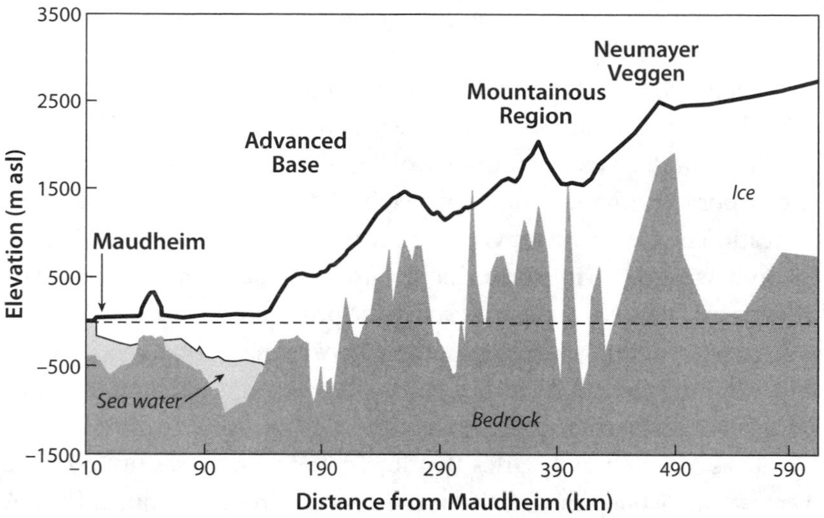

Figure 2.6. Seismic profile by Robin from Maudheim onto the East Antarctic Ice Sheet.

2.4 The International Geophysical Year and Its Aftermath

Antarctica was seen as a major focus for the IGY of 1957–58 and investigation of the ice sheet a primary goal. Many nations contributed to this task, which led a few years later to the continued scientific exploration of the continent and its surrounding seas. The IGY brought cooperation between East and West during some of the most difficult political times on the world stage; the challenges of the science and the polar conditions forged long-lasting collaborations. It was paradoxical that on the coldest continent on Earth the Cold War was thawed the most dramatically. The planning for the IGY called on several nations to undertake widespread seismic soundings to obtain as comprehensive a picture as possible of the ice thickness and the nature of the underlying terrain. Whilst the IGY was confined to a single year, it rapidly became clear that the scientific issues and problems could not be addressed in this intensive but short phase. With considerable investments in infrastructure, particularly the establishment of bases around the periphery of Antarctica, it was not surprising that many nations considered it good sense to continue scientific work at the close of the IGY, and they negotiated and adopted the Antarctic Treaty in 1959. By 1961 the Treaty had come into force with its signing by the various acceding parties. Putting aside national claims, the Treaty, uniquely, set the course for international science to be the determinant of what long-term activities would be undertaken south of 60°S. Nations already committed to the IGY were able to continue, indeed expand, their work under an international umbrella. The collaboration and coordination established in the IGY was followed by the creation of the Scientific Committee on Antarctic Research (SCAR).[29] This body with its various working groups and specialist committees was able to assist in agreeing priority scientific projects and the effective melding of national plans to achieve them. The Working Groups on Glaciology and Solid Earth Geophysics were the primary groups that encouraged the continued seismic work in later years.[30]

In all, as reported by Charles Bentley in 1965,[31] traverses up to 1962 had covered some 25,000 km with the addition of 23 aircraft landings. The countries that had participated to that date included Australia, Belgium, the British Commonwealth, France, Japan, the US, and the USSR. Maps of ice thickness, surface elevation, and the topography of the sub-ice surface were compiled at a coarse continental scale, albeit with major gaps and in some cases considerable interpolation. Bentley[32] undertook such enterprises in the US (Figure 2.7), as did Andrei Kapitza[33] in the USSR (Figure 2.8), and the mappings revealed the large-scale features. It should be noted that the American scientists were much more circumspect regarding large regions without data, leaving these tracts completely blank, compared with their Soviet colleagues, who were more prone to provide 'complete' geographical coverage involving considerable imaginative interpolation.

The surface shape of the ice sheet, and hence of the continent, was broadly laid out. A dome rising to an elevation of about 4000 m was depicted in the eastern, larger portion, and a lower dome characterised West Antarctica, flanked by two great floating ice shelves, the Ross Ice Shelf and the

[29] The Scientific Committee on Antarctic Research (SCAR) is a thematic organisation of the International Science Council (ISC). SCAR is charged with initiating, developing, and coordinating high-quality international scientific research in the Antarctic region (including the Southern Ocean).

[30] Walton, DWH; and Clarkson, P D (2011) *Science in the Snow*, Cambridge: SCAR, 258pp.

[31] Bentley, C R (1965) (footnote 20 above).

[32] Bentley, C R; Cameron, R L; Bull, C; Kojima, K; and Gow, A J (1964) *Physical characteristics of the Antarctic Ice Sheet*, Antarctic Map Folio Series, Folio 2, New York: American Geographical Society.

[33] Kapitza, A P (1967) 'Antarctic Glacial and Sub-glacial Topography', in Nagata, T (ed.), *Proceedings of the Symposium on Pacific-Antarctic Sciences*, Tokyo: Japanese Antarctic Research Expedition Scientific Reports, special issue no. 1, 82–91.

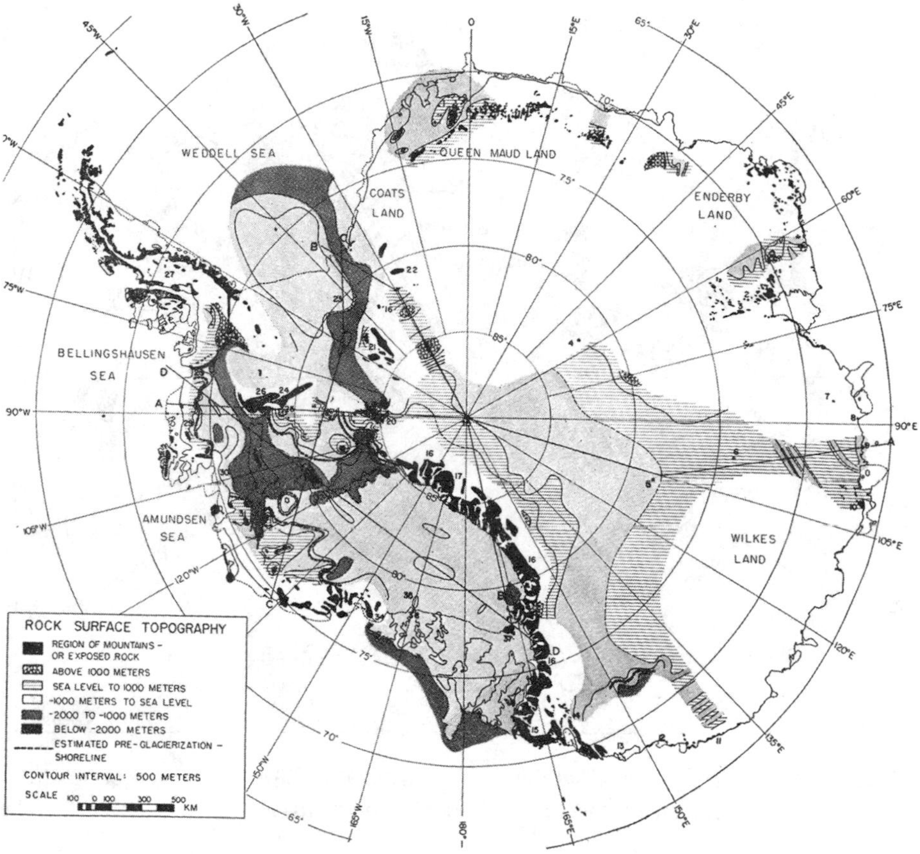

Figure 2.7. Subglacial map of Antarctica from Bentley (1965). (Courtesy New Zealand Antarctic Society).

Ronne-Filchner Ice Shelf. Beneath the ice the geophysical measurements indicated the central part of East Antarctica was dominated by two major relief features. The first was an extensive highland zone called the Gamburtsev-Vernadskii Mountains, stretching from about 79°S, 95°E to 72°S, 40°E and reaching elevations of 3900 m asl (above sea level), according to Kapitza. The mountains had first been identified by the Third Soviet Antarctic Expedition,[34] but difficulties of interpretation of their seismic records cast

[34] Sorokhtin et al. (1959) Rezul'taty opredeleniya moschnosti lednikovogo pokrovo v Vostochnoi Antarkide, *Informatsionnyy Byulleten' Sovetskoy Antarkticheskoy Ekspeditsii*. no.11, 9–13. The 'Vernadskii' portion of the mountains was later shown to be a diffuse area of highland not connected to the Gamburtsev chain, and the name was dropped.

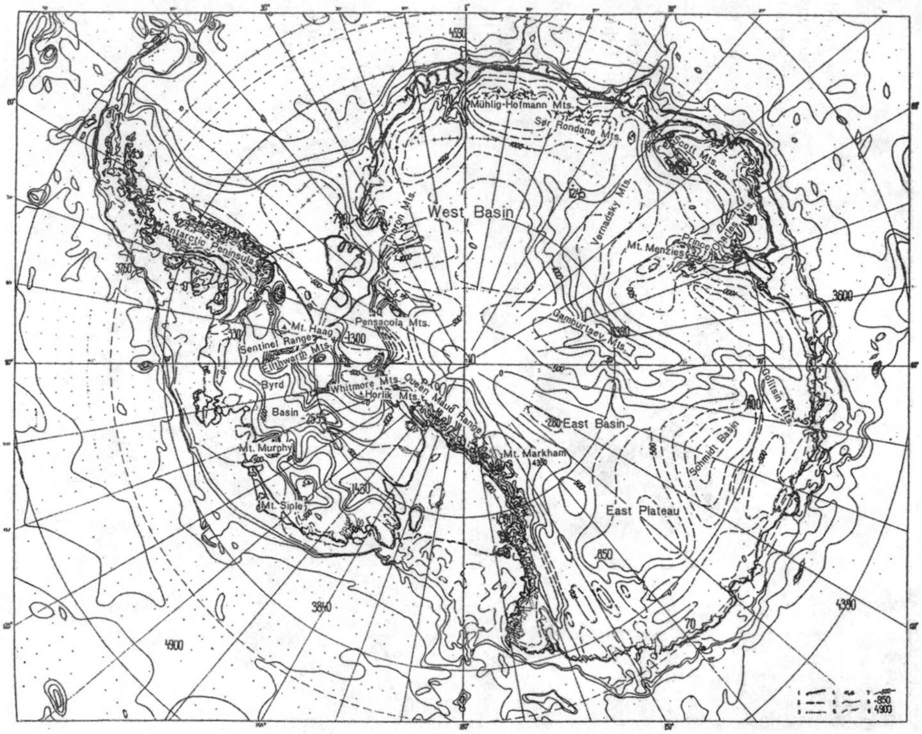

Figure 2.8. Subglacial map of Antarctica from Kapitza (1967) with contours at 500 m intervals. (Courtesy National Institute of Polar Research, Tokyo).

doubt on the true existence of this mountain range.[35] Later investigations confirmed the presence of this vast highland massif. The second major feature was a very deep basin (termed the 'Schmidt Plain' by Kapitza) lying between Vostok and Wilkes Stations in latitude 72°S and extending between 95° and 120°E. According to Kapitza, this basin reached 1500 m below sea level. The remainder of East Antarctica appeared to comprise large swells reaching 1000 m in elevation, whilst the plains that lay between varied from 0 m to 500 m below sea level. Undoubtedly, such generalised topography resulted from the smoothing between sparse data points many hundreds

[35] Kapitza, A P (1960) Novye dannye o moschnosti lednikovogo pokrova tsentral'nykh relonov Antarktidy. *Informatsionnyy Byulleten' Sovetskoy Antarkticheskoy Ekspeditsii* 19:10–14; Woollard, G (1961) *Crustal Structure in Antarctica*, Geophysical Monograph 7, Washington, DC: American Geophysical Union, 57–73.

of kilometres apart. The coastal margin was characterised by a series of blocky massifs.

There was better coverage by seismic data in West Antarctica, principally from US traverses, many radiating from Byrd Station. This region exhibited an altogether different landscape dominated by a very deep sub-ice basin lying below sea level over a substantial area stretching from the Ross Sea, beneath the Ross Ice Shelf, through the central region of West Antarctica to the Amundsen and Bellingshausen Seas (Figure 2.7). This extensive feature was termed the 'Byrd Basin', and its topography appeared quite rugged. Towards the interior and around 80°S,110°W the bedrock surface descended to the lowest elevation in Antarctica (2560 m below sea level), containing Antarctica's deepest ice at that time (4335 m). This became known as the Bentley Subglacial Trench. Another narrower trough appeared to run from the eastern Weddell Sea, beneath the Filchner Ice Shelf, extending towards the Thiel Mountains with a maximum elevation of 1500 below sea level. Between these deep depressions lay a highland area that looked to be the inland extension of the elevated Ellsworth Mountains, the highest range on the continent. Shallower and smoother terrain was found towards and beneath the Ross Ice Shelf.

The oversnow traverses undertaking seismic sounding and gravity observations continued into the late 1960s with a suite of US campaigns in East Antarctica (Queen Maud Land Traverses) and Soviet forays inland of its bases along the coast, adding further detail to the general picture that had emerged during and immediately after the IGY.[36] Other nations that made contributions included Japan, with a traverse from its station at Syowa to the South Pole between 1967 and 1969.[37] Most of these soundings were integrated in map form into the Soviet Atlas of Antarctica published in 1967.[38]

Impressive as these geophysical programmes and their hard-won results were, the outcome at this time could be likened to deducing the features of the landscape geography of the United States from some 1500 height determinations, not randomly scattered but along somewhat arbitrary lines across

[36] Beitzel, J E (1971) 'Geophysical Exploration in Queen Maud Land, Antarctica', in Crary, A P (ed.) *Antarctic Snow and Ice Studies II, Antarctic Research Series*, vol. 16, Washington, DC: American Geophysical Union, 39–87.

[37] Murayama, M (ed.) (1971) *Report of the Japanese Traverse Syowa–South Pole 1968–69*, Japanese Antarctic Research Expedition Scientific Reports, special issue no. 2, Tokyo: Polar Research Center/National Science Museum, 279pp.

[38] Tolstikov, Ye I; et al. (eds.) (1966–67) *Atlas Antarktiki I* (Atlas of the Antarctic I), Moscow: Glavnoye Upravleniye Geodezii i Kartografti.

the continental land mass. There were still many and enormous gaps and little specific detail. Traverses followed routes not necessarily governed by the objective of gaining a systematic coverage and yielded only spot determinations. In the case of seismic lines, the average spacing was 60 km (ranging between 10 and 150 km); for the much more rapid but less accurate gravity observations it was 8 km (with a range of 3–50 km).

The mid- to late 1960s marked a turning point in the geophysical and glaciological exploration of Antarctica. An exhilarating age was dawning that would mark a step change in our ability to map this last great continent—based upon the new and exciting technology of radar sounding.

3

The Advent of Radio-Echo Sounding

In this chapter we look back at the historical origins of radio-echo sounding (RES), particularly the advances made at the SPRI in Cambridge. The spur to develop new methods of sounding glacier ice came, in part, from seeking fresh and innovative ways to make measurements more quickly and continuously; undertaking seismic depth measurements across the Antarctic continent was a laborious, slow, and often dangerous process. There was also the possibility, as in other fields of science, that different technology would be developed. Eventually, several new findings and experimental activities converged, and a new era dawned.

The RES technique is comparable with seismic methods: a pulse of radio energy (rather than acoustic energy) is transmitted through the ice to be reflected at its base, and the returned signal is recorded; however, unlike a seismic source, which generates sound waves from a single explosion, a radio transmitter can send out thousands of pulses a second.

3.1 Experiments and Happenstance

The earliest venture into experimenting with radio waves was made by an academic at the University of Göttingen in Germany in the late 1920s. Stan Evans at the SPRI unearthed a reference to a paper by W. Stern that considered the theoretical background for 'electrodynamic' thickness measurements. Field trials were made on the Hochvernagtferner in Austria in 1927 and 1928 using an antenna laid out along the surface of the glacier and measuring its changing capacitance. Depths of 40 m were calculated despite considerable uncertainties in aspects of the technique.[39] This work came to the notice of William Pickering, then a professor at the California Institute of Technology, who encouraged one of the doctoral students, B O Steenson,

[39] Evans, S (1963) Radio techniques for the measurement of ice thickness, *Polar Record* 11 (75): 795.

Figure 3.1. Amory Waite (second from left) during the Second Byrd Expedition. (Photography by Joseph A Pelter in Byrd, R E (1935); see footnote 41).

to undertake radar experiments in Alaska in 1947. Robert Sharp, the expedition leader and early luminary in glaciology in the US reported: 'Bernard Steenson . . . made good progress in attempting to adapt radar to the determination of ice thickness in a glacier above its bedrock floor. He obtained a reasonable transverse profile of a valley glacier, and this method appears to have considerable promise.'[40] Steenson's work was presented in his thesis but was not taken any further.

The next and more extensive radio sounding determinations were by Amory ('Bud') H Waite Jr in Antarctica and Greenland in the mid- to late 1950s. Waite had been to Antarctica with Admiral Byrd on his Second Antarctic Expedition (1933–35) (Figure 3.1[41]) as a radio communications engineer, and he had noticed that the burial of cables and radio aerials made little difference to their performance. After WWII Waite was working for the

[40] I am grateful to an anonymous reviewer for drawing this early experiment to my attention. Sharp, R P (1948) Project 'Snow Cornice', *Engineering and Science* 12 (2): 6–10.

[41] Byrd, R E (1935) *Discovery: The Story of the Second Byrd Antarctic Expedition*, New York: G P Putnam's Sons, facing p. 281.

US Army Signal Research and Development Laboratory, Fort Monmouth, New Jersey, where he became involved in experiments to investigate the electrical properties of ice in order to determine whether an electromagnetic system could be used to measure ice thickness.

There was a highly practical reason for his work. Aircraft pilots flying over ice-covered land had noticed that the early onboard radio altimeters, which operated at megahertz frequencies, appeared to give false readings of the plane's height above the ground. The 'terrain clearance' from these instruments was much greater than expected, which implied penetration of the radio waves into the ice. The altimeters were so unreliable over snow and ice terrain they may well have been responsible for the crashes of several aircraft on the Greenland Ice Sheet towards the end of WWII. Pilots refused to use them, and the US Air Force issued a warning notice on their use in such circumstances. These occurrences were sufficiently worrisome to cause the US military to seek investigations into their operation, with which Waite and his colleagues were tasked in the early to mid-1950s. Waite recalled that 'it was . . . important that studies be made to quickly determine the optimum frequency and technique for reflecting radio waves from the surface of the ice-covered areas of the earth to low-flying aircraft so that fatal casualties of the type recently reported on the Greenland ice-cap would not occur in the future'.[42]

Waite proceeded to organise a series of experiments to measure ice depths and radio propagation in Antarctica at the commencement of US involvement in the IGY, designated Deep Freeze I. His equipment was landed at the Bay of Whales at the new base, Little America V, on the Ross Ice Shelf. Waite travelled a few kilometres inland to where early seismic determinations of ice thickness had been made by his former Byrd Expedition colleague Thomas Poulter. Waite used a standard aircraft radio altimeter operating at 440 MHz (SCR 718 pulsed ultra-high-frequency (UHF) unit). With suitably configured aerials he was successful in determining the dielectric constant of the upper layers of snow to a depth of approximately 6 m. This parameter was essential in converting time of propagation of the radio wave through ice (i.e. velocity) into an ice depth. Waite's first experiments at Little America were curtailed for logistics reasons, but his vessel, on its

[42] Waite, A H (1959) Ice depth soundings with ultra-high frequency radio waves in the Arctic and Antarctic, and some observed over-ice altimeter errors (DN Task NR 3A-99-20-001-04) *USASRDL Technical Report 2092*.

return journey north, called at the US Wilkes Station on the margin of the East Antarctic Ice Sheet at 66.25°S,110.5°E. He travelled out from the base about a mile up the ice ramp, where it was estimated the ice would be several hundreds of feet thick, and in February 1958 made the first unequivocal determination of depth of polar ice using radio waves. The value of 'slightly over 500 feet' was corroborated a few months later by a seismic shot which gave a thickness of 540 ft.

Following his return to the United States, Waite was encouraged to undertake further and more elaborate experiments by Henri Bader, the influential scientific director of the Snow, Ice and Permafrost Research Establishment (SIPRE) of the US Army Corps of Engineers. In the late summer of 1959 Waite travelled to the US military facility at Thule in northwest Greenland. His equipment (operating at 440 MHz), mounted on a 'Weasel' tracked vehicle, was driven along the 3-mile ramp at the edge of the ice sheet at Camp TUTO ('Thule Take-Off"). Furthermore, Bader had assigned the Swiss geophysicist Hans Röthlisberger to assist with seismic sounding to enable comparison with the radar data. Waite was thus able to comment on the rapidity and ease with which such soundings could be made compared with explosive methods. Waite's results agreed with the seismic data to within 20 ft (Figure 3.2).

The potential of radio frequencies for ice sounding had been demonstrated conclusively, but few of these interesting and pioneering activities were published in the international scientific literature. Waite was an employee of a military establishment with little time or inclination to explore the wider glaciological opportunities the technique promised. He wrote up his results in a technical report and only later did he distribute them more widely.

Meanwhile, various radio-frequency methods for ice sounding were developing in parallel in other parts of the world. Igor Bogorodsky and his colleagues were particularly active at the Arctic and Antarctic Research Institute (AARI) in Leningrad. In the book *Radioglaciology*[43] Bogorodsky describes his and his co-workers investigations in 1955 of the theory of radio wave propagation through glaciers, and their publication of a paper in 1960 entitled 'Problems of measuring of glacier thickness by electromagnetic

[43] Bogorodsky, V V; Bentley, C R; and Gudmandsen, P E (1985) *Radioglaciology*, Dordrecht: Reidel, 254pp.

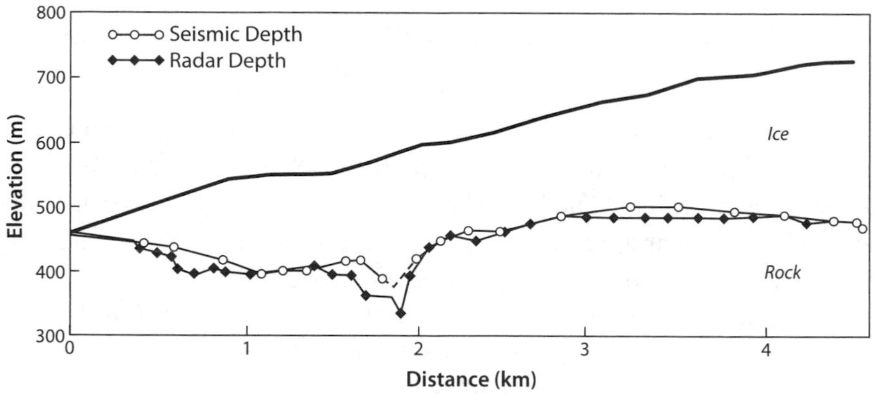

Figure 3.2. Waite's 1959 profile on the TUTO Ramp, northwest Greenland comparing seismic- and radio-determined ice depths. Seismic soundings by SIPRE in 1957. (Adapted from US Army Signal Research and Development Laboratory Technical Report 2092).

methods'.[44] The first field experiments by these Soviet scientists were in East Antarctica inland of Mirnyy Station during the 10th Soviet Antarctic Expedition in 1964. In February and March 1965 they made spot radar soundings at 32 locations during the traverse from Mirnyy to Pioneerskaya Station.

Evans had some interesting, if not extreme, views regarding the Soviet experimental work and had visited Bogorodsky's institute in 1966. In his interview for the British Library, he stated:

> Russians were under the impression that they'd started all this. . . . Their approach was, you know, get the biggest possible transmitter—just like their approach to satellites, get the biggest possible one and drive it as hard as you can and see what you get. But they used—they actually used some US Air Force equipment, which in the Russian literature you will find described as 'Giuss' . . . as the maker or manufacturer or type and brand. . . . That's actually a transliteration from the Roman alphabet. We call the firm Hughes, H-U-G-H-E-S, you see, which became Giuss.[45]

[44] Rodakov, V N; and Bogorodsky, V V (1960) *Zhurnal Tekhnicheskoi Fiziki* 80:82–89.

[45] Dr Stan Evans, National Life Stories, An Oral History of British Science, The British Library, C1379/51.

3.2 Developments in Cambridge and Antarctic Tests

Ideas on radio propagation through ice were also emerging in Cambridge but via an entirely different route, from the interpretation of anomalous ionosphere echoes on Antarctic floating ice shelves. Stan Evans returned from Antarctica in 1957 to Manchester University to work at the Jodrell Bank Radio Telescope and, recruited by Gordon Robin, moved to Cambridge University in 1959, to engage in a variety of physical projects.

Evans' work at Halley Bay, located on the Brunt Ice Shelf (on the east coast of the Weddell Sea (see Figures 5.21 and P.2), involved recording and studying auroral and geomagnetic storms. Ionospheric sounders were mounted on the surface of the floating ice shelf and beamed radiation upwards into the upper atmosphere. Echoes received from the plasma could provide information about the processes occurring at high altitude (more than 80–100 km) which affected radio communications.

Evans describes how in the early 1960s he was approached by Roy Piggott, the avuncular doyen of upper atmospheric physics in the UK, who was based at the Radio Research Station at Slough. Piggott showed Evans ionospheric records from Halley Bay which displayed some unusual features—a series of gaps in the frequency domain. Investigating these further, Piggott and Evans observed the gaps or weak echoes to be spaced regularly, corresponding to particular frequencies. They concluded that some of the signals had penetrated the ice shelf to a depth of several hundreds of metres and had been reflected upwards from the very strongly contrasting boundary between the base of the ice shelf and the seawater on which it was floating. The phase of some of the reflections was the opposite of the phase of the signals being transmitted, so they cancelled each other out—hence the gaps. This discovery led Evans to investigate ionospheric records from three other locations in Antarctica where similar measurements had been made. Evans published his results in 1961, interpreting the anomalous echoes in their glaciological context and showing that it was possible to deduce not only the ice thickness but the velocity of radio waves through the ice and their attenuation.[46]

Robin, with his considerable experience with seismic depth sounding in Antarctica, encouraged Evans to apply techniques of radio-echo sounding

[46] Evans, S (1961) Polar ionospheric spread echoes and radio frequency properties of ice shelves, *Journal of Geophysical Research* 66 (12): 4137–41.

similar to those with which he was familiar from Jodrell Bank to the study of the thickness of ice sheets—downwards-looking radars, not upwards-looking!

Evans commenced a thorough review of all the existing literature on the electrical (dielectric) properties of snow and, particularly, solid ice, both laboratory and naturally occurring glacier ice.[47] This was a prerequisite to understanding the frequency-related transmission of radio waves through ice and led to numerous questions: Did this phenomenon occur at many or only at selected frequencies? How much of the radio signal did the ice absorb? What properties of the ice controlled or influenced propagation—density, temperature, impurities, crystal orientation, water content? The list was long! It became obvious that for scientific purposes an explicitly designed radio-frequency sounder would need to be constructed rather than using some form of modified aircraft altimeter. This raised several new problems: Which frequency? How much power would be needed? How might the transmission aerials be best configured? How would the signals be recorded? To Evans' tidy and analytical mind these were exciting scientific challenges. The starting point, however, had to focus on the fact that both he and Waite had observed the real thing.[48]

Over the next year and a half Evans worked steadily at identifying suitable parameters for an RES system which could be mounted on an oversnow vehicle and was capable of deep ice penetration. Absorption rises steeply with frequency, so as low a frequency as possible was needed, but the trade-offs were that—depending on bandwidth—the lower the frequency, the lower the resolution and resolving power of the returned pulses. Scattering within the ice caused by air bubbles, ice lenses, pockets of water, and solid materials such as sand and gravel, as well as by cracks and fractures, needed to be considered. Antarctic ice is mostly very cold and thick and possesses fewer of these scattering centres, so a critical factor was going to be generation of sufficient power to penetrate to the greatest depth.

Then, there were practical matters if equipment was to be deployed in the aggressive environment of Antarctica and carried over hundreds of

[47] Evans, S (1965) Dielectric properties of ice and snow—a review, *Journal of Glaciology* 5 (42): 773–92.

[48] Turchetti, S; et al. (2008) Accidents and opportunities: A history of the radio-echo-sounding of Antarctica, 1958–1979, *The British Journal for the History of Science* 41 (3): 417–44. These authors commented that Waite and Evans had arrived at the idea of ice-depth sounding by radar from two different directions but both arising serendipitously—aircraft crashes and strange ionospheric observations.

Figure 3.3. Early radio-echo sounding on the Brunt Ice Shelf by Mike Walford (left) in January 1963 from the UK base at Halley Bay. A sledge of fuel and a live-in caboose (attended by Douglas Finlayson) are pulled by a Muskeg tractor. (Courtesy David Petrie).

kilometres of rough ice on sledges or in vehicles: How large would the equipment be? How much power would be required? Would the electronics be robust enough to withstand penetrating cold and physical shocks?

Evans tackled these questions progressively; he settled on a centre frequency of 32 MHz and later 35 MHz with an accuracy for the depth measurements of ±5 m and designed and constructed a sounder known as the SPRI MkI. By this time, he had secured funding from the Royal Society Paul Instrument Fund and been joined by a doctoral research student, Mike Walford. Walford undertook the first tests of the SPRI equipment in Antarctica in January 1963 during an oversnow traverse of 320 km over the Brunt Ice Shelf supported by the British Antarctic Survey (Figure 3.3).[49] Douglas Finlayson and David Petrie ably assisted in this activity. Petrie, a first-class radio technician, was to undertake airborne RES in the Antarctic Peninsula with BAS in 1967. He later joined the SPRI to work with Evans and participated in field operations out of McMurdo. Although Walford's records were discontinuous, the results validated the fundamental principles and demonstrated the efficacy of the equipment. Among the lessons learned from this early trial was that attention must be paid to the method of

[49] Walford, MER (1964) Radio-echo sounding through an ice shelf, *Nature* 204 (4956): 317–19.

recording the returning signals, as the use of a marine-style chart recorder proved unworkable.

3.3 Trials in Greenland

A further opportunity to test Evans' design came in April 1963, when the US Army Electronics Research and Development Laboratory (ELRDL, later the USAEL, US Army Electronics Laboratory) issued an invitation to several research organisations to participate in comparative trials of radar systems for measuring ice thickness. This initiative, headed by Bud Waite, had been encouraged by the US National Science Foundation's Office of Antarctic Programs under its chief scientist, Albert 'Bert' Crary, whose own Antarctic experience with seismic sounding was undoubtedly influential. The trials would be held in northwest Greenland at Camp TUTO, close to the large American Thule Air Base (76°31.5′N, 68°42.1′W). Waite was the project director for ELRDL, and the activity was also supported by the US Army Cold Regions Research and Engineering Laboratory (CRREL), the successor to SIPRE. Participating in addition to USAEL staff were A Bauer and M Andrieux from the Centre de Géophysique Appliquée, Paris, conducting electrical resistance experiments (also undertaken by Hans Röthlisberger); Charles Bentley and Manfred Hochstein from the Geophysical and Polar Research Center at the University of Wisconsin, undertaking seismic sounding and gravity measurements for comparison; and Stan Evans from the SPRI. The SPRI equipment (operating at 32 MHz) suffered some electronic problems but measured depths to 550 m and recorded widely fluctuating echo strengths, suggesting an irregular base of the ice.[50]

It was an encouraging but not fully successful experiment, and further and more exhaustive field trials were carried out in northwest Greenland between June and August the following year (1964), organised once more by USAEL and CRREL. Evans describes this work as 'the biggest leap forward in technique and analysis so far'.[51] USAEL deployed its own sounders at 440 MHz and 30 MHz; the SPRI equipment at 35 MHz constituted an improved version—the SPRI MkII. There were significant enhancements to

[50] Evans, S (1963) International co-operative field experiments in glacier sounding: Greenland, 1963 *Polar Record* 11 (75): 725–26.

[51] Evans, S (1967) Progress report on Radio-echo Sounding, *Polar Record* 13 (85): 413–20.

Figure 3.4. Oversnow vehicles used for the Greenland trials. In the foreground a snow core is being extracted. (Courtesy S. Evans).

the electronics, and the Z-scope recording system was introduced, as well (described in Appendix 1).

The field operation in Greenland involved an oversnow traverse from Camp TUTO climbing the ramp onto the main ice sheet and thence along a steadily rising ridge to Camp Century some 220 km inland at 1930 m asl (Figure 3.4).

From there two journeys were made to the north and west (90 km) and south (60 km), followed by a return journey to Camp TUTO (Figure 3.5). The SPRI team comprised Gordon Robin, Stan Evans, Jeremy Bailey (a new research student), and Terry Randall (a technical consultant). Garry Clarke, from Toronto University (later at the University of British Columbia), undertook a series of seismic soundings along the main trail for detailed comparison and provided the 'first real check of the validity and accuracy of all our values for the thickness of polar ice sheets'.[52] The average difference in

[52] Robin, G de Q; Evans, S; and Bailey, J T (1969) Interpretation of radio-echo sounding in polar ice sheets, *Philosophical Transactions of the Royal Society A* (London) 265 (1166): 437–505.

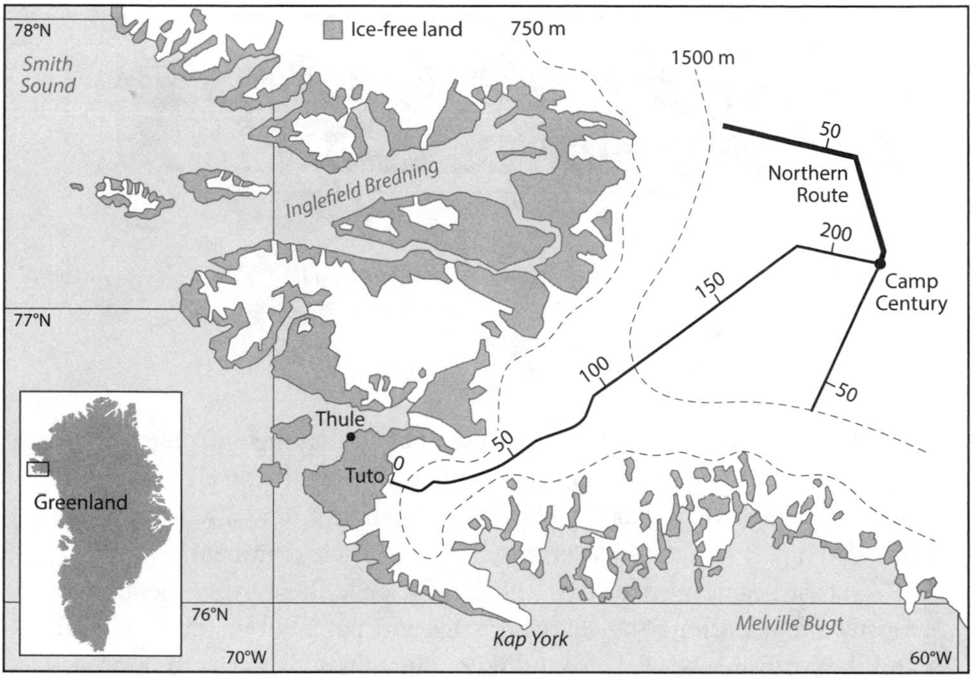

Figure 3.5. Sketch of the route of soundings during the trials at Camp TUTO, 1964. The RES record obtained along the northern leg is shown in Figure 3.6. Distances are in kilometres.

ice thickness for 27 common points was ±15 m (without regard to sign), giving a good indication of the uncertainty of a measurement.

At Camp Century, CRREL had drilled a core through the ice to its bed in 1966. The bottom of the core showed dirty bands in the lowest 17 m and encountered frozen subglacial sediment known as *till* at 1387.4 m. The radio-echo sounding depth, made on the surface 300 m from the core hole, gave a value of 1370 m, some 17 m shallower than the directly measured depth to the till layer.

The work undertaken in Greenland proved seminal; it was the first continuous profiling of the ice sheet in Greenland (Figure 3.6). Careful experiments with the SPRI equipment revealed a wealth of information on many of the parameters that had been of concern or uncertain: the effect of temperature and absorption on the strength and duration of reflections from the bottom surface of the ice, refraction effects due to the height of the aerials and the changing density of the upper layers, and determination of the speed of transmission of radio waves in the ice, necessary for converting

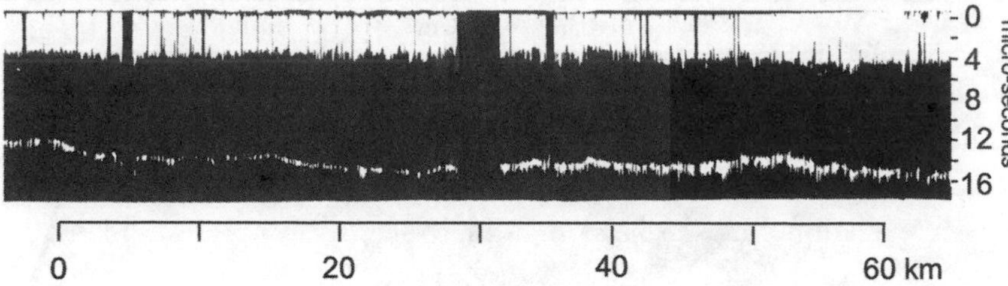

Figure 3.6. A section of RES profile along the northern leg from Camp Century (see Figure 3.5). The scale on the right is delay time; 16 μs is equivalent to about 1350 m in ice. (Courtesy SPRI).

travel times to ice depths. An unexpected but fascinating outcome was the recording of some reflecting horizons within the ice that were to become significant later when radar records from Antarctica (and also Greenland) revealed highly structured layering across the whole continent.

As at the beginning of any novel field of science, these experiments were mostly about learning the characteristics and potential of the technique and the performance of the sounding equipment with a view to improvements and ancillary features. It was a heady and exciting time, and Robin and Evans worked steadily through the data they had collected on the traverse. The material came together in one of the most influential papers in RES and which placed the technique and its value for geophysical exploration of the great ice sheets on a rigorous and intellectually sustainable footing.[53] This lengthy report not only dealt with the theoretical background and methodology but began to explore the glaciological information it provided—its raison d'être.

It is interesting to compare the outcomes from the other systems being tested in Greenland, which, to this author, reveal why the SPRI system, along with Evans and Robin, gained considerable reputation. Bud Waite's principal objective, it would seem, was to establish the effectiveness of radar sounding in ice and thus to establish some basic concepts and principles. This he achieved effectively. He did not, however, pursue the technique's wider application and the detailed glaciological investigation of the features it exposed; it was not, I suspect, his primary interest. This task was left to the focused, academic approach at Cambridge, where considerable knowl-

[53] Robin, G de Q; Evans, S; and Bailey, J T (1969) (footnote 52).

edge of ice sheet behaviour and radio wave propagation could be melded with the new and exciting radar data. An entirely new field of research was opening up, in which concepts and ideas could be challenged and new hypotheses developed.

Robin used the continuous profiles of ice thickness along with detailed surface elevation data, independently surveyed by Stephen Mock of the CRREL, to examine aspects of the flow of large ice sheets that were of interest to glaciologists at that time, namely, the relationship between the shape of the ice sheet as reflected in the local and regional surface slopes and the underlying bedrock topography. Did the bottom surface of hills and valleys govern the surface outline of the ice sheet? Did it determine its speed, and how far above the bed was the ice still affected by the bottom bumps and hollows?

Robin's contribution to these discussions demonstrated the value of the continuous RES ice-thickness data. He was able to show that variations in stress in the direction of ice flow had to be considered and that along a band of ice outflow the surface topography mirrors the underlying bedrock topography but at a reduced or 'dampened' scale. He also confirmed the relationship, derived principally from valley glaciers, that the speed of the ice is proportional to either the square or the cube power of the shear stresses at the bed (equivalent to friction), and that these latter quantities in turn show no clear dependency on temperatures in the lowest layers of the ice.

The readily identifiable but weak reflections from within the ice sheet were a surprise. Robin described them as having 'the characteristic of weak specular reflexions [*sic*] of slowly varying intensity and appear to be caused by layers of wide areal extent approximately parallel to the surface'.[54] The question of their origin was discussed, and density variations were thought to be the most likely explanation, giving rise to a change in the dielectric constant. This was later to be seen as just one explanation and possibly relevant only in the upper layers. Based on measurements of snow accumulation and information from the deep Camp Century ice core, the radar layers at several hundred metres depth were calculated to have been formed at the surface of the ice sheet approximately 1000 years previously, possibly

[54] Robin, G de Q; Evans, S; and Bailey, J T (1969) (footnote 52).

by a warming event that had melted the surface, which subsequently refroze, leading to a change in snow/ice density.

These studies were the first ripples of what later become a wave of fresh insights that RES provided into the investigation of these vast ice masses and whose link to climate would later become of crucial global concern. Like many other new techniques, radio-echo sounding was opening up avenues of new and exciting research.

The SPRI MkII sounder became something of a standard system following its successful operation in Greenland and Antarctica and, surprisingly, went into semi-commercial production by Terry Randall, who had accompanied Robin and Evans to Greenland. Randall was ex-BAS and had established his own small but financially successful electronics company manufacturing central heating controllers at a plant near Harpenden, north of London. Evans contacted Randall on the recommendation of Sir Vivian Fuchs, the director of BAS, to see whether he knew of suitable electronic construction facilities for the echo sounders; at the time the SPRI had virtually no laboratory capabilities. Randall was enthusiastic about the project and for several years provided expertise and funding in kind to assist in building the apparatus and underwriting experimental trips to Norway and the European Alps. Responding to enquiries via Evans, Randall commenced constructing the MkII for overseas sale and at a reasonable cost. Units were sold to investigators in the US, Australia, South Africa, and Belgium, and loaned to Norwegian and other researchers. Charles Bentley commented: 'It [the SPRI MkII] has remained the basic unit employed in the airborne systems that produced most of the soundings of the Antarctic Ice Sheet by the USA, Great Britain, and other western countries up until about 1970. . . . It is still widely used to this day' [i.e. the date of publication, 1985].[55] The later and improved MkIV sounder has been described fully by Evans and Beverley Ewen-Smith.[56]

[55] In Bogorodsky, V V; et al. (1985) (footnote 43).

[56] Evans, S; and Smith, BME (1969) A radio-echo equipment for depth sounding in polar ice sheets, *Journal of Scientific Instruments* (*Journal of Physics*) ser. 2 (2): 131–36.

The basic SPRI units comprised a pulse-modulated transmitter, logarithmic receiver unit with attenuation facility, power supply, and a system control unit. A crystal-controlled clock provided the overall time base, and calibration marks were made at specified intervals. Michael Gorman provided an excellent commentary on the details of the MkIV in his audio and video recordings in 2019. (Gorman, M (2019) University of Cambridge, SPRI, Oral History collection (RS202 1.04 .MOV to 1.09. MOV)).

3.4 Field Activities Elsewhere

By the mid-1960s the pace of development and field experimentation was gathering speed amongst the research groups involved in pioneering this new technique. Evans recorded several activities in his 'progress reports' throughout the period, to 1966. This included work by the AARI in Antarctica during the austral summers of 1964–65 and 1965–66[57] along the route from Mirnyy Station at the coast to Pioneerskaya Station using 211 MHz and a higher-frequency system operating at 440 MHz. The team obtained almost fully continuous ice-thickness recordings with a maximum depth of 2000 m and detected internal ice sheet layering with the latter sounder to a depth of 700 m. George Jiracek of the Geophysical and Polar Research Center at the University of Wisconsin and one of Charles Bentley's doctoral students, using a USAEL MkII 30 MHz device, made radio wave velocity determinations on the Ross Ice Shelf and at Skelton Glacier.[58] The following Antarctic season (1965–66) the same equipment[59] was deployed on the second leg of an oversnow traverse from the South Pole towards the unexplored interior of Dronning Maud Land (SPQMLT II). The traverse was from the Pole of Relative Inaccessibility (PoI)—the point most distant from the coast of the continent—where a Soviet station had been established for a few days during the IGY (see Figure 5.28), towards the Greenwich Meridian then eastwards to Plateau Station. John Beitzel obtained signals from the base of the ice over 80% of the track. The somewhat rudimentary recording system consisted of photographing an A-scope display every 1 NM (1.85 km), as well as using a 'sequence' camera to record a profile. Depths of 3500 m were recorded.[60] The SPQMLT III continued the traverse work from Plateau Station to a point 320 km northeast of the Shackleton Range, and John Clough obtained radio-echo soundings along the first third of the traverse; beyond there only very occasional returns were recorded in thin ice.[61]

[57] Bogorodsky, V V; and Fedorov, B A (1967) Radar measurements of glaciers, *Zhurnal Tekhnicheskoi Fiziki* 37 (4): 781–88.

[58] Jiracek, G R (1967) Radio sounding of Antarctic ice, University of Wisconsin, Geophysical & Polar Research Center, *Research Report Series 67-1*, 127pp.

[59] In the report by John Beitzel (footnote 36) he states he used the USAEL 30 MHz sounder. In Bogorodsky et al. (footnote 43) Bentley states that the SPRI MkII was used on this and the subsequent traverse (QMLT III), p. 67).

[60] Beitzel, J E (footnote 36).

[61] Clough, J W; Bentley, C R; and Poster, C K (1968) Ice-thickness investigations on SPQMLT III, *Antarctic Journal of the United States* 3 (4): 96–97.

3.5 Radio-Echo Sounding Goes Airborne

It was a logical development to move from operating RES equipment in oversnow vehicles to using aircraft—an approach that was both quicker and safer. Oversnow traverses in severely disturbed, crevassed areas had resulted in bad experiences. In September and October 1965 a journey from Halley Bay on the Brunt Ice Shelf to Tottanfjella, a group of mountains in western Dronning Maud Land, had resulted in tragedy. The Muskeg tracked vehicle containing three persons, D P Wild, J K Wilson from BAS, and Jeremy Bailey from SPRI, broke through the ice into a crevasse, and all three were killed. Their loss was a severe and distressing blow. Bailey's death was felt acutely at the SPRI, as he had contributed substantially to the early electronic developments and fieldwork experiments in Greenland. Close encounters also were experienced by American and French teams conducting oversnow work.

In addition to the advantage of speed, the ability to better delimit the lines of sounding from the air made the progression indisputable, notwithstanding the irony that faulty readings on altimeters mounted in aircraft had spawned RES. Bud Waite had experimented using the SCR-718 altimeter from helicopters along a coastal section of Antarctica, and in February 1966 the Soviets flew their 211 MHz and 440 MHz radars in an IL-14 in two experimental, low-level flights (approximately 200 m above the surface) out of Mirnyy. The former was used for the ice depths and the latter to measure the aircraft terrain clearance. The radar travel times were sampled every 1 to 1.5 km, and a maximum ice depth of 2875 m was recorded. Fedorov, head of the group, reported enthusiastically that 'the possibilities afforded by the radar sounding in the measurement of ice thickness . . . are tremendous'.[62]

At the SPRI serious consideration was being given to an airborne system. Robin sought high-level support from the UK Ministry of Defence to see whether it would be possible to operate the equipment from aircraft flying over the Greenland Ice Sheet. This proposal failed to gain acceptance, but an opportunity materialised from a completely different direction—Canada.

[62] Fedorov, B A (1967) Primeneniye aklivnoy radiolokatsii dlya izucheniya antarkticheskikh lednikov (Radar sounding of Antarctic ice), *Informatsionnyy Byulleten' Sovetskoy Antarkticheskoy Ekspeditsii* 62:19–24.

Geoffrey Hattersley-Smith, a glaciologist working for the Defence Research Board of Canada (DRB) was a good friend of Robin and Evans and was a regular visitor to the SPRI. He indicated that Canada was looking to reinforce its political presence in the Northwest Territories and was seeking suitable projects to do so. Undertaking a trial airborne survey to investigate the glaciers in northern Ellesmere Island would, he considered, be highly appropriate (Figure P.3). Hattersley-Smith had been with the Falklands Islands Dependencies Survey (FIDS—the forerunner of the BAS) at Admiralty Bay on King George Island as base leader before moving to Canada. Besides considering the political angle, he was keen to obtain ice-thickness information from a range of Canadian Arctic ice caps and glaciers, as well as the Ward Hunt Ice Shelf. The project was agreed, with the DRB providing a DHC-3 Otter aircraft for the purpose in the spring of 1966.

A major concern was to rig aerials beneath the wings of the aircraft which would have good electrical behaviour and not compromise the mechanical performance of the aircraft through drag or by inducing instability. This would be a recurring issue for all future airborne operations.

The basic principles of antenna construction and use focus on the frequency, the bandwidth (i.e. the frequency range over which the antenna will radiate), gain (the maximum power radiated in a particular direction), polarisation, and impedance (the voltage to current ratio at the antenna). Importantly, the antenna has to be 'matched' to the transmission line or cable that feeds it. For the Otter the arrangement was relatively straightforward: one aerial was used for transmission and the other for reception of radar energy. A good deal was learned concerning the mounting and operation of aerials, and improvements were introduced into subsequent designs.

Just over 18 hours of missions were flown between 10 and 20 April from the air facility at Tanqueray Fjord, the DRB field station on Ellesmere Island. The team comprised Hattersley-Smith and Harald Serson (from the DRB), Gordon Robin, and Stan Evans. Also from the SPRI was Julian Paren, who had joined as a doctoral research student in October 1965. Paren was working on the electrical properties of ice that would be pertinent to understanding better the propagation of radio waves. He was to conduct experiments on the White Glacier on neighbouring Axel Heiberg Island in a separate project but joined Robin and Evans for the first flight.[63] Navigation for the airborne work was relatively crude, depending on line-of-sight

[63] J. Paren, personal communication, December 2020.

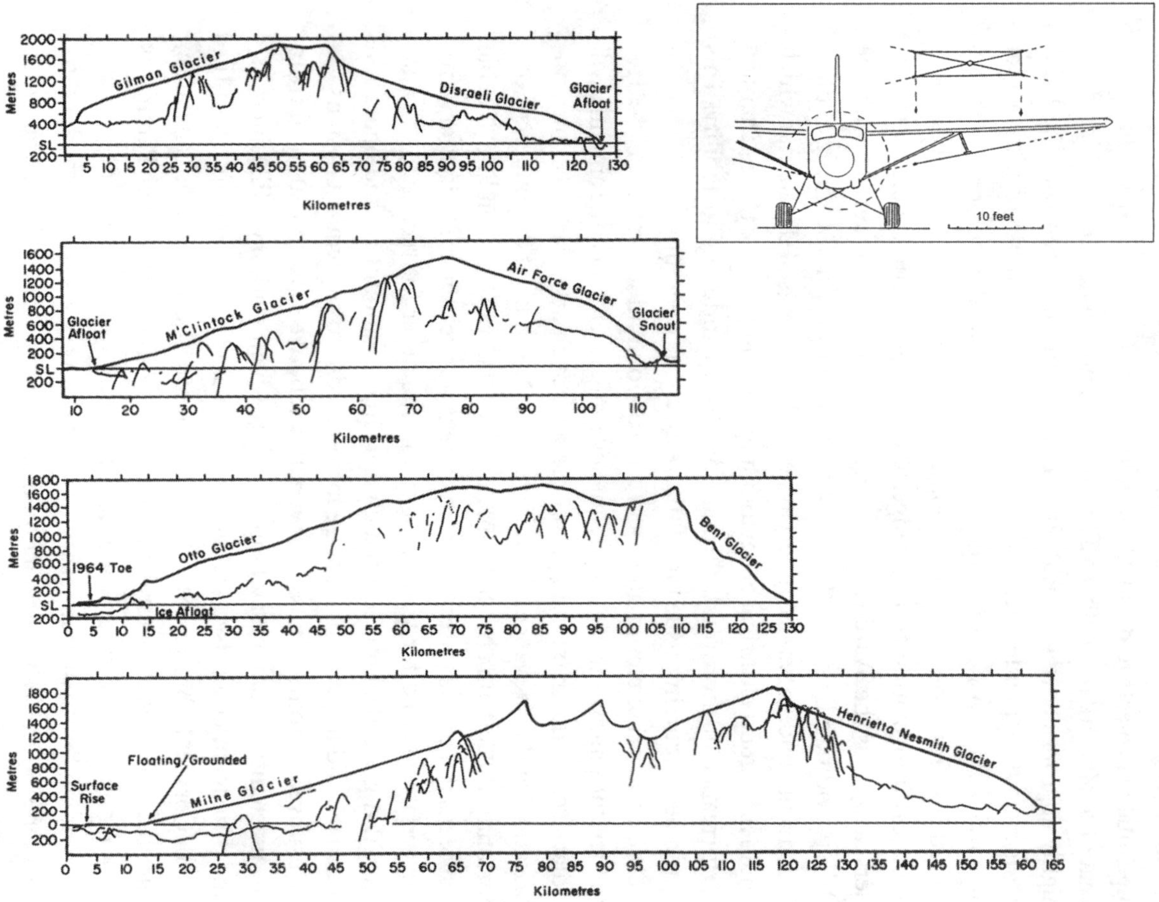

Figure 3.7. Glacier ice thickness profiles from the first airborne testing of the SPRI equipment in Ellesmere Island. Top right: antenna configuration on a single-engine Otter aircraft (From Hattersley-Smith et al. (1969); see footnote 64).

or compass bearings using features on maps and aerial-photo composites. The flights were conducted mostly at 100 knots, and the surveys were 'draped'; that is, they were flown at relatively constant height above the ice surface, in this case 500 m.

The results were an excellent demonstration of the capability of the SPRI MkII sounder to penetrate cold, high-latitude glaciers and small ice caps (Figure 3.7[64]). The primary drawback was the reflections generated by adjacent mountain sides and other rock features. Such oblique returns arise from the relatively wide beam of the radio sounder. The lower the frequency, the wider the beam and the greater the illuminated area. Over the upper areas of the glaciers, where the valleys were narrower and the adjacent mountain sides steeper and higher, there was a good deal of interference from these features. Some of these echoes could be identified readily and thus eliminated; others could be accounted for by examining the geometrical setting—aircraft height, the echo-delay time, and comparing transmission in air (300 m μs^{-1}) versus in ice (~169 m μs^{-1}). Much later a digital method of recognising problematic echoes from valley sides was developed by researchers at the SPRI.[65]

This was an extremely effective experimental season and a milestone in the development of airborne RES. Evans remarked: 'We flew down the Gilman Glacier. I had a steady echo down the centre of the glacier towards the snout, . . . and I saw the depth decreasing, just exactly as you would expect. . . . And I saw it peter out on the snout and then we just got the rocky ground below. And I think I can honestly say that was the first time I really believed in my heart that what we were recording was the depth of the ice'.[66]

Evans quickly wrote up the research in Canada before returning to Cambridge, where he immediately typed up the draft and fixed the diagrams and photographs. Then, he describes, 'I dropped this paper addressed to . . . the editor of *Nature*, who was John Maddox at the time, into a letterbox in Cambridge. . . . [I]t appeared in *Nature* with a notice on the front cover

[64] Hattersley-Smith, G; Fuzesy, A; and Evans, S (1969) Glacier depths in Ellesmere Island: Airborne radio-echo sounding in 1966, *Operation Hazen Report No. 36*, Ottawa: Defence Research Board of Canada, 23pp.

[65] Benham, T J; and Dowdeswell, J A (2003) A simple visualization method for distinguishing between subglacial-bed and side-wall returns in radio-echo records from outlet and valley glaciers, *Journal of Glaciology* 49:463–68.

[66] Dr Stan Evans (footnote 45).

within two weeks.[67] I think he [Maddox] subsequently told me he didn't send it to any referees; he just put it straight into print. So that's probably the most exciting publication I can lay claim to and, you know, this was a real announcement'.[68] RES had officially become airborne.[69]

Hattersley-Smith, Evans, and research assistant Anne Fuzesy published a more extensive analysis of their findings a couple of years later. They reported that beneath the ice caps, which reached a maximum depth of 900 m, the topography was very irregular.[70] The main glaciers draining the highland ice had maximum depths of between 600 and 700 m. For the tidewater glaciers, it was possible to identify the grounding line—where the glacier front on entering the sea comes afloat—from a significant change in echo strength. On the Ward Hunt Ice Shelf, small ice domes (ice rises) were shown to be grounded below sea level, and thicknesses of between 70 and 80 m were recorded to the east of Ward Hunt Island.

In July 1966 a USAEL team performed trials in Greenland using their 30 MHz system mounted onboard a Super Constellation aircraft and measured depths in excess of 3300 m.[71] They conducted four 8-hourflights out of Thule and Søndrestrøm Fjord extending from the south to the north of the ice sheet, as well as transecting from east to west. One flight resulted in identifying the main source of error as uncertainty in the dielectric constant and hence the electromagnetic velocity in ice.[72] These operations proved to be very promising; they were conducted from a much larger aircraft and ranged more extensively than the SPRI campaign in Canada, so it was surprising that USAEL did not pursue this research further nor did it publish any additional information or analyses. Because USAEL is a military organization, it can only be presumed that priorities changed, and staff moved on to other projects. The responsibility in the West for the development of RES now lay almost exclusively with the SPRI.

[67] Evans, S; and Robin, G de Q (1966) Glacier depth sounding from the air, *Nature* 210 (5039): 883–85.

[68] Dr Stan Evans (footnote 45).

[69] When this article was published in May 1966 the Soviet experimental flights in Antarctica in February 1966 had not yet been reported.

[70] Hattersley-Smith, G; et al. (footnote 64).

[71] Walker, J W; Pearce, D C; and Zanella, A H (1968) Airborne radar sounding of the Greenland ice cap: Flight 1, *Geological Society of America Bulletin* 79:1639–46.

[72] Pearce, D C; and Walker, J W (1967) An empirical determination of the relative dielectric constant of the Greenland Ice Cap, *Journal of Geophysical Research* 72 (22): 5743–47.

At the other end of the world preparations were being made to undertake the first airborne trials of the SPRI MkII equipment in Antarctica by the British Antarctic Survey (BAS). Charles Swithinbank had returned from the University of Michigan to accept a 'split' position at the BAS (paying his salary) but physically located within the SPRI in Cambridge. He was keen to use the new RES system and organised for it to be flown in another DHC-3 Otter and in a Pilatus Porter. Swithinbank and David Petrie, who had assisted Michael Walford at Halley Bay, flew some 75 hours during the summer season between December 1966 and February 1967 from the BAS base at Adelaide on the west coast of the Antarctic Peninsula (Figure 3.8). Swithinbank has given a detailed personal account of these flying operations and the day-to-day circumstances of operating in the Antarctic in his autobiographical book, *Forty Years on Ice*[73].

Flights were made over a range of glacial phenomena—the extensive ice sheet of the Antarctic Peninsula and smaller adjacent ice caps, as well as outlet glaciers flowing from them, valley glaciers, and ice-covered islands. Attention was paid to floating ice shelves and their ice rises. Swithinbank discovered that the warmer ice at low levels in the north of the Peninsula increased the signal absorption, and few reliable bottom echoes were obtained. However, results were much better at higher elevations (2000 m asl) and further to the south (beyond 72.5°S) with colder ice conditions. The thickest ice on the ice sheet was 600 m, along the crest of the Peninsula. Excellent results were obtained on the ice shelves, with continuous and strong returns, and a thickness of 700 m was recorded at the inner zone of the Larsen Ice Shelf on the western side of the Weddell Sea. A continuous longitudinal profile was made of the entire George VI Ice Shelf, 450 km long and 475 m thick some 80 km from its southern extremity.[74]

These various airborne forays were are a prelude to the future direction of ice-thickness sounding. However, the aircraft were small, slow, and had little range. In contrast, the Antarctic Ice Sheet was of continental dimensions (14 M km^2)—somewhat larger than the whole of Europe ((10.2 M km^2) or the conterminous United States (8.08 M km^2)—and demanded a commensurate scale in the size and capability of airplanes that could be used. The

[73] Swithinbank, CWM (1998) *Forty Years on Ice*, Lewes, East Sussex: The Book Guild, 228pp.

[74] Swithinbank, CWM (1967) Radio-echo sounding of Antarctic glaciers from light aircraft, International Union of Geodesy and Geophysics; IASH, General Assembly Bern. Commission for Snow and Ice. IASH Publication 79:405–14.

Figure 3.8. BAS radio-echo sounding operations in the Antarctic Peninsula, 1966–67. Left to right: David Petrie, Bob Vere (pilot), and Charles Swithinbank stand in front of the BAS single-engine Otter aircraft. (Courtesy SPRI).

challenge then, certainly for the SPRI, was how to make that leap forward. It was clear that the RES system flown from fixed-wing aircraft had the potential to rapidly gather information of ice thickness and map the base in unprecedented detail over vast tracts of Antarctica. The question was, how would this tantalising scientific goal be achieved and by whom?

4

Flight into the Unknown

Long-Range Antarctic Campaigns Commence

Progress in the mid- to late 1960s with the design and testing of the new radio-echo sounding equipment at the SPRI was brisk and exciting. The results obtained from ground-based and airborne operations and experiments conducted in Greenland, Canada, and Antarctica were tantalising. Robin recognised that their research was now at a turning point. The challenge lay in the interior of the great ice sheet of Antarctica. To date, the experimentation had only pecked at its margins and not in very thick ice. Robin was convinced that flying the radio-echo sounder across Antarctica would transform our understanding of this ice mass and the continent beneath. He recalled his own tribulations with seismic sounding and the experiences of others on oversnow traverses during and since the IGY. The exhilaration of obtaining good results was tempered by mechanical breakdown of vehicles, poor weather, and the continuous threats posed by crevasses, blue ice swept bare of snow, and sastrugi (wind-eroded wavelike ridges of snow). Only a few hundred depth soundings had been obtained by seismic methods after 10 years of international activity on a vast continent, so the prospect of gathering orders of magnitude more data on a single flight was a scientific prize of the greatest order. But there was a problem. Swithinbank's reconnaissance work on the Antarctic Peninsula had signalled the limitations of small, short-range aircraft, and the BAS did not operate any long-range airplanes with which to support Robin's vision of a continental scale survey.

4.1 International Cooperation

At that time, in the 1960s, the only countries possessing aircraft in Antarctica with a capability of flying over large tracts of the continent were the Soviet Union and the United States. With the new Antarctic Treaty in place there was, perhaps, the prospect of international collaboration to achieve this goal. Nevertheless, Robin was under no illusions that developing a joint research project with the Soviets would be easy. Their aircraft were not reliable, their operational bases could be accessed only by long sea voyages, and, at the height of the Cold War, despite the Treaty, technical collaboration would be problematic, especially back in the USSR. Besides that, language would be a considerable handicap in operations. Furthermore, under Bogorodsky the Soviets were moving ahead with their own radar plans, and given the location of their bases, perhaps it was as well to let them make whatever progress they could—the Antarctic was big enough for several groups to work in.

This left the Americans. Through his SCAR connections and background as a geophysicist, Robin had good access into that arm of the US science system that was coordinating American activities in the Antarctic. The Office of Antarctic Programs (OAP) was located in a number of offices in midtown Washington—at the corner of 19th and G Streets—and was a component but distinct part of the National Science Foundation (NSF) and at that time was headed by Dr T O Jones. The chief of the science staff was Dr Albert 'Bert' Crary.

Crary, a highly respected and influential scientist, had been responsible for some of the first seismic sounding work by the US during and immediately after the IGY (Figure 4.1).[75] He was aware of the radar experiments conducted by Bud Waite and had been partly responsible for stimulating the international experiments in northwest Greenland. By 1966 he had conceived of an 'airborne laboratory' that would support a range of remote-sensing systems, including a radio-echo sounder. Naturally, the USAEL was identified as the principal organisation to undertake the ice-thickness measurements, some funds were earmarked, and aircraft were allocated via the US Navy's Squadron VXE-6 that operated in Antarctica.[76]

[75] Crary, A P (1963) Results of United States traverses in East Antarctica, 1958–1961, *IGY Glaciological Report No. 7*, IGY World Data Center A: Glaciology, New York: American Geographical Society, 143pp.

[76] Turchetti, S; et al. (2008) (footnote 48).

Figure 4.1. Bert Crary examining a seismograph record on a traverse in 1959. (Courtesy National Academy of Sciences Archives).

In his autobiography Charles Swithinbank recounts his return from the Antarctic Peninsula following the airborne RES survey and his trip to Washington DC, in the northern spring of 1967, where he shared the new and exciting results of his RES work with Crary at the NSF. Crary was impressed by the clear visual representation in Swithinbank's photographic cross sections. These must have counteracted some of the negative criticism by one or two US investigators of the SPRI system and of Evans' insistence on having a visual record.[77]

Crary reported to Swithinbank that he had been discussing with Gordon Robin a possible collaboration which would involve mounting the SPRI equipment in a US aircraft the next season. Robin and Crary were good friends—both experienced polar geophysicists. In 1966 Robin had approached Crary to enquire about the availability of airplanes operating from the main American base at McMurdo Sound in the Ross Sea and shared his ideas and plans with a receptive colleague. Over the ensuing months an agreement was reached for a pilot season of airborne echo sounding with

[77] Turchetti, S; et al. (2008) (footnote 48).

the SPRI mounting its apparatus in one of the US Navy's C-121 Super-Constellation aircraft, a development of the greatest significance.

Soon, SPRI personnel were liaising with the US Navy on the necessary modifications to the aircraft. The generosity of the NSF was extraordinary, as it met the costs not only of operating this large airplane in Antarctica but also of making the necessary structural modifications. Robin, in the meantime, applied successfully for a substantial research grant from the UK Natural Environment Research Council to support this developing collaborative activity.[78] In late August the US Navy Special Development Squadron (VXE-6) that undertook the airborne support activities in Antarctica for the OAP flew one of its two remaining Super Constellation C-121 aircraft (named 'Phoenix') across the Atlantic to RAF Lakenheath, east of Cambridge, to liaise with Evans and the SPRI group on the necessary modifications. At the same time the aircraft commander, Lt Cmdr J 'Jake' K Morrison, and the crew that would operate in Antarctica would get to know the plans for the coming season, which were significantly different from their usual flying tasks.

The Americans were operating the C-121 (J model) primarily as a passenger transport aircraft, between Christchurch in New Zealand, their staging base, and McMurdo Station. The aircraft was old and slow, with piston engines, and did not have ski capability, so its schedule was restricted to the period when the smoothed sea-ice runway near McMurdo was operational—the first half of the season—up to the beginning of January. These factors placed the C-121J—unlike the ski- equipped C-130 Hercules the US Navy had available—at greater risk flying over the ice sheet; indeed, very few missions by this aircraft had been undertaken over the inland ice. If there was an emergency, the plane would be unable to land at one of the remote interior stations or on the ice sheet; it would have to return to McMurdo. Robin accepted all these caveats and risks and proceeded with preparations for a field programme in the 1967–68 austral season.

4.2 Plans for Antarctic Season 1967

The radio-echo sounding technical details were under the direction of Evans, who was assisted by Beverley Ewen-Smith. 'Bev' had joined SPRI as a research student filling the gap left by the untimely and tragic death of

[78] NERC Research grant (1 August 1967), £30,600.

Jeremy Bailey. He had recently graduated in physics from Manchester University and was quickly immersed in the planning for the Antarctic operation. Robin drafted preliminary flight plans so as to test the equipment in a variety of glaciological situations. From the results of earlier seismic sounding it was clear that very deep ice existed in West Antarctica—in the Byrd Subglacial Basin—and would need to be investigated. The Ross Ice Shelf (the size of France) spreading out from Marie Byrd Land to float over the Ross Sea would be a fascinating part of Antarctica to examine. East Antarctica, the highest and most remote area, with only a handful of seismic depths, many of dubious validity, would undoubtedly reveal much new information. The opportunities were boundless, the excitement palpable. Trials would also need to be undertaken of the technical performance of the antenna configuration and the recording systems of the radio-echo sounding equipment. It would be a challenging and hugely important field season.

One critical matter was seen to be navigation, firstly ensuring the aircraft followed planned lines as closely as possible and, secondly, precisely reconstructing the flight track afterwards. In the late 1960s the technical means of achieving these aspects were fairly rudimentary, as inertial and satellite positioning systems were not yet widely available. For the first requirement, the normal mode was dead reckoning. The navigator and the pilot would have a feed of various in-flight parameters by which to adjust the course of the flight in the face of variable atmospheric conditions such as cross winds and pressure changes that might cause the plane to drift off course or alter its groundspeed.

One navigational aid was a Doppler system that integrated the speed derived from measuring the Doppler effect of echoes from three or four beams directed downwards from the aircraft. Comparing the returns from left- and right-hand beams would give the drift angle, and the shift between forward- and aft-facing beams would give the groundspeed. In practice, over several years of flying campaigns in Antarctica the Doppler system was frequently inoperative, as it typically required a demanding setup routine. Also, the system would lose 'lock' quite regularly over the ice surface, as it was predominantly smooth.

For much of the time navigation was reliant on more basic dead reckoning supplemented by astronomical observations using a sextant. The navigator would make regular sightings and calculate the position from a standard book of tables. The problem in the Antarctic was that during the austral

Figure 4.2. A sample of a trimetrogon horizon-to-horizon image array from three large-format (9″×9″) cameras installed in C-121 and later in C-130 aircraft. (Courtesy US Geological Survey).

summer, with 24-hour daylight, it was impossible to make star sightings, so only 'sun shots' could be made, which yielded less accurate position lines. Such observations could be supported by flying over features of known location—*fixes*—to pin or 'fix' the track. There were very few geographically known points to be observed on flights over the ice sheet, so the number of fixes was extremely small.

However, the C-121J had been configured for aerial photographic mapping, which meant that accurate time-based images could be taken of known features (mainly in coastal and mountainous locations) for use as fixes. Furthermore, the high-quality imagery produced was extremely valuable for study of glaciological features of ice shelves, ice rises, and outlet glaciers. The photographic arrangement was known as *trimetrogon* and comprised three cameras—two oblique-looking and one vertical-looking (Figure 4.2).

The resulting images overlapped, so there was horizon-to-horizon coverage. The high-quality 9″×9″ negatives and positive prints allowed precise fixes to be obtained. In the case of the verticals this was simply a matter of identifying the feature on the image and its corresponding coordinates on the relevant maps. The oblique photographs necessitated a different technique: *resectioning*, which requires the height of the aircraft, the angle of incidence, and the focal length of the camera lens. Rays are drawn from identified features to determine the nadir (the point vertically below the observer/camera/aircraft) and thus the navigation fix. Bev Ewen-Smith developed a routine for calculating the aircraft position by this method, and it worked well when the plane was flying in mountainous terrain.[79]

[79] Ewen-Smith, B M (1972) Automatic plotting of airborne geophysical survey data, *Journal of Navigation* 25 (2): 162–75.

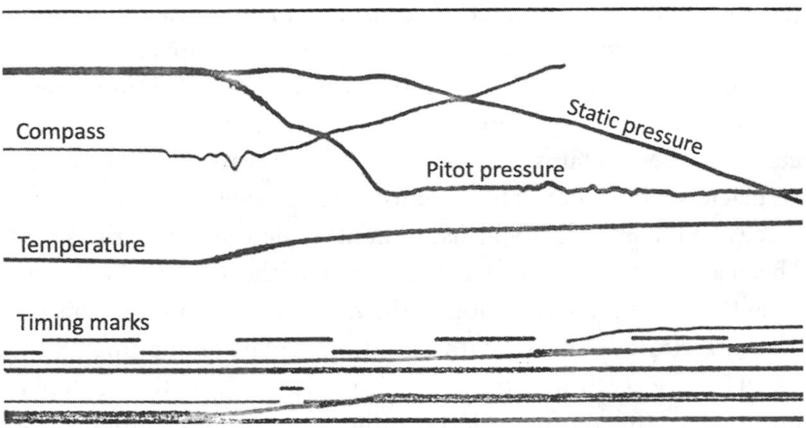

Figure 4.3. Example of a section of the SFIM (flight-recorder) trace. (Courtesy B M Ewen-Smith).

Navigation fixing of this sort required a collection of the best maps for the areas of Antarctica over which flights were to operate. The US Geological Survey had produced an excellent series of 1:250 000 scale maps of the Transantarctic Mountains and coastal ice shelf and the mountain areas in West Antarctica. The trimetrogon system was later installed in a C-130 Hercules aircraft that the SPRI used in two further seasons and proved equally valuable.

For the later reassembly of the flight route, the data logged on flight recorders could be used, but at the time, these systems were primitive. The SPRI had chosen to install its own flights recorders (a French-manufactured model, SFIM), which required inputs from all the important aircraft flight instruments (airspeed, static and pitot pressure, outside air temperature, compass heading, terrain clearance). These were recorded as traces on reels of 60mm photographic paper alongside the same time marks and event marks on the radio-echo films (Figure 4.3[80]). Back in Cambridge, following processing, the traces could be digitised and the track re-constituted.

The SPRI team for this important programme comprised Gordon Robin, Stan Evans, Bev Ewen-Smith, and Charles Swithinbank. Swithinbank had been asked to join the group because his experience of undertaking RES

[80] Ewen-Smith, B M (1971) 'Radio-echo Studies of Glaciers', PhD thesis, University of Cambridge, 157pp.

from light aircraft in the Antarctic Peninsula, with the same equipment, was relevant. A few years earlier, whilst working at the University of Michigan, Swithinbank had undertaken ground-based observations of the giant outlet glaciers streaming from the East Antarctic Ice Sheet through the Transantarctic Mountains (Figure P.4). He had measured the flow rates of these glaciers along transects close to mouths where they issue into the Ross Ice Shelf. Using gravity measurements to determine ice thickness, he had been able to calculate the ice discharge into the Ross Ice Shelf, demonstrating the glaciers' contribution to the flow and behaviour of this largest of the Antarctic ice shelves. Swithinbank wanted better ice-thickness data along his transects to improve his research, so one of the objectives of Robin's programme was to undertake flights across these outlet glaciers.

The 1967–68 season was undoubtedly the turning point in the development of airborne RES and, therefore, to a large extent in our understanding of the Antarctic Ice Sheet and its hidden continent. The analyses of the Greenland oversnow campaign had generated a body of research insights and deepened our knowledge and appreciation of what this new technique could offer; it had set a new and rigorous standard for future work. Likewise, this first Antarctic long-range sounding season triggered an explosion of new studies, an ever-expanding intellectual horizon, and a plethora of scientific publications. At this point RES went viral! Of course, at the time, for those immersed deeply in the fieldwork, technical developments, and the analysis of the data, it did not seem like that. Rather, it was a continuous and exciting series of discoveries with leaps of faith, testing out new hypotheses, puzzling over curious records, finding new problems, discovering dead ends and confronting frustrations, piecing together new ideas and new insights, and attempting to write up these developments in a consistent and robust fashion.

It was clear that new avenues were opening up—characterising the electromagnetic properties of polar ice, radio wave propagation phenomena, radio systems design including antenna performance, internal layering, basal reflection types, subglacial morphology, ice sheet dynamics, the study of ice shelves, navigation problems, data recording; the list was varied, long, and getting longer. These aspects and the programme's many results will be discussed in later chapters, examining their importance and role in shaping future research and operations. In the meantime, the first major Antarctic season beckoned. Swithinbank gave a personal perspective of airborne RES within the continental interior of Antarctica in a chapter of his book

An Alien in Antarctica.[81] Similarly, Bev Ewen-Smith recorded his impressions during an audio interview.[82]

4.3 Antarctic Operations

In Christchurch, New Zealand, the US base located at Harewood Airport was the preparation and forwarding station for the US Antarctic Research Program (USARP). The first arrival there of the SPRI team was Ewen-Smith, who had flown via Washington and Hawaii on a US military transport.[83] Here he met up with the aircrew and its commander, Lt Cmdr J K Morrison of the Super Constellation C-121J 'Phoenix' and spent much of his time installing the equipment. A rack and seating were provided for the RES units and the operator. The antenna was manufactured by the New Zealand National Aircraft Corporation at its facility in Christchurch and comprised two terminated three-quarter-wavelength folded dipoles. These were mounted beneath the tail of the aircraft, and each just fitted between the vertical fins but were positioned on struts to ensure adequate separation of a quarter wavelength from the tailplane surface (Figure 4.4). Robin, Evans, and Swithinbank arrived about four weeks later by commercial air.

Testing of the RES equipment and the antenna followed by flying out over Christchurch to the Pacific coastal waters. On the first occasion the drag caused buckling of the struts and loss of cable and insulators somewhere over the city! (Figure 4.5). Fortunately, there was no report of damage.

Shortly afterwards, the SPRI team, minus Evans[84], flew to McMurdo Sound on 5 December 1967. This meant the team lost a significant part of

[81] Swithinbank, CWM (1997) *An Alien in Antarctica*, Blacksburg, VA: McDonald & Woodward, 214pp.

[82] Ewen-Smith, B M (2019) University of Cambridge, SPRI, Oral History collection, RS2020_3_EwenSmith-Bev_ACCESS.mp3.

[83] We shall discover in later chapters that this could be a challenging and eventful journey in its own right.

[84] Stan Evans told the author that the reason he did not continue down to 'the ice' was because of a rather silly bureaucratic faux pas. The NSF required that all Antarctic personnel undergo and pass a full medical examination. The US Navy medical forms were comprehensive but also appeared somewhat illogical and incongruous. Evans completed his form in a 'cavalier' manner, and consequently, he was rejected. By the time he learned of this, it was too late to re-submit a more moderated medical form. It should be said that in the next season (1969–70) he had no medical issues to prevent his participation. This incident demonstrates the consequences of not following military procedures, however absurd they appear to the academic mind!

When I was preparing for my first Antarctic season, in 1969, Evans sent the requisite medical forms with a covering note that read: 'Dear Drewry. You will be tempted to mock some of the questions on these forms. We are just as tempted, but we must appear to take them seriously.' (27 May 1969)

Figure 4.4. Evans (left) and Swithinbank with Lt Cmdr Morrison at Harewood Airport. Above them is the RES 35 MHz antenna arrangement beneath the tailplane of the C-121J 'Phoenix'. (Courtesy B M Ewen-Smith).

Figure 4.5. Robin, Ewen-Smith (centre), and Evans (right) discuss damage to the antenna beneath the C-121J 'Phoenix' port tailplane at Harewood Airport, Christchurch. Note the bent strut below the outboard fin, and collapsed cabling. (Courtesy SPRI).

Figure 4.6. Robin (left) and Ewen-Smith stand in front of C-121J Super Constellation 'Phoenix' showing RES aerials mounted beneath the tailplane. (Courtesy SPRI).

its electronic expertise. Ewen-Smith commented: 'I was the only person present with any technical knowledge of the RES hardware and felt a heavy burden of responsibility for ensuring the season was a success'.[85] Because of the limited period of operation of the sea-ice runway there was little time to undertake the range of sounding that had been planned to thoroughly test the equipment (Figure 4.6). The SPRI group were ably supported by Jerry Huffman—the NSF representative in McMurdo, as well as by the air wing of the US Navy in Antarctica, VXE-6.

At that time the US Navy was responsible for flying fixed wing aircraft and a number of helicopters in support of not only the American flying programme but also much of the US logistical infrastructure in Antarctica. This function included aspects of ground transportation, berthing, and provision of a range of services such as catering and the limited entertainment in the several 'clubs' positioned around McMurdo Station.

This last feature was an interesting cultural component, as each of these clubs related to the various ranks of the Navy personnel—the Enlisted Men's Club, the Chiefs' Club, and the Officers' Club. The clubs provided relaxation for off-duty personnel—they were mainly drinking establishments and places for showing films, playing pool and shuffleboard, and other pastimes.

[85] Ewen-Smith, B M (footnote 82).

Figure 4.7. SPRI MkII radio-echo sounding system installed in C-121J aircraft 'Phoenix'. Ewen-Smith operates an oscilloscope. (Courtesy SPRI).

The bars were a unique feature of the Antarctic operation, since elsewhere in the world the US Navy was and remains 'dry', along with the rest of the US armed forces. Special dispensation was given to the Antarctic operation owing to its mix of civilian and military participants, and that under the Antarctic Treaty, strictly military activities are forbidden except in support of scientific programmes. Scientists were in an interesting position on these bases that were run almost exclusively along service lines. Treated as 'honorary officers', they could move freely among the various activities and ranks and thus helped ensure that bases and programmes performed in as harmonious and integrated a fashion as possible.

The US Antarctic support programme was called COMNAVSUPPFORANTARCTICA (Command Naval Support Force for Antarctica) and was headed in 1967 by Navy Rear Admiral J L Abbott. His role was to run the operations in McMurdo and at other stations in liaison with the NSF Representative. This resulted, at times, in an uneasy relationship between civilian and naval procedures and priorities—science was paramount, but Navy tradition and procedures pervaded all operations.

For the SPRI group arriving on 5 December no time was to be lost, and the first flight took place only hours after landing on 6 December, on a

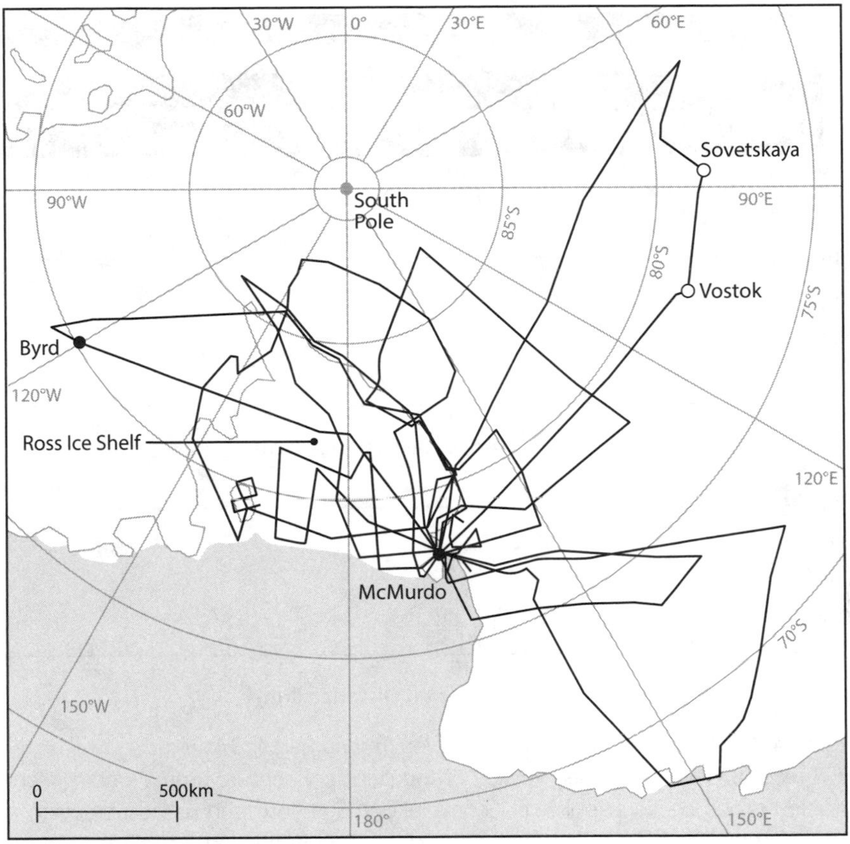

Figure 4.8. RES flight lines conducted in the 1967 season.

routing out to the west of McMurdo. Robin and Swithinbank variously directed the flight from the co-pilot seat on the flight deck, whilst Bev Ewen-Smith ran the radar and recording systems, both radar and navigational, back in the belly of the aircraft, sharing the confined space with a large rubber bladder that provided additional fuel for extended sorties (Figure 4.7).

Immediately on their return, the RES 35mm film records were taken to the Navy Photographic Laboratory, where staff were on standby to process them. The team members were well pleased with these first results over a variety of terrain. Figure 4.8 is a map of flight lines.[86]

The next two flights were to explore the radar performance over the Ross Ice Shelf and to seek opportunities to measure ice-depth cross sections of

[86] Robin, G de Q; Swithinbank, CWM; and Smith, BME (1970) Radio-echo exploration of the Antarctic Ice Sheet, in Gow, A J; et al. (eds.) *International Symposium on Antarctic Glaciological Exploration (ISAGE)*, IASH Publication 96:97–115.

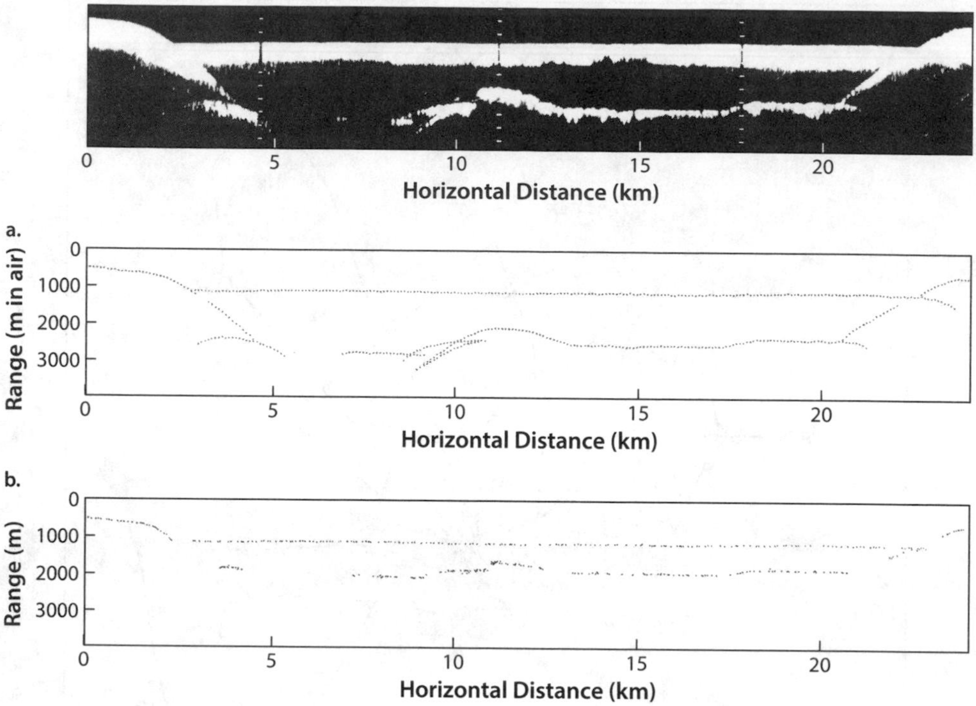

Figure 4.9. Cross section of Nimrod Glacier, Transantarctic Mountains. Top: original RES record. Centre: digitised version of the principal echo returns. Bottom: reconstructed cross section applying deconvolution/migration techniques (removing effects of the wide beam). The method is described in section 6.5. (From Harrison (1970); see footnote 149).

the major outlet glaciers flowing through the Transantarctic Mountains into the ice shelf, to contribute to Swithinbank's work (Figure P.4). In both regards the operation was successful if somewhat testing for the pilots, as the cross sections required the aircraft to fly at low level, about 300 m above the ice surface, between the steep cliffs and mountainous edges of the glaciers. The flight took in the Scott, Amundsen, Liv, Beardmore, and Nimrod Glaciers, and the team were gratified with the results; they obtained more or less continuous records on the ice shelf and some very good valley cross sections (Figure 4.9). The next day the aircraft flew south to the great defile occupied by Byrd Glacier—the fastest moving in the Transantarctic Mountains and with the greatest discharge of ice.[87] The pilot flew an exciting transect,

[87] Measured velocities have varied over several decades from 850 m a^{-1} (1978–79) (Brecher, H (1982) Photogrammetric determination of surface velocities and elevations on Byrd Glacier, *Antarctic*

banking steeply only at the last minute at Horney Bluff, a sheer 3700 ft cliff, to the consternation of many onboard! Later, the team flew down and completed cross sections of the Mulock and Skelton Glaciers before returning to McMurdo Sound.

On 19 December the SPRI team undertook one of their longest flights, penetrating deep into the interior of East Antarctica. The route was to make soundings in the thick, cold ice of the central plateau, first by way of the Soviet station Vostok. This base lies at an altitude of 3700 m asl and some 1400 km from the coast at Mirnyy; it has the dubious honour of being the coldest place on Earth, with temperatures having been measured at −89°C (−128°F) by intrepid Soviet meteorologists. Thick ice of the order of 3000–3500 m had been encountered sporadically during the flight en route to Vostok, and weak echoes were discerned at the base between 3760 m and 3810 m, which compared favourably with a seismic determination by Andrei Kapitza of 3700 m made during the IGY.[88] Further west beyond Vostok the ice thinned considerably, and the subglacial relief became so distinctively mountainous that Robin described it as 'alpine' in character.[89] This rugged highland terrain with peaks at 2600 m above sea level confirmed the presence of the Gamburtsev Mountains, which had been inferred on an earlier Soviet oversnow traverse principally from gravity measurements, their seismic results being generally unreliable. The flight extended further inland to overfly the abandoned Soviet Sovetskaya Station, temporarily occupied during the IGY, where exceptionally deep ice of 4200 m was measured by the RES equipment (Figure 4.10). In comparison, the earlier Soviet seismic results gave only 1830 m.[90] Furthermore, the very strong echo at this depth was unexpected and should have required an additional 10–15 dB of system sensitivity (termed *system performance* in earlier RES

Journal of the United States 17 (5): 79–81) to 650 m a^{-1} in 2000–2001 (Stearns, L A; and Hamilton, G S (2005) A new velocity map for Byrd Glacier, East Antarctica, from sequential ASTER satellite imagery, *Annals of Glaciology* 41:71–76). The total discharge has been estimated at 22.32 ± 1.72 km^3 a^{-1}. Stearns and colleagues re-measured the velocity between 2005 and 2007, revealing a 10% increase they attributed to discharge of 1.7 km^3 of subglacial water some 200 km upstream of the grounding line (Stearns, L; Smith, B; and Hamilton, G (2008) Increased flow speed on a large East Antarctic outlet glacier caused by subglacial floods. *Nature Geoscience* 1:827–31).

88 Kapitza, A P (1960) New data on ice thickness in the central regions of Antarctica, *Informatsionnyy Byulleten' Sovetskoy Antarkticheskoy Ekspeditsii* 19:10–14.

89 Robin, G de Q (1969) Long-range radio-echo sounding flights over the Antarctic Ice Sheet, *Geographical Journal* 35 (4): 557–59.

90 Sorokhtin, O G; Avsiuk, Yu N; and Koptev, V I (1959) Rezul'taty opredeleniya moshchnosti lednikovogo pokrova Vostochnoy Antarktide, *Informatsionnyy Byulleten' Sovetskoy Antarkticheskoy Ekspeditsii* 11:9–13.

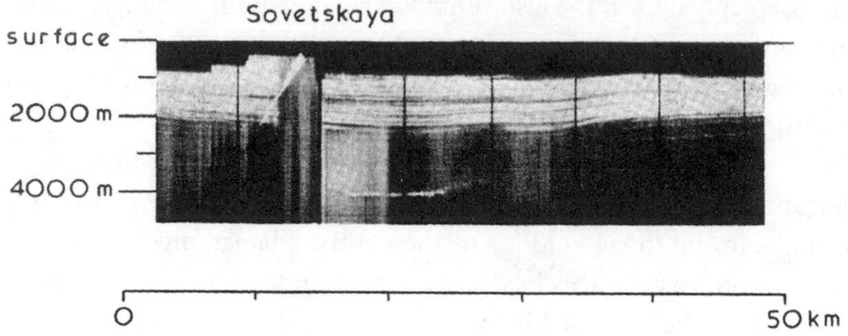

Figure 4.10. RES profile of the 'lake' at Sovetskaya recorded in December 1967. Note internal layers to a depth of 2200 m and a strong, continuous, and near-horizontal return from the base of the ice, interpreted as a thick water layer. The oblique echo is the return from the surface building.

terminology).[91] A very high reflection coefficient at the ice base (i.e. with minimal reflection loss) was the only way to explain this paradox—and that implied an ice/water interface similar to that found beneath the floating ice shelves. Robin and his colleagues argued that Sovetskaya must be floating on a water layer at least 1 m thick.[92] The surprising presence of water at the bed of the East Antarctic Ice Sheet is a topic we shall return to as one of the many discoveries of RES in chapters 7 and 16.[93]

The flight continued further into East Antarctica, across the highest part of the great dome of the ice sheet, following the line of an earlier Soviet over-snow traverse that led to another of their temporary camps located at the Pole of Relative Inaccessibility. Once more, comparisons indicated considerable underestimation by the seismic sounding depths. It is worth repeating that the continuity of the radio-echo records, and time integration of the slow-moving photographic film, provide a powerful tool for the identification of bottom signals even in areas of patchy or weak returns.

This, the lengthiest of all the season's RES flights, resulted in a near emergency being declared by the air controllers in McMurdo when the aircraft failed to keep its regular radio contact for more than three hours owing to equipment failure. Swithinbank provides a vivid account of this situation

[91] The system sensitivity (SNR) describes the ability of a radar sounder to detect echoes and is defined as the ratio of the transmitted power to the minimum detectable received power at the antenna.

[92] Robin, G de Q; Swithinbank, CWM; and Smith, BME (1970) (footnote 86).

[93] Bell, R E; Studinger, M; Fahnestock, M A; and Shuman, C A (2006), Tectonically controlled subglacial lakes on the flanks of the Gamburtsev Subglacial Mountains, East Antarctica, *Geophysical Research Letters* 33, no. 2, https://doi.org/10.1029/2005GL025207.

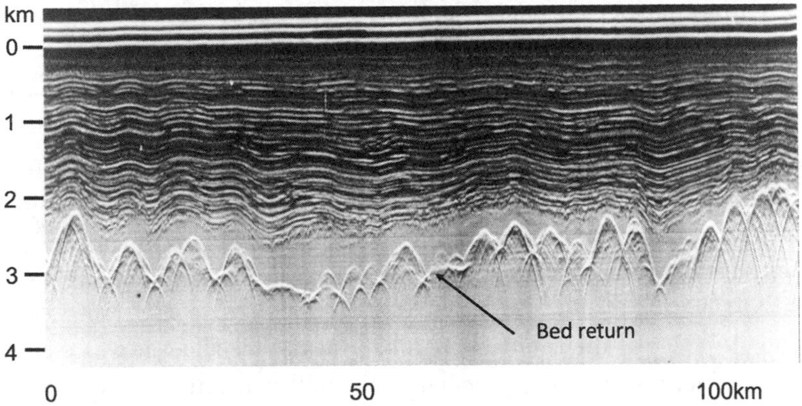

Figure 4.11. RES image (TUD 60 MHz) acquired in the 1974–75 Antarctic season from central East Antarctica showing internal reflecting horizons (layering).

and the somewhat tense aftermath for the plane commander who had pushed the scientific mission to, and perhaps beyond, the regulation limits.[94] Bev Ewen-Smith describes briefly that as it 'was six hours or more back to McMurdo, . . . by the time we came within VHF range, they had assumed the worst and there were Hercules' and helicopters searching our track for the wreckage. As soon as we landed, the Captain was whisked off to an interview with the base commander who had assumed, since we were so long without comms, that the Captain had not followed "turn-back rules". The matter was sorted out and no blame apportioned'.[95]

An important feature identified on many of the radar records from East Antarctica in particular but also in parts of Marie Byrd Land was the presence of internally reflecting horizons (Figure 4.11). Observed along several tens of kilometres of the Tuto Trail in northwest Greenland, and now in Antarctica from an airborne platform, they appeared as stacked 'strata' extending through several hundreds if not thousands of metres of ice and seen to extend over 100 km or more. In some places where the sub-ice surface was relatively level the internal horizons would display a similar subdued pattern. Over terrain with greater relief the layers showed larger variation, bending and dipping with the bedrock but still retaining their horizontal continuity.

[94] Swithinbank, CWM (1997) (footnote 81, pp. 114–15).
[95] Ewen-Smith, B M (2019) (footnote 82).

Over mountainous relief, however, their continuity was broken by steeply dipping reflecting surfaces. Robin considered these were due to shear with possible vertical slumping within the ice sheet. To explain the deep, well-developed layers the same mechanism as invoked in Greenland could not, unfortunately, be applied in Antarctica—the virtual lack of surface melting made it impossible to generate melt layers of ice of higher density. Furthermore, their considerable lateral extent strongly suggested they were features arising from deposition on the ice sheet surface which could give rise to small dielectric variations and hence reflections. From such early and enticing glimpses of the internal structure of the ice sheet it was clear that much further research was necessary. We shall return to this topic in chapter 14 and to the use of these layers in recent times to study the stability of the ice sheet (in chapter 17).

More flights followed. On 22 December 1967 the team flew once more across the East Antarctic Ice Sheet to about 130°E and then north to intersect the coast at the Ninnis Glacier, a powerful outlet of the ice sheet, returning south more or less along the 150°E meridian over Victoria Land. On Christmas Eve a flight was directed westwards over the central Ross Ice Shelf and thence inland across the grounded ice of Marie Byrd Land to pass above US Byrd Station at 120°W and 80°S. In this latter location a team of ice core drillers from the CRREL in Hanover, New Hampshire, under the brilliant leadership of Lyle Hansen were drilling through the ice sheet and reached the bottom on 29 January 1968 at a depth of 2164 m.[96] The RES flight line was within 3 km of the drill site and recorded weak echoes at about 2200 m[97]. This flight proved important and instructive in other ways. To the east of Byrd Station the ice was very deep, as indicated by seismic soundings, but also 'warm' owing to the lower elevation of the ice sheet. The SPRI equipment did not have the system sensitivity to penetrate this combination of depth and higher ice temperatures. The region was to prove difficult, a bête noire, for several years to come.

Following a well-earned break over Christmas Day for the aircrew and SPRI group alike, the team undertook a final and long, 12-hour flight arranged for 26 December. This was to further explore the Ross Ice Shelf and its boundary with the grounded ice of Marie Byrd Land where, as we later

[96] Ueda, H T; and Garfield, D E (1970) 'Deep Core Drilling at Byrd Station, Antarctica', in Gow, A J; et al. (eds.) *International Symposium on Antarctic Glaciological Exploration (ISAGE)*, IASH Publication 96, 53–68.

[97] Robin, G de Q; Swithinbank, CWM; and Smith, BME (1970) (footnote 86).

Figure 4.12. C-121J 'Phoenix' and aircrew at McMurdo (Robin (left) and Swithinbank on the extreme right of back row; Ewen-Smith on the extreme right in front row). (Courtesy SPRI).

discovered, several fast-moving streams of ice debouch into the ice shelf. The line of soundings commenced at the coast to the east of Roosevelt Island, a large ice rise where the shelf is grounded on a submarine bank, and then south on the 150°W meridian to the Transantarctic Mountains. The RES picked out an interesting pattern of ice rises and flatter features where strong bottom echoes were encountered, leading Robin and his colleagues to interpret the pattern as a series of trapped 'lakes' forming 'pseudo' ice shelves. It was the beginning of an intense study of this boundary region between the Ross Ice Shelf and the inland ice of Marie Byrd Land by the SPRI and later a number of US-led research groups both on the ground and from the air.

On completion of the RES flights the aircraft and crew returned to Christchurch, and all the SPRI personnel except Ewen-Smith flew back to the UK (Figure 4.12).

Bev Ewen-Smith was left, responsible for the demobilisation and return of the RES equipment. He comments:

> I uninstalled the gear and packed it up for shipping back to Cambridge. I drove it over to Lyttleton Harbour and asked about any ship going to the UK. Someone pointed towards one of the cargo ships and I

> drove alongside. A crane was lowering a net to the dockside so with no word to anyone I piled the crates into the net, and it disappeared into the hold. There was no paperwork and no conversation. A month or two later, we got a message to collect it from the docks in England. I hand carried the processed SFIM and 35mm films on the commercial flight back to UK.[98]

This was quite extraordinary and reflected a different era with a more casual attitude to security and bureaucracy.

4.4 Review of the Season

The 1967 reconnaissance season of Antarctic airborne RES had been an outstanding success; the team were justifiably enthusiastic and in high spirits with their scientific achievements. The SPRI system had proven itself reliable and fully capable of operating in long-range aircraft for long periods of time; its performance had met expectations, and future improvements to antenna, power, and recording were easy to identify. The greatest weakness was seen to be accurate knowledge of the aircraft location and hence positioning of the ice thickness and related data. Use of the SFIM recorder was only partly a solution, and there was a high reliance on using known ground points as fixes, of which there were few in Antarctica. Years later this positional uncertainty would be a limiting factor on the continued use of these early RES soundings as a baseline for subsequent changes in ice sheet dimensions and flow. In the meantime, however, it was clear that the flights had covered significant tracts of terrain and sampled a range of glaciological conditions. Much had been learned about the ice sheet and its subglacial environment. It was the dawn of a new era in which aircraft fitted with increasingly sophisticated scientific sounding, recording, and navigational equipment would criss-cross the ice sheet, revealing in unparalleled detail the physical nature of this, the last of Earth's continents to be explored.

[98] Ewen-Smith, B M, personal communication, January 2021.

5

The Second Antarctic Season 1969–70

A Task for Hercules

The 1967 fieldwork had achieved its goals. The SPRI MkII sounder proved to be a capable instrument and had yielded promising results over a variety of glaciological situations. During 1968 the SPRI team worked up much of the data into preliminary publications so as to communicate their exciting findings at several scientific conferences; a hint of their results has already been outlined in the previous chapter. A timely meeting of the Royal Geographical Society on 17 February 1969 was on the subject of 'glacier sounding'. The meeting had been organised in collaboration with SPRI in order to present the new discoveries from radio-echo sounding, and Robin, Evans, and Swithinbank, as well as Hattersley-Smith from Canada, gave papers. Professor Preben Gudmandsen from the Technical University of Denmark (TUD) also gave a talk summarising the results of a similar Danish programme of radar sounding in Greenland.[99] The opportunity for collaboration between Gudmandsen's group and the SPRI seemed logical; it would lead to a highly effective and beneficial programme of joint operations in subsequent years.

The trial season had heralded a scientific breakthrough in investigating the Antarctic Ice Sheet. It held out the promise of surveying huge tracts of Antarctica on a routine basis to compile a detailed picture of the shape of the ice sheet surface, the ice thickness, the topography of the land surface beneath, and myriad details of the internal structure of the ice. But a

[99] Gudmandsen, P (1969) Airborne radio-echo sounding of the Greenland Ice Sheet, *Geographical Journal* 135 (4): 548–51.

follow-up campaign would require several ingredients. First, only the US could provide the necessary logistics, as there still was no suitable long-range aircraft support in the UK. This would mean negotiating a longer-term plan with the NSF and the committing of significant US resources to a primarily British research programme. Second, should the Americans agree, the C-121J aircraft used in 1967 was inadequate for future work, and its lack of skis posed serious risks. In any case, the US Navy was expecting to phase out this model in the near future. Third, any replacement aircraft would require substantial mechanical modifications to carry redesigned aerials and once more would cost a great deal of money to meet scientific and, importantly, airworthiness criteria. Who would pay? Fourth, the SPRI would need a larger team not only to undertake the fieldwork but to prepare equipment and later analyse and publish the results as the momentum built. Last, but certainly not least, such a research programme would require substantial financing.

All these were daunting issues facing Robin and Evans. Sir Vivian Fuchs, then the director of the British Antarctic Survey, had attended the Royal Geographical Society meeting, and he made some remarks in the discussion following the main presentations. In part there was a tinge of envy, as an old rivalry between the BAS and the SPRI surfaced when he commented: '[I]t is also important to have a strong back-up team at home who will help to work it all out. Because of all the traverse miles that Dr Robin proposes to cover in the next two years, he is certainly going to have his time very full working up the results.'[100] By some fortunate alchemy or alignment of the planets, as 1968 and 1969 rolled forward many of the jigsaw pieces fell miraculously into place.

The effective collaboration with the NSF in 1967 proved to be open to further development. Phil Smith, a perceptive and astute senior government official, was director of field requirements at OAP and committed to international collaboration. In accord with his director, Thomas Jones, he viewed the SPRI-NSF work as a primary demonstration of the intent of the Antarctic Treaty, which had principally been a US-inspired international agreement. Robin outlined the success of the reconnaissance season and the outstanding support that the SPRI group had received and discussed the options for the future. The NSF agreed to provide the aircraft platform and the necessary modifications in time to meet the 1969–70 field

[100] Fuchs, Sir Vivian (1969) Discussion, *Geographical Journal* 135 (4): 559.

season with a generous allocation of 350 flight hours (representing a significantly expanded campaign compared with that of 1967). The agency proposed using one of its C-130 Hercules transport aircraft, which the US Navy had been operating in Antarctica for some years. These planes had several advantages over the Super Constellation, the principal one being that they were ski-equipped, thus massively reducing the hazards of flying at great distance from bases over the ice sheet. Second, the skis enabled them to operate throughout the Antarctic season when the sea-ice runway at McMurdo ceased to be operable for wheel-only airplanes. Third, they were able to land at stations on the ice sheet such as the US bases of South Pole and Byrd to refuel, thus considerably extending the range of RES flights.

5.1 Cambridge Preparations

Evans and David Petrie immediately commenced designing a new antenna system, liaising with VXE-6 and Lockheed Aircraft Corporation in Marietta, Georgia, the manufacturers of the C-130.[101] Petrie, it should be recalled, had worked closely with Charles Swithinbank during the BAS radio-echo season in the Antarctic Peninsula during 1966–67, where he had been responsible for the installation and operation of the equipment. He had now joined the SPRI team.

The particular aircraft designated for the radio-echo sounding work was an early 'F' model with the bureau number (BuNo) 148320—henceforward designated #320 and had been operating in Antarctica for many years. The new aerial array, comprising two terminated dipole antennas was to be mounted beneath the tail using the two tail planes as reflectors, an arrangement similar to that adopted for the C-121J. The mounting necessitated opening up the tail and installing hardpoints for the struts that would hold the antenna cables. The struts and related fixtures had to be manufactured and the whole system assembled and flight-tested. Lockheed, under contract to the NSF, undertook this work, and the materials were then packaged and shipped to Christchurch, New Zealand, where they would be reassembled on #320.

[101] In the first two SPRI seasons the antenna arrangements were similar. Two terminated dipoles were positioned a quarter wavelength beneath the tail plane of the respective aircraft.

In the laboratory at SPRI work also went ahead on improvements to the RES equipment.[102] These comprised an automatic annotation of the film record every minute with date, time, and receiver gain. This enabled rigorous correlation with the flight recorder (SFIM) and the trimetrogon photography system that was installed by the US Navy. Chris Harrison, who began his PhD investigations in the autumn of 1968, designed an electronic device that would 'flip' the radar beam from side to side, which he hoped could be used in examining the source of certain echoes in the records. This device was operated by switching 90° phase shift into one of the two loaded dipole elements under the tail of the aircraft. In all other respects the now redesignated sounder, SPRI MkIV, was the same as the MkII.[103]

Attention was also focused on recording the aircraft navigation data, which was a tricky matter. The SFIM system had been used in the 1967 season, but considerable resources were required to digitise the traces and convert them into a track once they were returned to the UK. Furthermore, inherent drift in many of the instruments meant there would be significant closure errors—which could be distributed back along the flight but would leave residual errors of the order of many kilometres. As there appeared to be little prospect of a breakthrough in technology, it was decided to run the SFIM again in 1969–70, supplemented when in mountainous or coastal terrain with the 9″×9″ high-resolution aerial photographs taken by a trimetrogon system similar to that which had been installed in the C-121J and described in section 4.2.

5.2 The Team Assembles

Two of the three persons who participated in 1967 were not to be on the team for the next season. Swithinbank was increasingly committed to BAS activities, and Ewen-Smith[104] had completed his PhD and moved to work in a local computer-aided design company. Robin had applied for research studentships from the NERC in several of the years since commencing the RES work, and new students had joined the SPRI. Julian Paren had arrived

[102] See also footnote 56.

[103] Evans, S (1970) 'SPRI-NSF Radio-echo Flights in Antarctica 1969–70', in Gudmandsen, P (ed.), *Proceedings of the International Meeting on Radioglaciology*, Lyngby, Denmark, 113–14.

[104] Note that Beverley Ewen-Smith's 'true' surname is not hyphenated. He added the hyphen because of problems with his name being abbreviated to BME Smith, for example, as in the publications referenced in footnotes 56 and 86.

in 1965 and was now heavily engaged in completing his research on the electrical properties of ice and would submit his thesis in 1970. Chris Harrison was using the 1967 RES data to better understand the physics of reflections from the ice sheet and was developing a series of studies that he hoped to continue with new data from 1969–70. In August 1969 David Drewry, the author, arrived at the SPRI to take up a NERC PhD studentship to investigate the subglacial morphology and geology of Antarctica from the RES records under Robin's supervision.

I had recently graduated from Queen Mary College, University of London, with a degree in physical geography and geology and had the good fortune to have spent the summer months of 1968 in the Stauning Alps of East Greenland undertaking a glaciological project. I later commented, 'It was that experience that ensnared me with an enthusiasm for the polar regions and led me to the opportunity at SPRI and a career in glaciology'.

This meant that the fieldwork team for Antarctica would comprise Robin and Evans, Petrie, and two PhD students—Harrison and Drewry. Tasks were apportioned: Robin looked after the negotiations with NSF and the US Navy and worked closely with Evans on the progress of the radar systems. Evans and Petrie busied themselves on the detailed preparations of RES equipment: SPRI MkIV and the recording systems for the data and navigational information. Harrison assisted them from time to time and designed the electronic 'beam-flipper'. I was designated to assist Robin with flight planning—compiling potential areas for sounding, as well as preparing all the maps and charts. I had to learn as much as possible about air navigation, since I was to be involved in monitoring the flights, as well as recording any significant glaciological or geological feature that would be observed during each mission. It was a steep learning curve that included a stack of reading about RES techniques and earlier operations before deploying south—in just over three months' time.

During the summer of 1969 the various strands of the operation gradually merged. In October all the kit had to be assembled, packed, and shipped to New Zealand. This was quite worrisome, given the extensive season of flights that lay ahead; there would be little chance to retrieve or replace any items should something be forgotten, broken, or lost—so thorough checklists were prepared, and spare components, tools, testing equipment, maps, logbooks, reference manuals, and the like were carefully packed away in large cabin trunks. Then, personal arrangements for departure had to be organised. The group was likely to be away for at least four to five months, spending time first in Washington, DC, at the NSF and the US Geological Survey to make face-to-face contact with American colleagues whose program [*sic*] Operation Deep Freeze we were to join. The plan was to fly to Christchurch using the US Military Airlift Command (MAC) to rendezvous with the C-130 aircraft and commence installation and testing of equipment. Only after that would the team deploy south to McMurdo and Antarctica.

5.3 Washington, DC

In October all the personnel were ready to leave for Washington, DC, but the best laid plans can go awry. At the check-in counter at Heathrow there was a message awaiting Harrison and Drewry. It simply said, 'Do not fly! Robin'. This was disconcerting, and they immediately contacted the SPRI by telephone. Within the previous 24 hours Robin had been informed the airborne programme had been delayed at the US end and that departure was put on hold for just over a week. Harrison and I, along with Evans, were back at Heathrow on 3 November and en route to Washington. Gordon Robin would follow later via Australia, where he had a number of engagements.

David Petrie, it had been decided, would fly to Atlanta and visit Lockheed's Marietta facility. There he met the personnel who had designed the RES console, to discuss its configuration and was shown the assembled unit, the wiring diagrams, and associated installation details for interfacing with the aircraft airframe and systems. He then flew to Washington to meet up with the SPRI team at the Office of Antarctic Programs (OAP)[105] of the NSF. Evans knew some of the staff there, and we were given a warm welcome even by the formidable Helen Gerasimou, the director's personal assistant, who appeared to control pretty much everything that went on in the office (and beyond—possibly the director, as well)[106].

This visit provided the opportunity to meet a number of the OAP staff whom the group would get to know much better in subsequent years. Several would become good colleagues and friends—Bob Dale, Walt Seelig, and

[105] On 27 October 1969, a major reorganization took effect at the NSF. The US Antarctic Research Programs became a part of National and International Programs, and Dr T O Jones, formerly Division Director, Environmental Sciences, was appointed Acting Deputy Assistant Director.

[106] An interview with Helen Gerasimou provides interesting background (Interview of Helen Gerasimou by Brian Shoemaker, Polar Oral History Program, Byrd Polar Research Center Archival Program), 18-08-2005. http://hdl.handle.net/1811/6059.

Phil Smith. Bert Crary, the former chief scientist and now deputy director of the Division of Environmental Sciences in NSF, also looked in on meetings. It was evident that the programme was seen as an exciting scientific project and its international dimension important; Robin's influence had ensured much assistance. There was a briefing on operational aspects of the US Antarctic Research Program (USARP), relations between the NSF and the US Navy, and much anecdotal information regarding life and work in McMurdo.

The next day a call was made on the US Geological Survey Branch of Special Maps and its head, Bill McDonald. MacDonald was a large, genial individual who had been responsible for the highly successful programme of topographic mapping in Antarctica. Back in Cambridge I had contacted him by mail to request certain maps, and he was only too ready to oblige. He drove the group to the branch office at Silver Springs and for more than two hours provided fascinating details of the reduction, compilation, and manuscript drawing of topographic maps from aerial photographs. The maps required for the RES missions were then identified and collected. The considerable quantity was provided gratis. Combined with the personal attention of McDonald, such treatment reinforced the generosity and helpfulness of these American polar colleagues.

Departure from Washington was scheduled for 6 November, and our party was driven to Andrews Air Force Base to be processed and to join several Navy personnel and other 'SARPS'—(short for USARPs) a somewhat pejorative military sobriquet for those scientists working in the American programme. Owing to strong crosswinds, the flight was diverted to Dulles Airport, and personnel were ferried across Washington to board the aircraft that took off at 4.30 a.m.

As a novice in military matters and one of the junior SPRI members, I was soon to learn that operations could be called at any time of the day or night, and usually included cancellations, endless waiting time, and mandatory doses of uncertainty. The air bridge to and from New Zealand was operated by MAC using large-transport aircraft, the C-141 StarLifter (Figure 5.1). This was a cavernous monster whose interior was bare metal, webbing, straps, valves, cables, nozzles, and the like. Our RES team were directed to rows of seats, all facing rearwards, that had been placed in the centre of the hold to be occupied by some 60 passengers. Behind and in front were piles of cargo lashed down with webbing, and two portable toilets had also been installed on a pallet. There were no windows—only a few small

Figure 5.1. C-141 StarLifter at Harewood International Airport, Christchurch. This transport aircraft was used extensively in carrying USARP personnel and cargo from the continental US to New Zealand. Note: virtually no windows!

'portholes'. The route was across the country to California and Travis Air Force base outside San Francisco, for refuelling, and on over the Pacific to Hickam Airfield adjacent to Pearl Harbor on Oahu, Hawaii. Here there was to be a crew break for 16 hours. We were checked into the officers' billet but left immediately—to explore Honolulu and its exotic surroundings in a rented car. Somewhat exhausted, we flew on to American Samoa, landing to refuel at the small equatorial island of Tutuila at the airbase serving the capital, Pago Pago. The final leg was to Christchurch, New Zealand, crossing North Island and its central volcanic terrain of cones and lakes to descend over the Kaikoura coast and landing at 4 p.m. on 7 November.

5.4 New Zealand Activities

Considerable activity followed during the next three weeks. After the arrival in Christchurch (CHCH)[107] we made contact with the NSF/USARP representative, the personnel of VXE-6 Squadron, as well as with the New Zealand Antarctic Research Programme (NZARP), including Robert (Bob) Thompson, the director. The USARP representative, based at the airport,

[107] Sometimes also shortened to CHC.

Figure 5.2. NSF/USARP Offices, part of the US Navy facility at Harewood Airport, Christchurch, New Zealand. (Courtesy NSF).

maintained a close link with McMurdo, the centre for the American operation, by radio, and with the Office of Polar Programs (OPP)[108] in Washington. At that time the representative was Bill Austin. The reception was, to great surprise and consternation, chilly and indifferent; Austin offered nothing in the way of support or cooperation. This was very different from the attitude encountered in Washington and in all the dealings to date with the OPP and the US Navy fraternity. It gave rise to concern and worry in preparing for a major Antarctic season.

The office was run and organised by a very energetic, highly efficient, and friendly Kiwi, Margaret Lanyon (Figures 5.2 and 5.3). It was through her that quite often information and assistance would be elicited. Over several years her welcoming, 'can-do' attitude was appreciated by the SPRI teams and many other transient scientists. Some years later she took over the role of the NSF representative herself, a confirmation of her expertise and dedication.

[108] In 1970 the Office of Antarctic Programs changed its name to Office of Polar Programs to recognize increasing involvement in Arctic research.

Figure 5.3. Margaret Lanyon, assistant to the USARP Representative in Christchurch, New Zealand. (Courtesy The Antarctican Society).

The following few days were occupied with unpacking and checking the RES equipment in one of the aircraft hangers rented to the US Navy by the Christchurch airport authority at Harewood. There was a fully equipped avionics laboratory for the maintenance of aircraft navigation and other instruments, and the SPRI team was fortunately able to utilise part of this space. An important aspect of the programme necessitated processing of the RES photographic film following each airborne mission. In 1967 the Navy Photographic Laboratory in McMurdo had undertaken the task and was again commissioned for this work. A 'photo-mate', Mike Gottlieb, was assigned to the group by the Navy, a very helpful gesture, to undertake the processing. He was in CHCH to familiarise himself with the requirements and would fly down with the team to McMurdo. Unfortunately, his skills did not match either the SPRI's or the Navy's expectations, and much of the processing necessitated considerable supervision or even being undertaken by someone from the SPRI! Part of the problem was the switch to a different, more rapid technique termed 'Bimat Colortrace' that involved minimal fluid development. The 35mm film lengths were wound in contact with a pre-soaked strip of similar size—quite a difficult task to physically handle. The fluid processed the film, which became the negative, and the other strip the positive. It proved much less satisfactory than expected.

On Friday 28 November Hercules aircraft #320 flew in from McMurdo, and thereafter an intense period followed of fitting out the tailplane with antenna fixtures and installing the specially constructed internal equipment

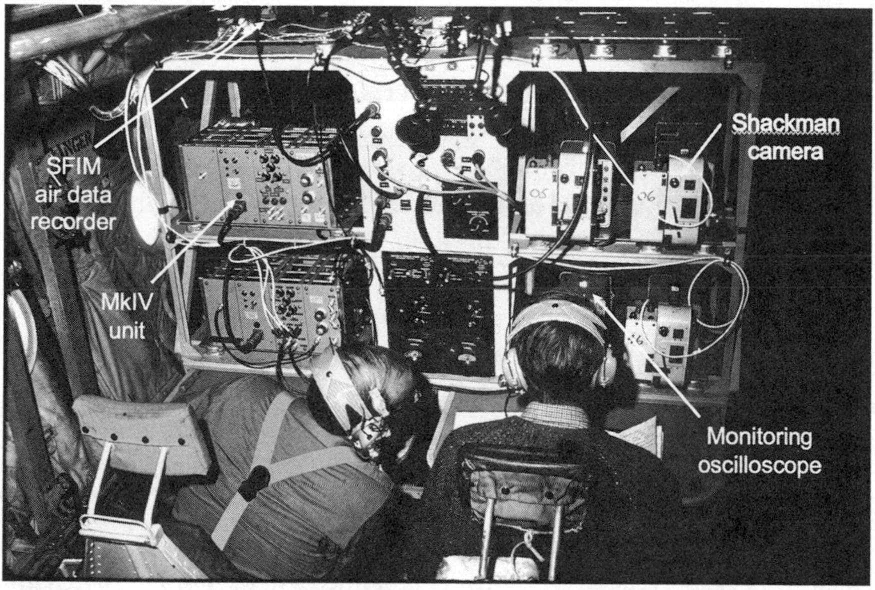

Figure 5.4. Main units of the SPRI MkIV system rack-mounted and installed in C-130 #320, 1969–70.

rack for the radio-echo sounding and navigational logging gear. Two sets of the SPRI MkIV system were mounted and three recording cameras linked together with a spaghetti-like maze of cables and wires (Figure 5.4).

Lt Cmdr Richard Hamilton, the aircraft and squadron senior navigator, had already revealed that several of the plane's navigation instruments were inoperative; the Doppler had burnt itself out, and the drift sight had jammed in its 'up', parked position with its gyro broken. Most serious of all was that the Bendix polar path compass, which it was hoped would provide a substantial improvement at high magnetic latitudes over the normal magnetic compasses, was also defunct. It was not possible to repair or replace any of these. All this news was depressing, particularly as one of the reasons for delaying the second Antarctic season had been to make improvements to navigation. It was also annoying that the information about the failure of some of the instruments had been available in June but had not been passed to the SPRI.

During this time the SPRI team took accommodation in two large bed-and-breakfast establishments in the Merivale area: "Merivale" (where Evans and Petrie stayed) and 'Highway Lodge' on the Papanui Road (Harrison and Drewry). The latter was run by the colourful and warm Mrs Scarlett, an Italian émigrée, who mothered us and provided ample but seriously unhealthy breakfasts.

Luckily, the team was able to meet several of the Hercules flight crew who would be undertaking flying the RES missions in Antarctica. From the outset the importance of establishing a positive relationship was recognized, as crew and scientists would be working closely for several months, and success would depend a great deal on their cooperativeness and enthusiasm for the work. The plane commander was Major R Cantrell (USMC), a Marine flyer but assigned to VXE-6. After a couple of months in Antarctica some of the SPRI team concluded that Major Cantrell had a wristwatch fetish. Every week or so he would come onboard the aircraft having purchased a new watch in the US Navy Exchange or mini-store at McMurdo. The author recalls that on one occasion the major proudly displayed two watches on one arm and three on the other.

There were to be two crews so flight operations could be undertaken back-to-back; the second crew was commanded by Lt Cmdr Harvey Heinz, a jovial former anti-submarine pilot. It should be recalled that in 1969 the Vietnam War was in one of its most intense phases, and many of the aircrew had seen service in the Vietnam theatre of operations. It was jokingly said that the US Navy air support personnel were given the devil's choice of posting—Vietnam or Antarctica (hot metal or freezing ice)!

Over the next few weeks, scientists and aircrew laboured together. On 2 December Robin, who had arrived from Melbourne on 21 November, gave a presentation to the aircrews, briefing them on the nature of the work—outlining the general scientific details, the crucial nature of their support, and the sort of operating procedures he wished to develop—and getting their input on details of the aircraft procedures. This discussion was followed by a long, separate session on a range of navigation and related issues with Hamilton and the other navigator, Lt Springate (accomplished at radar navigation and poker!), that included range and endurance at the flying heights considered to be optimum for the radio-echo sounding work.

On 6 December the aircraft was ready for a test flight, which was conducted over the Pacific Ocean to the east of the Canterbury Plains; it was an important step. The newly constructed aerials (Figure 5.5), two terminated dipoles, supplied to SPRI specification by Lockheed, mounted beneath the tailplane had been tested twice before in the US, and on both occasions they had failed mechanically. And on one of them the loose cable,

Figure 5.5. SPRI 35 MHz aerials (two terminated dipole antennas) on C-130 #320.

oscillating in the slipstream, had beaten some serious holes in the rear port elevator!

There were two test flights, the first to check the integrity of the aerials which survived the experience and the second, longer flight of about three hours to evaluate the RES gear. Critical manoeuvres were undertaken at about 10,000 ft to evaluate the characteristics of the antenna and power gain by flying at fixed bank angles and measuring the power of the returned signal from as smooth a reflecting surface as possible—in this case a calm sea. The resulting beam pattern or *polar diagram* would assist in modelling the likely performance of the radar system overall.

Stan Evans describes such a calibration flight in 1967 when the Super Constellation aircraft made a series of steep banks: 'I do remember, when they got to—I think . . . seventy degrees bank angle they said, (i.e. the plane commander) "How far do you want us to go?" And we said, "How far can you go?" And they said, "We can go to ninety degrees." And I said, "No thank you, no."'[109]

The antenna results for the C-130 were very satisfactory, including the operation of Harrison's new facility to switch the RES beam by 15° to either

[109] Dr Stan Evans (footnote 45).

left or right in order to better isolate side echoes. However, there appeared to be 'ringing' in the antennas,[110] and the system sensitivity did not meet expectations; this was disappointing news.

Meanwhile, Robin had been contacted by Dr. Alton Wade, professor of geology at Texas Technological University, in Lubbock, Texas. Wade had been with Admiral Richard Byrd on his Second Expedition (1933–35) and was a veteran Antarctic scientist. His group was working on the geology of western Marie Byrd Land on the eastern side of the Ross Ice Shelf. Wade asked Robin whether it would be possible for the SPRI team to sound the glaciers and ice sheet in this area, as the ice thickness and subglacial bedrock topography data would be very helpful to their interpretations. Robin agreed and also invited Wade to send a couple of his people to join the SPRI for part of the season to help direct the flights where they needed the soundings and to provide some additional support for the SPRI generally. In short order John Wilbanks and Larry Osborne joined us in Christchurch, and Wade himself participated in the early flights in Byrd Land until about 2 January 1970.

The teams were now assembled. SPRI, Texas Tech, and the aircrews were getting the feel for the operation, and the aircraft was ready—the time had arrived to launch for Antarctica.

5.5 Antarctic Sounding Commences

On 7 December the SPRI contingent of five and 21 aircrew with all the kit loaded aboard #320 took off for McMurdo in a spectacularly colourful evening sky of crimson-infused clouds. After a few hours of crossing the Southern Ocean, we approached and crossed the Antarctic Circle as the sky brightened. For several of the team it was an exciting moment—our first encounter with Antarctica. I quote from my diary:

> I went onto the flight deck and there was my first view of Antarctica! A faint series of dark mountains set amidst a sea of cloud. Nearer and nearer they came and gradually the mountains of northern Victoria Land, snowy and now gleaming in the sun, became distinct. And then the pack ice! Soon we were over the high peaks of the Admiralty Range. The radio-echo recorder was switched on and soundings con-

[110] The input impedance will vary widely with the length of the transmission line, and if the input impedance is not well matched to the source impedance, little power will be delivered to the antenna. This power ends up being reflected back to the generator-'ringing'.

Figure 5.6. Arrival at McMurdo. Left to right: Stan Evans, Gordon Robin, David Petrie, and Chris Harrison.

> tinued over Tucker Glacier. . . . I could pick out Mount Melbourne, Terra Nova Bay, where Priestley and Scott's Northern Party wintered in an ice cave more than half a century ago. And then there were only the mountains fading in the background as we headed down towards McMurdo.

During 8 December the team established itself on the base (Figure 5.7).[111] Personal accommodation was basic (Figure 5.8), but two spacious workrooms were made available in the Earth Sciences Laboratory halfway up Observation Hill, where electronics kit, maps and other materials could be organised.

The first RES flight, primarily to test equipment over different ice conditions, was on 9 December. The aircraft operated from the skiway at Williams Field on the McMurdo Ice Shelf with all the SPRI team aboard (Figures 5.9 and 5.10). The route was south to the Byrd Glacier and then

[111] It is worth recording that season 1969–70 coincided with the US programme hosting its first women scientists, a group of four geologists from The Ohio State University, led by Dr. Lois Jones. It was probably this situation that impressed on the author, as a young graduate student, the ludicrousness of earlier exclusion of women from Antarctic field science and resulted in my easy decision in later years to allow women scientists to overwinter at British bases when I was director of the British Antarctic Survey. (See also Seag, M (2015) 'Equal Opportunities on Ice: Examining General and Institutional Change at the British Antarctic Survey 1975–1996', Master of Philosophy thesis, Scott Polar Research Institute, University of Cambridge, 95pp.)

Figure 5.7. McMurdo Station from Observation Hill, circa 1977. View across sea ice in McMurdo Sound to the Royal Society Range.

Figure 5.8. Building 108, in which most of the SPRI team were housed for the field season, with basic bunk rooms only, and ablutions in an adjacent block.

Figure 5.9. Buried Jamesway huts, part of the air facilities at Williams Field.

Figures 5.10. SPRI MkIV RES rack-mounted in C-130 #320. Stan Evans (right) with Larry Osborne (Texas Tech). Evans is making an entry in the RES logbook; a similar logbook was kept by the SPRI 'glaciologist' on the flight deck. These logbooks are retained in the SPRI Archives.

Figure 5.11. David Petrie undertaking installation of RES aerials beneath the tailplane.

ascending this giant outlet onto the polar plateau to 140°E. The flight altitude was altered in stages to observe the effects on performance, starting at 20,000 ft and then coming down progressively to 3000 ft, 1000 ft, and then 500 ft of terrain clearance—the aircraft skimming the ice surface with its veneer of wind-aligned sastrugi. The aircraft re-gained height and flew out to the Ross Ice Shelf. descending again and passing over the ice front, a glistening white cliff standing some 30–50 m high out of the blue and ice-flecked waters of the Ross Sea. The flight returned to McMurdo after 6½ hours, and the data were evaluated during the next day. The results confirmed the evidence from the Christchurch test that the system sensitivity did not meet expectations, and no reflections from the base of the ice sheet were recorded over much of the inland plateau sections. On the ice shelf the ice/water interface was detected easily, but the ringing arising from the possible mismatch of the antennas generated a mass of surface echoes. Based on these facts, further work commenced on the RES equipment, the plan of the season was re-considered to prioritise regions within the capabilities of the system.

It was decided that operations were to concentrate on the following regions: (i) the Ross Ice Shelf and the ill-defined margin between the 'grounded' Marie Byrd Land ice and the floating ice shelf which had been the subject of some reconnaissance soundings and speculative ideas from the 1967 flights and was of particular interest to Robin; (ii) the inland side of the Transantarctic Mountains buried by the ice sheet to explore their extent, character, and glacial drainage, which was to be studied by Drewry; (iii) Harrison was to investigate internal reflections, especially complex patterns of overlapping echoes; (iv) the little-investigated Filchner Ice Shelf; and (v) local glaciers in the coastal area of Marie Byrd Land and Ellsworth Land for Texas Technological University. In addition, Robin had received a request from the British Antarctic Survey to make soundings of the Brunt Ice Shelf on which the BAS Station Halley Bay was located and over adjacent inland areas. This was to assist a glaciological programme examining the dynamics and mass balance of the ice shelf being undertaken by Robert (Bob) H Thomas. There would be opportunities to attempt soundings in other areas, to investigate further the major outlet glaciers of the Transantarctic Mountains and inland regions. Indeed, several flights were made deep into the interior of East Antarctica as far as the PoI. Regardless of the change in plans, it was going to be a full and busy season (Figure 5.12).

An interesting footnote was the presence of Dr Anthony Michaelis, science correspondent for the British newspaper the *Daily Telegraph* during the first part of the SPRI field season. He had visited the SPRI earlier in the year following his acceptance by the US Antarctic Program as a visiting journalist. Large in body and spirit, self-confident and shrewd, Michaelis joined several of the RES flights and sent well-written reports back to London that gave the Cambridge team first-class exposure in the British press.[112]

[112] 'U.S. Navy to aid Cambridge team plot Polar cap', *Daily Telegraph*, 13 October 1969; 'Science in the Antarctic: British survey of the ice cap', *Daily Telegraph*, February 1970.

Figure 5.12. The aircrew greeted us one morning with this version of the use of the RES aerials.

5.6 Personal Experiences

Before providing an outline of the scientific results from this season I have paused for some reflections, as recorded in my diary, of the many flights that were made; they were memorable and exhilarating.

> In the Queen Maud Mountains (Figure P.4), the flying was at low level, typically no more than about 1000 m (3000 ft)[113] above the ice surface. The terrain was spectacular—inland the vastness of the ice sheet was spread out before us flowing inexorably from great central domes towards the coast and there encountering the first sentinels of the Transantarctic Mountains, themselves stretching as far as the eye could see. Nunataks and isolated peaks gave way to continuous belts of higher rugged terrain. The mountains were clearly layered, formed of an enormous thickness of sedimentary strata called the Beacon Supergroup intruded with intervening sills or bands of dark igneous dolerite. These created spectacular cliffs with extensive ledges and steep

[113] Most of the soundings were made at this altitude above the surface and in limited areas at 300 m (1000 ft) in order to reduce the obscuring surface returns in crevassed areas.

Figure 5.13. View of the lower section of the Beardmore Glacier, looking south; a large crevasse field is evident.

faces, similar to those that Scott and his party had described on the arduous journeys up and down the Beardmore Glacier on their ill-fated yet heroic polar mission in 1911 and 1912 (Figure 5.13). Camped below Buckley Island at the head of that glacier they inspected moraines and found rocks that when broken-open contained the imprints of fossils plants attesting to the change of climate over many hundreds of millions of years. Now in our comfortable aircraft we flew slowly by these scenes marvelling at their grandeur and perfect isolation.

Where the inland ice encountered the mountains, it had breached the highland barrier in giant glaciers to reach the sea two hundred miles further on and ten thousand feet lower. These are some of the greatest glaciers in the world, assembled one-by-one along the huge elongate mountain chain stalking the continent from Cape Adare at the entrance of the Ross Sea to the several massifs extending out towards the Weddell Sea on the Atlantic Ocean side. Their names are redolent of the early history of exploration of this part Antarctica; of men and ships and sponsors: Priestley, David, Mackay in Northern Victoria Land, the Ferrar and Taylor in what Scott termed the "Western Mountains" of Southern Victoria Land and then Skelton, Byrd, Nimrod, Beardmore, Shackleton, Liv, Axel Heiberg, Scott, Amundsen,

Figure 5.14. Thiel Mountains, looking east.

Reedy—an endless succession of outlets to the south feeding the Ross Ice Shelf.

In the following diary entry, I recall one remarkable day in January 1970.

We started our flight from McMurdo climbing out over the Ross Ice Shelf, its ice glistening in the sunlight, with the frontal ranges of the Transantarctic Mountains on our starboard side, heading for the entrance to the Ramsey Glacier. Now at 300 metres above the surface we turned right to climb steadily up the glacier over stupendous ice falls cascading into the main channel from higher icefields. The plane commander, Harvey Heinz who had conducted low-level antisubmarine duties in P-3 aircraft was in his element. The weather was unbelievably kind, with bright blue skies and virtually no clouds to obscure the glacial landscapes. . . . Passing over the Thiel Mountains (Figure 5.14) we headed . . . to the top of Scott Glacier, which we descended over the next hour with its harsh, impressive scenery of granite peaks reaching to 10,000 feet, etched into jagged needles and serrated ridges, and later on precipitous, smooth ice-truncated cliffs. This area, I thought, would be a climber's paradise (Figure 5.15).

Back on the Ross Ice Shelf we headed east, parallel to the mountain front to ascend the Axel Heiberg Glacier. This was the route that Amundsen had taken on his journey to claim precedence at the Pole

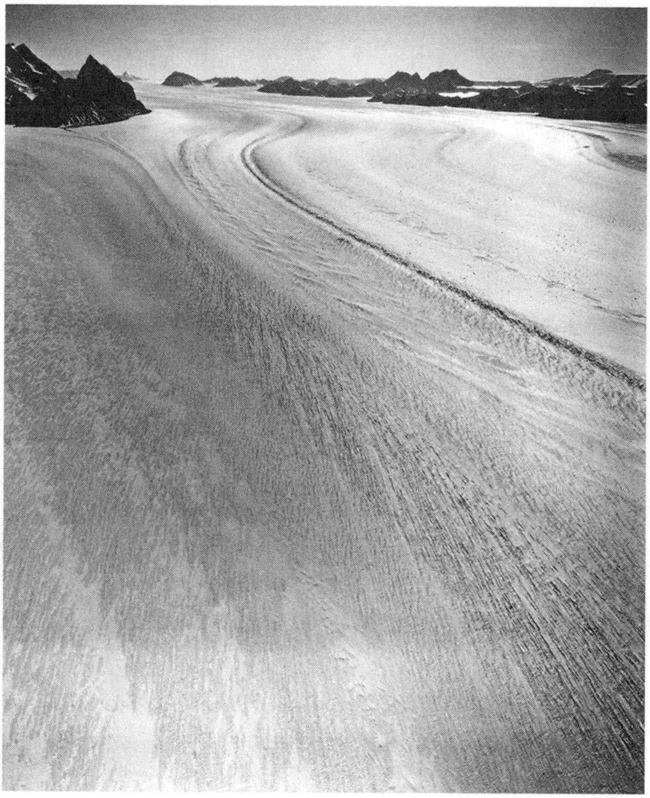

Figure 5.15. Scott Glacier. (Courtesy US Navy, US Antarctic Program).

in 1911. It is a steep and heavily crevassed glacier, a continuous cascade of ice falls. Stan Evans, who was on the flight, recorded in the log-book—'a frothy extravaganza of crevasses' (Figure 5.16).

Then we were in the ice basin that Amundsen called the Devil's Ballroom, where the constant winds off the plateau have stripped away the surface snow creating a field of glazed blue ice. Now once more on the polar wasteland we reached the very last, the most southerly of all exposed land in the world, the nunatak at Mount Howe. Beyond here lay only the vastness of the ice sheet, an interminable dome of cold and isolation stretching to the continent's other extremity, 2500km away in the Indian Ocean. We turned north to the top of Mill Glacier a large tributary of the Beardmore Glacier. Flying close to the mountain sides our aircraft passed impressive hanging valleys, cirques, side glaciers and chaotic ice falls. For the next hour we steadily descended

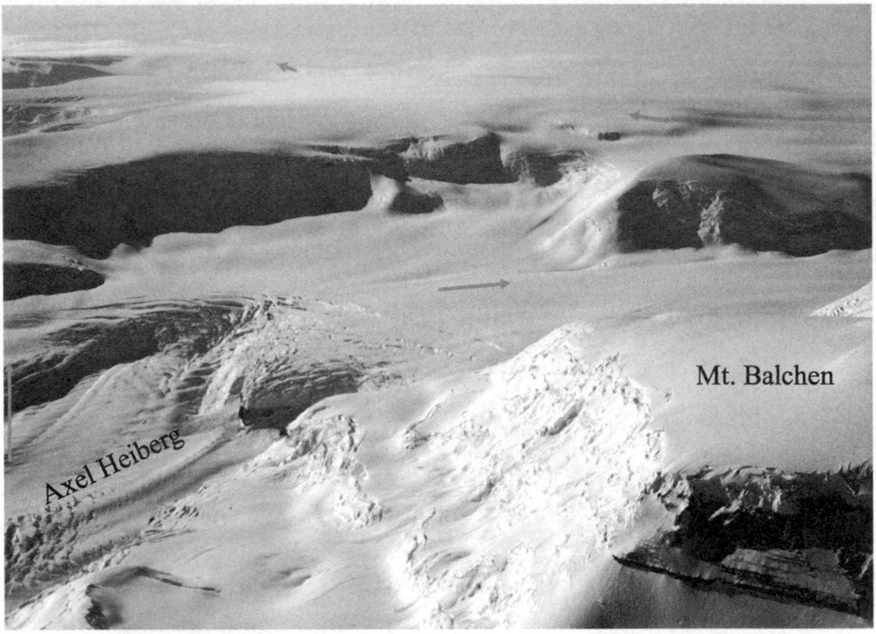

Figure 5.16. Head of the Axel Heiberg Glacier showing its crevassed ice falls, as well as the route taken by Amundsen en route to the South Pole in December 1911. Mount Balchen is formed of Beacon Supergroup rocks (see Figure 8.7).

the immensity of the Beardmore, zones of crevasses and past the beacons to Scott and Shackleton—the Dominion Range, the looming bulk of The Cloudmaker, the prominent peak of Mount Kyffin, and then at Mt. Hope we were at the ice shelf once more, where we climbed back to altitude and returned to McMurdo Sound (Figure 5.17)—what an exhilarating day—flying and viewing the two iconic glaciers of the Heroic Age. All that and a RES record rich in soundings to aid our analyses.

A few years later I had the good fortune to meet Ed Stump, from Arizona State University, a hard-rock geologist researching the basement sequences of these mountains. Over many years he worked along almost all the ranges and in 2011 produced a superb book describing the early exploration of this spectacular mountain chain, which he called aptly the *Roof at the Bottom of the World*.[114]

[114] Stump, W (2011) *The Roof at the Bottom of the World: Discovering the Transantarctic Mountains*, New Haven: Yale University Press, 254pp.

Figure 5.17. Mount Erebus (3794 m) the ever-present volcanic mountain overlooking McMurdo Sound, with a gentle plume of gas emanating from its summit crater.

5.7 Western Marie Byrd Land and the Ross Ice Shelf

Some 25 flights were made across the Ross Ice Shelf and into Marie Byrd Land. In order not to eat into the number of hours allocated for the RES project, some soundings were combined with fuel supply flights headed to Byrd and South Pole Stations. Whilst the interior of the C-130 was almost filled with a large fuel tank for these rotations, there was just enough room to operate the RES console. The flights were mostly made at high altitude, at 15,000–28,000 ft (~4500–8500 m), but still gave good penetration of the ice shelf. Requests for small lateral deviations from the direct flight path allowed progressive extension of the coverage. The navigation used the trimetrogon photographs over geodetically well-positioned features, as well as solar fixes. Robin estimated that final navigational errors were unlikely to exceed 10 NM (18.5 km).[115] Given the slowly changing thicknesses of the ice shelf, such errors were adequate for reconnaissance studies but of great magnitude when compared with contemporary GPS accuracies.

[115] Robin, G de Q (1980) Glaciology of the Ross Sea Sector: Contributions from the SPRI, *Journal of the Royal Society of New Zealand* 11(4): 349–53.

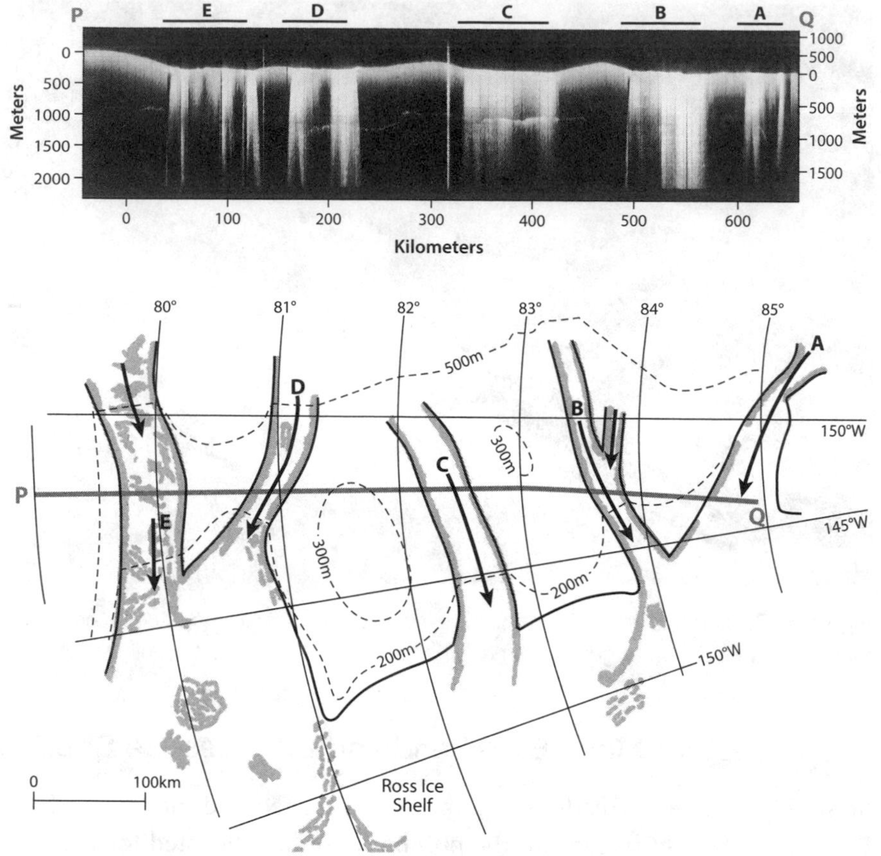

Figure 5.18. Ice streams (labelled A to E) feeding into the Ross Ice Shelf along the Siple Coast recorded in 1969–70. Top: RES (35 MHz) record along the line P-Q. Note that the two scales relating to range and depth are different in air and ice owing to different velocities. Strong, deep returns are from crevassing at the margins and across the ice streams. The map depicts the ice streams as determined from this season. (Courtesy NSF).

The outline which emerged of the ice flow from Marie Byrd Land was to be of considerable significance and the focus of a number of later important glaciological studies by other research groups. The radio-echo records identified five major streams of ice flowing into the Ross Ice Shelf (Figure 5.18[116]). A pattern of ridges and domes characterised the regions

[116] Robin, G de Q; Evans, S; Drewry, D J; Harrison, C H; and Petrie, DL (1970) Radio-echo sounding of the Antarctic Ice Sheet, *Antarctic Journal of the United States* 6: 229–32.

Figure 5.19. Complex boundary of an ice stream close to its entry into the Ross Ice Shelf. The ice beyond the crevassing in the distance is moving at an order of magnitude faster than that to the left, giving rise to shearing and rupture.

between the ice streams. On a flight crossing all five more or less orthogonally it was possible to detect their boundaries by extensive scattering of the radar signal from surface crevassing due to the faster ice shearing past the more stagnant ridges (Figure 5.19).

The ice streams are between 20 and 60 km wide and extend as distinct features, with low surface slopes, several hundreds of kilometres within the interior of Marie Byrd Land. They occupy shallow bedrock depressions between the adjacent ice rises.

They were designated, with great 'sophistication', from south to north, A,B,C,D, and E. Ice Stream A is an extension of Reedy Glacier, with some ice discharging from East Antarctica through the Transantarctic Mountains, whilst B to D evacuate ice from Marie Byrd Land and the West Antarctic Ice Sheet alone. Ice Stream C appeared to be less crevassed at the surface but still had scattering from buried fractures. Ice Stream E drains ice from the area around the Rockefeller Plateau. Robin surmised the low surface gradients of the five ice streams could be accounted for if a water layer was present at the base.[117] Many scientific questions arose from these findings—where exactly was the grounding line between the floating ice of the Ross Ice Shelf and the interior ice lodged on solid bedrock, be it ice stream or ice rise? What was the contribution of Marie Byrd Land and the Siple Coast to the mass balance of the Ross Ice Shelf, and, therefore, how fast was the ice flowing, and what was the total discharge of the ice streams? What was the nature of the sub-ice terrain and its effect on ice flow, and what was the coverage and role of any water at the bed? Perhaps the immediate substrate was water-saturated moraine or till from the erosion of rocks further inland. Were the ice streams all active and stable, and what was their history? These many matters were to be tackled by future SPRI RES campaigns, which are explored in section 9.3. In addition, an ambitious programme was being devised in the United States—the Ross Ice Shelf Project (RISP),[118] focused on a multidisciplinary plan of drilling through the ice, and the Ross Ice Shelf Geophysical and Glaciological Survey (RIGGS), a comprehensive geophysical and glaciological undertaking to study the whole of the ice shelf.[119]

Robin assembled the RES data from the radiating array of flight lines across the Ross Ice Shelf (Figure 5.20) in part to better understand the ice-thickness pattern and, by derivation, the flow characteristics of this enormous floating plate of ice and, second, to provide detailed input to the planning of the RISP. The results are elaborated further in the next chapter.

5.8 Halley Bay—Visit to the Brits!

Several lines of sounding were planned on the Weddell Sea side of Antarctica from the UK base Halley Bay (Figure 5.21) to assist Bob Thomas with his glaciological work. The BAS had agreed to position 50 barrels of aviation fuel at the base to extend the flying in the area. This fuel had come in by ship the previous season and then been transported inland on sledges

[117] Robin, G de Q; et al. (1970) (footnote 116).

[118] Zumberge, J H (1971) Ross Ice Shelf Project, *Antarctic Journal of the United States*, 6 (6): 258–63.

[119] Bentley, C R (1990) 'The Ross Ice Shelf Geophysical and Glaciological Survey (RIGGS): Introduction and Summary of Measurements Performed', in *The Ross Ice Shelf: Glaciology and Geophysics*, vol. 42 of *American Geophysical Union Antarctic Research Series*, 1–20.

Figure 5.20. The Ross Ice Shelf, looking north. In the distance (left) is Ross Island. The ice cliffs are 30–35 m high.

towed by Muskeg tractors. Arrangements had been made for the RES team and the aircrew to live on the base during the operations. Although plans were already laid out for the flights, it was not until the third week in January that a suitable opportunity surfaced to head for Halley Bay.

Evans and three others departed in the afternoon of 21 January for Williams Field to make necessary preparations with the RES system. Robin and I left with Fred Brownworth[120] of the USGS at 1630. Food and drink for the team at Halley Bay was collected—two cases of steaks, a crate of fresh eggs, sacks of oranges, and two mailbags full of beer—American largesse.

The team left in glorious sunny weather at 18:00, flying more or less directly for the South Pole Station to refuel and then continued towards the Weddell Sea. The route passed over the Shackleton Range on the edge of the Filchner Ice Shelf before descending towards the Brunt Ice Shelf on the Coats Land coast. There were low clouds, but dipping beneath it, the pilot was able to follow some tractor tracks towards the base. Radio con-

[120] Brownworth, a good-humoured and helpful colleague, was in charge of the USGS mapping program in Antarctica and had requested to accompany the SPRI to Halley Bay.

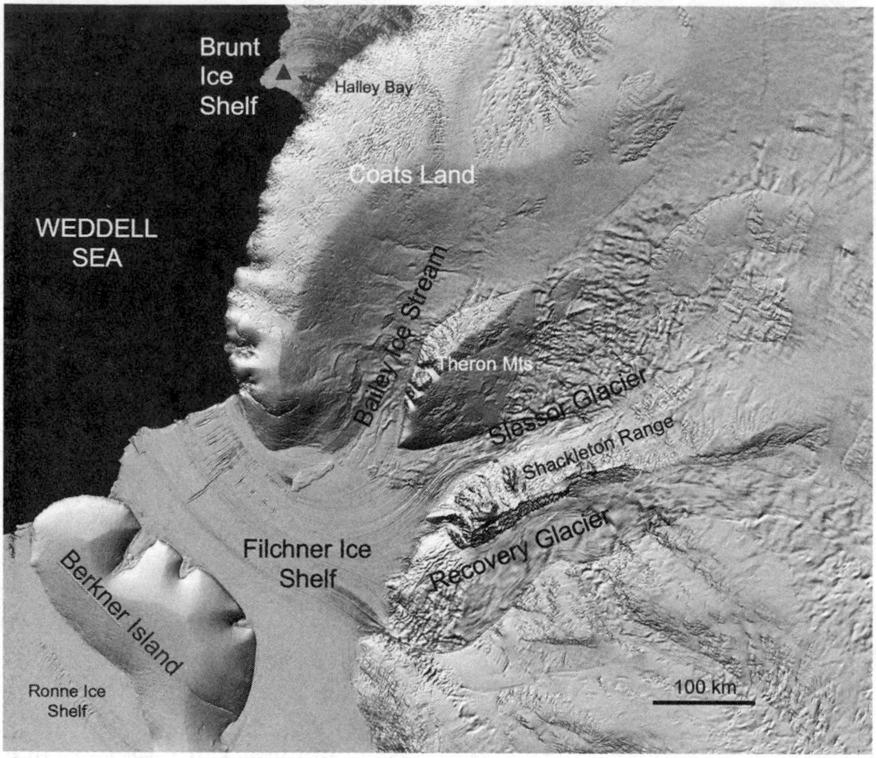

Figure 5.21. Satellite image of the outer zone of the Filchner Ice Shelf, Coats Land, and the Brunt Ice Shelf (location of Halley Bay). (From NASA MODIS instrument, courtesy NASA NSIDC DAAC).

Figure 5.22. Fuel dump at Halley Bay. These barrels will soon be covered by several metres of snow; the stakes and flags are there to indicate their location for later retrieval.

Figure 5.23. Refuelling at Halley Bay from 50-gallon barrels! No. 3 engine propeller is feathered, that is, shut down owing to an oil leak.

tact had not been established, but buzzing the base twice identified a suitable place to land. The aircraft was greeted by some 15 base members, and the team was driven by tractor to the station. In the meantime, some of the crew had commenced refuelling the C-130 (Figure 5.23). The supply from BAS was in 50-gallon barrels, and a barrel was emptied in seconds with a high-speed pump, but it took quite some time handling the many barrels to fill the aircraft tanks!

Halley Bay[121] is situated at 73°30′S, and the base was mostly underground, consisting of six huts joined by interconnecting passageways. Two of the units comprised bunkrooms, and two others were mess and kitchen, lounge, and bar. Further huts had generators, workshops, and scientific offices. The buildings were constructed on the ice shelf surface, but with a high snowfall of about 1 m per year, the base quickly became buried, and at the time of our visit was 4.5 to 6.0 m (15 to 20 ft) below ground. Entry was by means

[121] 'Halley Bay'—the name given to the inlet at the front of the Brunt Ice Shelf and the station established in 1956 (after Sir Edmund Halley the astronomer). When the inlet disappeared, the name was changed to Halley, in 1977.

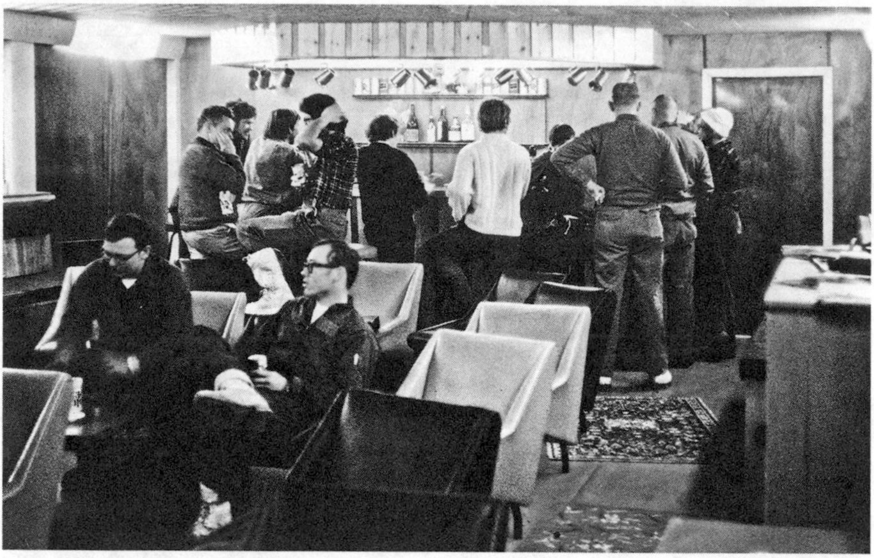

Figure 5.24. Interior of Halley III base. Aircrew and BAS base personnel enjoying an evening around the bar in the lounge area.

of a series of ladders in wooden shafts extending to the surface, where other buildings were located for the scientific programmes—ionospherics, meteorology, geomagnetism, and auroral observations. It should be recalled that it was the careful long-term observations of stratospheric ozone at Halley that by the late 1980s revealed the dramatic depletion of the ozone layer owing to its destruction by chlorine atoms derived principally from CFCs.[122] The base had teams of dogs that were kept outside along with sledges and tractors.

The interior presented an extremely warm, welcoming, and comfortable environment (Figure 5.24). Cups of tea were served on arrival, and a short tour of the base inside and outside was provided. A British bicycle parked against a rather fine signpost made an interesting vignette.[123] After a couple of hours' rest the RES team and the aircrew returned to the aircraft for the local sounding flight. Norris Riley from Halley Bay joined the

[122] Farman, J C; Gardiner, B G; and Shanklin, J D (1985) Large losses of total ozone in Antarctica reveal seasonal ClO*x*/NO*x* interaction, *Nature* 315:207–10.

[123] Almost all Antarctic bases have erected a signpost, most indicating the direction to their capital but also the home cities of the base members.

operation. He had been undertaking glaciological work and knew the terrain. Unfortunately, after less than an hour of sounding, an oil leak in no. 3 propeller was detected, and the mission was aborted. The plane commander decided it was necessary to return to McMurdo but not before a 9-hour crew rest.

Petrie, Harrison, the author, and four of the base personnel (known as FIDS, for the Falkland Island Dependencies Survey) took the opportunity to travel to Emperor Bay transported in a Lansing Snowmobile—basically a covered sledge with an aircraft engine and propeller on the rear. It was very speedy taking us to a steep valley in the front of the ice shelf leading down to the sea, with ice cliffs on either side some 15–20 m high. The party stopped at the top and walked down to the ice edge, encountering a frisky leopard seal along the way. The horizon was ice-choked with many large icebergs in the distance, gleaming in the sunlight. Back at the base we were served a wonderful meal—steak, eggs, potatoes, cans of British Whitbread beer, followed by a huge helping of gooseberry pie and custard—extraordinary given the all-American diet at McMurdo.

The first base at Halley Bay had been constructed during the IGY. The snow overburden had crushed the buildings as it was buried progressively and was eventually abandoned, and the activities moved to a new base in 1967. It, too, was snowed over and subsequently closed in 1973, being replaced, in turn, by Halley III.

Evans had worked at the old IGY station, and Petrie at its replacement, and when offered a visit to the old site, several of our team accepted readily. We travelled by Muskeg tractor to a tall metal-frame tower and a few huts that marked the location. It required digging some 6 ft down to locate the entrance to the shaft leading to the old huts and thence the IGY base. A metal caving or climbing ladder was thrown down into the dark void, and then the party descended almost 20 m (about 60 ft) carrying paraffin lamps amongst a rain of snow and ice crystals from the sides and top of the shaft (Figure 5.25). At the base everyone assembled and made their way along a tunnel to the hut constructed in 1961. Inside, the lamps illuminated a score of splintered beams and a half-fallen ceiling—all resulting from the overburden pressure of the snow. In the kitchen there was still a splendid stove looking quite forlorn. We now edged our way to the IGY station along a passage and descended further on a rickety metal ladder. Although now well

Figure 5.25. Entrance to Halley Bay IGY Station—20 m down the metal ladder!

down into the ice shelf, the IGY hut had stood up to the great pressures far better than its successor.

In my diary I wrote:

> Evans and Petrie showed us where they had slept and worked—it must have been quite emotional for them both. Evans had been here between 1956 and 1957, some thirteen years previously. Stan poked around, peered into a room and with a plaintive little sigh exclaimed, 'This is the dark room' and commented, 'it is all rather sad'. And it was. In the old generator hut, the well-preserved roof was a mass of ice crystals, suspended in their prismatic splendour and were even more fantastic, glinting in the glow of the lamps taking on weird proportions, shapes and colours—a ceiling studded with icy jewels. It was a little claustrophobic beneath those thousands of tons of ice pressing down relentlessly on these fragments of past human effort—gradually being crushed out of existence—memories to be lost forever, sealed away, frozen into layers of ice below a wilderness of a continent.

The group returned to the main shaft and ascended the spindly metal ladder to the surface. Petrie, Harrison, and I walked back to the base, about 2 miles, to regain some warmth from the chill of the icy depths.

The RES team managed four hours sleep before being awakened for breakfast—hot muffins, jam, cornflakes, and lashings of tea. Aircraft departure was at 0810 (GMT), and #320 flew out across the shelf, the inner part of the Weddell Sea with its many tabular icebergs, to the Theron and Shackleton Range, across the Transantarctic Mountains and Ross Ice Shelf to McMurdo. Two warning lights came on, indicating critical oil levels in the no. 3 propeller and one other. The landing was at 1415 (GMT), 0215 local time. It had been quite an experience, but the mission had not been completed. A return was planned to be at the first opportunity, which, remarkably, was two days later!

On the second Halley Bay mission the routing was via Byrd Station, enabling the sounding of a section of the Filchner Ice Shelf. The plane landed back at Halley at 0900 (GMT) and unloaded further stores for the base, even more than last time—the food and alcohol just poured off the back ramp. On the base the team enjoyed the hospitality of the bar before a rest. The first flight was a photo run along the coast to the Totten Mountains. The aircraft experienced a further oil warning light and shut down the no. 3 engine returning to Halley. Nevertheless, it was deemed possible and safe for a relatively short flight on the Brunt Ice Shelf, with Norris Riley once more assisting the navigation. It was a tight circuit for a large aircraft, but the base staff had assisted us earlier by running a tractor along the route, leaving distinctive tracks, and placing large patches of cocoa at various points, which showed up extremely well on the snow surface. When I asked if this had deprived them of their beverage, the reply was that all the FIDS hated this cocoa and were pleased to put it to good use! Following a few hours' sleep, the party returned to McMurdo—once again the RES work had been thwarted.

It is curious how circumstances can work out. One of the surveyors at Halley Bay during our visits was Alan Clayton. A few years later he joined the staff at the SPRI to work on the compilation of maps from the radio-echo sounding work. Norris Riley, sometime later, and following his PhD at Newcastle University, was employed by British Petroleum (the company changed its name to BP—Beyond Petroleum—in July 2000) as an exploration engineer, and the SPRI collaborated with him in undertaking RES in Svalbard in the early 1980s.

5.9 Inland Flank of the Transantarctic Mountains

The lower-than-expected RES system sensitivity resulted in focusing some of the sounding operations to the thinner ice covering the terrain inland of the exposed chain of the Transantarctic Mountains. This was of interest to me and formed a significant part of my PhD thesis.[124] Flights were centred on a network between the Queen Maud Mountains and the South Pole, yielding some 5500 line-kilometres of continuous or semi-continuous soundings. The well-mapped exposed mountains provided the opportunity to obtain good photographic fixes for the flight tracks, so that the errors in navigation were reduced to a few hundreds of metres at best and about 4 km at worst. The RES suggested a pattern of faulted blocks extending along the plateau-ward side of the mountains, tilted inland and reaching as far as latitude 89°S. Local outliers reach elevations of greater than 1000 m above sea level at this latitude. From these results it was possible to posit that much of the subglacial topography is controlled by the regional sub-horizontal inclination, or dip, of the Beacon Supergroup rocks and the associated dolerite sills. Beyond about 350 km from the mountains is a gently undulating interior lowland lying near sea level. Later airborne geophysical work by Michael Studinger and colleagues combining RES with magnetic and gravity measurements provided more detail and further explanations for the tectonic setting of the inland extension of the Transantarctic Mountains towards the South Pole. These suggest the terrain is dominated by early Palaeozoic granitic rocks with an absence of magnetic signatures of Jurassic Ferrar magmatism typical of the exposed Transantarctic Mountains elsewhere. Studinger and colleagues recognize a pronounced magnetic lineament extending from the South Pole towards the Shackleton Glacier along the zone where the author suggested a significant fault line, which they interpret as a possible 'lithospheric-scale structure' perhaps related to Gondwana continental break-up.[125]

A noteworthy discovery was the existence of a network of valleys with inland orientations. Their direction of drainage is thus opposite the present-day direction of flow of the ice sheet from its interior to and then through

[124] Drewry, D J (1973) 'Sub-ice Relief and Geology of East Antarctica', PhD thesis, University of Cambridge, 217pp.

[125] Studinger, M; et al. (2006) Crustal architecture of the Transantarctic Mountains between the Scott and Reedy Glacier region of the South Pole from aerogeophysical data, *Earth and Planetary Science Letters* 250:182–199.

the mountains. The form of these valleys suggested strongly that they were glacially eroded but now submerged by ice flowing the opposite way. In two studies resulting from this RES work[126] it was possible to speculate these inland-trending valleys constituted an active phase of erosion by local, temperate glaciers which descended the inland flank of the mountains at a time pre-dating the formation of the extensive ice sheet. This early, mountain phase of glaciation was estimated from other evidence to have been between 40 M and 20 M years ago. The later build-up of the East Antarctic Ice Sheet reversed the ice flow in the valleys and submerged them beneath the thick seaward-directed ice. There would also have been a change in thermal regime resulting in little modification by the later *cold-based* ice (i.e. frozen to the bed.) This research contributed to elucidating the early stages of the glaciation of Antarctica. Some years later it was recognised that the mountains would have been at a much lower elevation during this period, having been uplifted steadily during the last several tens of millions of years.[127]

5.10 To the Interior of East Antarctica and Vostok, the Coldest Place on Earth

Despite the reduced sensitivity of the RES system, it was considered worthwhile to undertake reconnaissance flights to the centre of East Antarctica. The opportunity arose on several occasions, especially when there were poor flying conditions, including low clouds, in West Antarctica. One of the first of these was just before Christmas 1969 with a flight to the Soviet station of Vostok (Figure 5.26). At that time Vostok was the most remote permanently occupied base in Antarctica, lying in the central part of the East Antarctic Ice Sheet (78°28′S, 106°48′E) at an elevation of 3470 m asl. It had been established during the IGY and operated continuously since that time despite logistic difficulties and its harsh conditions. It is the coldest occupied place on Earth[128] and one of the remotest spots on our planet—minus 89.2 degrees Celsius or, more impressively in Fahrenheit, minus 128.6 degrees! The

[126] Drewry, D J (1971) 'Subglacial Morphology between the Transantarctic Mountains and the South Pole', in Adie, R J (ed.) *Antarctic Geology and Geophysics*, Oslo:, Universitetsforlaget, 693–703. Drewry, D J (1972) The contribution of radio-echo sounding to the investigation of Cenozoic tectonics and glaciation in Antarctica, Institute of British Geographers, special publication, no. 4, 43–57.

[127] Smith, A G; and Drewry, D J (1984) Delayed phase change due to hot asthenosphere causes Transantarctic uplift? *Nature* 309 (5968): 536–38.

[128] As of 21 July 1983.

Figure 5.26. Aerial view of Vostok Station on the ice sheet showing the cluster of buildings and the skiway. The thin white line to the right marks the trail made by heavy tracked snow vehicles to Mirnyy.

base was in many ways the Soviet equivalent of South Pole Station of the Americans. It was located close to the south geomagnetic pole—the south pole of an ideal dipole model of Earth's magnetic field that most closely approximates Earth's actual magnetic field—so its scientific importance was considerable. Geophysical and glaciological work had been undertaken by the Soviet Antarctic Expeditions for many years. These had included seismic and gravity ice-thickness determinations on traverses that extended beyond Vostok to the South Pole and the Pole of Relative Inaccessibility. Andrei Kapitza's seismic measurement of ice thickness at the Vostok station was 3700 m and seemed reliable. During the 1967 RES the SPRI party had obtained echoes close by, so this was an opportunity to gain additional data in the area. For some of the current SPRI team there was also the added excitement of landing at the base and meeting Soviet colleagues!

In my diary for 23 December, I recorded: '[T]he plane circled a smudge on the featureless ice sheet that gradually resolved itself into a few snow-covered buildings. We landed on our huge skis and bumped across the ice close to the huts. Greeted by their commander sporting a splendid fur hat and high leather boots we were taken across to the Base comprising just 18 souls.' The buildings were somewhat rundown, probably not because of lack of money but because all the stores were brought up the 1400 km from Mirnyy on the coast at 94°E by tractor train—an 8-week journey. The interior was cosy, and the SPRI party and some of the flight crew were steered

Figure 5.27. Left: Author standing outside the entrance to Vostok Station. The East German and US flags denote overwintering guest scientists. The sign reads 'Welcome'. Right: welcome reception by Soviet colleagues—bottles of vodka and Russian champagne. Lt Elliot, the pilot, stands on the left.

into a rather cramped room in which the most noticeable feature was flowery wallpaper—it, too, brought all those kilometres from the coast. The Soviets were extremely hospitable and had laid out a table in the centre of the room for us (Figure 5.27).

I recall, from my diary, looking across the plastic tablecloth strewn with glasses, sweets, salmon, cheap Russian champagne, and more serious bottles of vodka,

> We were the first visitors the Russians had had since the beginning of their winter exile, half a year before. The Base Chief, a swarthy genial physicist, proposed a welcome toast, 'To our British and American colleagues'. Despite the depths of the Cold War, here in the coldest of all cold places, the atmosphere was warmed by the comradeship of international science nurtured by that remarkable Antarctic Treaty.
>
> We downed our vodka in one—fire flickered down my throat! It had only just recovered from the anaesthetising temperatures outside

where it was –50°C—so cold ice crystals form inside your nostrils as you breathe, and a cup of boiling water if thrown in the air explodes freezing instantly, falling in a myriad of glittering iridescent shards.

The leader of our party, Gordon Robin, responded, 'to our Soviet Antarctic friends'. We downed another vodka already a little heady from the altitude. The Soviet Chief stood once more, getting into his social stride, 'to international cooperation'. More Stolichnaya. Had this to be met with yet another reply? I remember cautiously glancing around at our air crew—the pilots—were they drinking? Someone had to fly us home sober! I spied Lt Elliot. Not a drop had passed his lips, thankfully! We made our farewells, hugs and photos, and we flew off. It was an experience of warm scientific friendship between people flung out across the cold Antarctic—poignant and moving.

It was not until over a month later that there was a chance to return to the Antarctic interior, when the RES programme commenced the first of three long flights termed Plateau I, II, and III. The first was 12 hours long, which in a C-130, given its incessant noise and vibration, was quite an endurance trip. The route was to head for the abandoned Soviet station Sovetskaya, to tie in with the infamous flight in the Super Constellation to this location in 1967. It was interesting to recall that the snow accumulation is so low in the interior (approximately 50 mm of water equivalent per year) that despite this station having been abandoned some 10 years previously, it had not been buried, and the buildings were easily distinguishable, including antennas and other masts. We circled the station a couple of times before flying towards but not over the PoI. Along this section the RES recorded continuous and clear reflections from the base of the ice sheet—rugged terrain of the subglacial Gamburtsev Mountains. The flight continued to the highest known point of the Antarctic Ice Sheet, at over 4000 m. Few if any had been to this spot, which must be even colder and more hostile than Vostok. The height was measured at 4100 m above sea level. The aircraft then turned south to head for the South Pole Station. After refuelling there, the flight carried on north along the 140°E meridian bound for McMurdo.

In 2009 the Chinese National Research Expedition established a new station, Kunlun, at this highest point of the ice sheet, on Dome Argus or A as it later became known (see chapter 14), and at a precisely determined elevation of 4091 m[129]; satellite radar altimeter measurements have subsequently mapped this region of the ice sheet in great detail. It was interesting to compare the contours resulting from the early SPRI flights with the latest surveys in a paper on this topic in 2016.[130] Even all those years ago SPRI hadn't done badly!

The next plateau mission was on 2 February, when just two RES operatives—David Petrie and I—completed a 13-hour marathon. This flight returned to Vostok at altitude, then descended to head to a most northerly point, turning to return to Sovetskaya, and then set a course to the South Pole to refuel before heading to undertake a number of sounding lines in the Queen Maud Mountains.

The final flight, Plateau III, was two days later and the longest, at 14 hours, which Stan Evans and I undertook (perhaps the correct term would be endured!). Once more, Vostok was used as the starting point for low-level sounding, and then a course took the aircraft over the highest part of the ice sheet and into the upper reaches of the immense Lambert Glacier Basin—this was the farthest from McMurdo—2200 km in a straight line. The flight then turned to pass over the PoI.

As described in section 4.3, the PoI is defined as the location furthest from any part of the Antarctic coastline. A tiny Soviet station was established and occupied at this spot for just a couple of weeks during the IGY and had a curious origin which is worthwhile recounting. (Figure 5.28[131]).

In the early planning stages of the IGY in 1955 a meeting was held in Paris to decide where national research stations should be located. For some time prior to this, there had been tacit agreement amongst several of the Western powers that the United States would establish a base at the geographic South Pole. At the meeting Vladimir Beloussov, the chief Russian delegate,

[129] Zhang, S K; Wang, Z M; et al. (2007) Surface topography around the summit of Dome A, Antarctica, from real-time kinematic GPS, *Journal of Glaciology* 53(180): 159–160.

[130] Gran, I; Drewry, D J; Allison, I; and Kotlyakov, V (2016) Science and exploration in the high interior of East Antarctica in the twentieth century, *Advances in Polar Science* 27(2): 1–13.

[131] In a personal communication (November 2020) Olav Orheim recalled an incident on the US Queen Maud Land Traverse that occurred at the Pole of Relative Inaccessibility in 1964: 'If you compare with USSR photos, you will see that the bust is rotated 90° to face west, and given a black scarf. . . . [T]his was a prank that happened during the night between 5 and 6 Dec'.

Figure 5.28. Bust of Lenin at the temporary Soviet station, Pole of Relative Inaccessibility, December 1964. (Courtesy Olav Orheim).

declared to much consternation that the Soviets, besides building a forward station on the coast in Australian Antarctic Territory, would also construct another at the South Pole. Both 'Tony' Fogg[132] and Dian Olsen Belanger[133] in their respective and excellent historical treatments of science policy of this period consider that it was virtually impossible for the Soviets not to be aware of American plans for the South Pole. Georges Laclavère of France had been elected chairman of the conference. He masterfully suggested to Beloussov that whilst the geographic South Pole might be of interest and had already been 'allocated', there was a much greater challenge lying in East Antarctica—the Pole of Relative Inaccessibility—and given the Soviet plan to create a base at the geomagnetic pole (that was Vostok), they would gain "two Poles"! Fait accompli.

From my diary:

[132] Fogg, G E (1992) *A History of Antarctic Science*, Cambridge: CUP, 483pp.

[133] Belanger, D O (2006) *Deep Freeze. The United States, the International Geophysical Year and the Origins of Antarctica's Age of Science*, Boulder, CO: University Press of Colorado, 494pp.

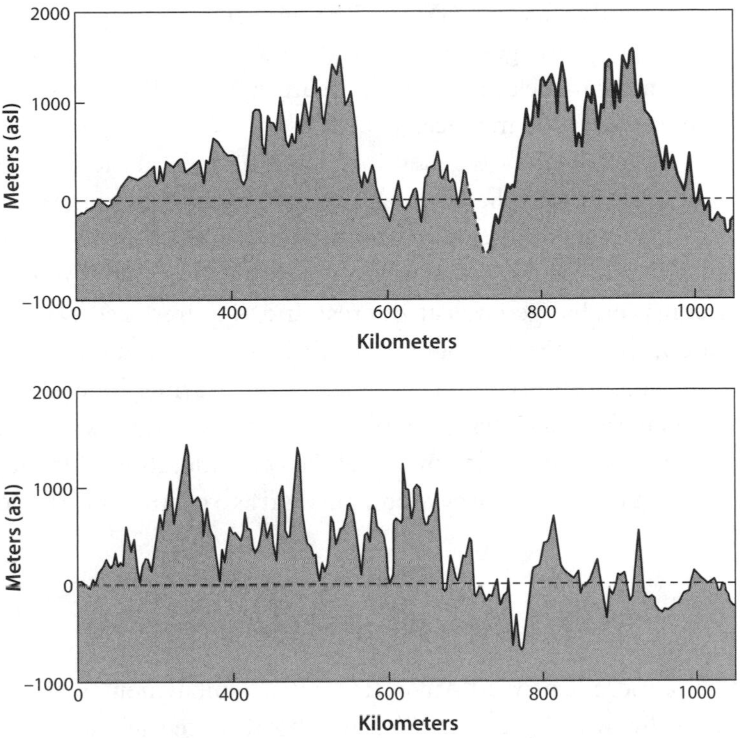

Figure 5.29. Two profiles of the bed topography across the subglacial Gamburtsev Mountains digitised from RES records. The vertical scale is in metres above sea level (0). With isostatic compensation these ranges would be elevated further by several hundreds of metres.

> [W]e planned to pass over the Pole of Relative Inaccessibility. Hamilton with his superb navigational skills brought us bang over this most desolate and isolated of IGY bases. We began to make out some detail of three huts and a number of antennae, as yet uncovered by the small snowfall, but adjacent and dwarfing all else a monstrous bust of Vladimir Ilych Lenin! In the IGY the Soviets had dragged this huge monolith some 1500 miles across endless wastes to abandon it in this uttermost of all places—extraordinary.

The results of these flights combined with those of the 1967 foray were to reveal for the first time some of the detail of the intriguing Gamburtsev range of subglacial mountains (Figure 5.29). These, it will be recalled, were first inferred and named by Kapitza from seismic and gravity measurements during and shortly after the IGY. The RES lines allowed only a very general sketch of the sub-ice ranges, which are extremely rugged west of Sovetskaya. The highest measured elevation for the bedrock was 2940 m—and still covered by more than 1000 m of ice.

Several large troughs were identified opening toward the coast and interior, and I speculated that some may date from an early phase of glaciation similar to that which I had already recognised in the subglacial Transantarctic Mountains.[134] It was clear that this whole remote region was of considerable geological interest and required dedicated future exploration. An extensive programme of RES surveys and other geophysical measurements was conducted in later years during the Fourth International Polar Year 2007–08—the AGAP project (Antarctica's Gamburtsev Province Project)—and led to more detailed consideration of the tectonic architecture of the Gamburtsev Mountains and as a centre for early ice sheet growth.[135]

5.11 The Filchner Ice Shelf

On the Atlantic side of West Antarctica, a great embayment, similar to the Ross Sea, is likewise filled with a gigantic floating ice shelf and in the late 1960s was called the Filchner Ice Shelf (see Figures P.1, 5.21, and 10.4).[136] Some geophysical work had been conducted in the IGY, for example, from the British base at Shackleton, and more intensively by oversnow sorties from the US Ellsworth Station. The main traverse from this latter base undertook a programme of seismic depth soundings that has been described in detail by John Behrendt.[137] The western region (the Ronne Ice Shelf) was virtually unknown owing to its difficulty of access across the ice-infested Weddell Sea and remoteness from existing research stations. In the US the principal interest on ice shelves had shifted focus to the logistically more feasible Ross Ice Shelf. Robin considered it would be highly advantageous

[134] Drewry, D J (1975) Radio-echo sounding map of Antarctica (~90°E –180°), *Polar Record* 17 (109): 359–74.

[135] Ferraccioli, F; Finn, C A; Jordan, T C; Bell, R E; Anderson, L M; and Damaske, D (2011) East Antarctic rifting triggers uplift of the Gamburtsev Mountains, *Nature* 479:388–92; Rose, K C; et al. (2013) Early East Antarctic Ice Sheet growth recorded in the landscape of the Gamburtsev subglacial Mountains, *Earth and Planetary Science Letters* 375:1–12.

[136] At a later date the larger portion of the ice shelf lying to the west of the grounded area called Berkner Island was named Ronne Ice Shelf, and the eastern portion, the Filchner Ice Shelf.

[137] Behrendt, J C (1998) (footnote 23).

to undertake a series of reconnaissance flights over this more distant region. It was not the highest priority for RES, so it was not until towards the end of the flying season in early February that a number of sorties were made to this area, and all these involved refuelling at either the Byrd or the South Pole Station.

The weather over the ice shelf and the adjacent Ellsworth Mountains, the continent's highest—4,892 m (16,050 ft) at Vinson Massif—was consistently poor and not helped by the lateness of the season. Nevertheless, flight lines were able to be fixed with reasonable accuracy using known features as control points in the mountains and, later, locating surface features on satellite imagery. The flights were, therefore, relatively tedious in their execution but gathered important new information about the thickness and flow of this least known of the great ice shelves. It was not until sometime after the 1969–70 season that a scientific story was constructed from these operations.[138]

5.12 Meanwhile in the Antarctic Peninsula

It will be recalled that shortly after the trials in Ellesmere Island, Charles Swithinbank and David Petrie undertook a first season (1966–67) of airborne RES in the Antarctic Peninsula. The thinner ice had been relatively easy to sound, although higher temperatures had affected some of the results. In BAS it had been agreed that a further programme of sounding should proceed as soon as possible using the SPRI MkIV apparatus. Swithinbank was unavailable to participate, and Beverley Ewen-Smith was in charge of the operations. He travelled with the SPRI equipment by ship, on the RRS *John Biscoe*, via the Falkland Islands to US Palmer Station. Ferrying his equipment onto the Piedmont above the base, he was air-lifted by a BAS Twin Otter, albeit with some searching by the pilot, and flown down to the BAS base at Adelaide. After the RES equipment and the antennas were installed, some 10,000 km of track were flown out of Adelaide and the small field station of Fossil Bluff with Andrew Wager, a BAS glaciologist, as the navigator (Figure 5.30[139]). The main areas covered were the Larsen, Wordie, and Wilkins Ice Shelves, George VI Sound, and on the plateau of the Ant-

[138] Robin, G de Q; Doake, CSM; Kohnen, H; Crabtree, R D; Jordan S R; and Moller, D (1983) Regime of the Filchner-Ronne ice shelves, Antarctica, *Nature* 302 (5909): 582–86.

[139] Ewen-Smith, B M (1972) Airborne radio-echo sounding of the glacier in the Antarctic Peninsula, *British Antarctic Survey Scientific Reports* no. 72, 11pp.

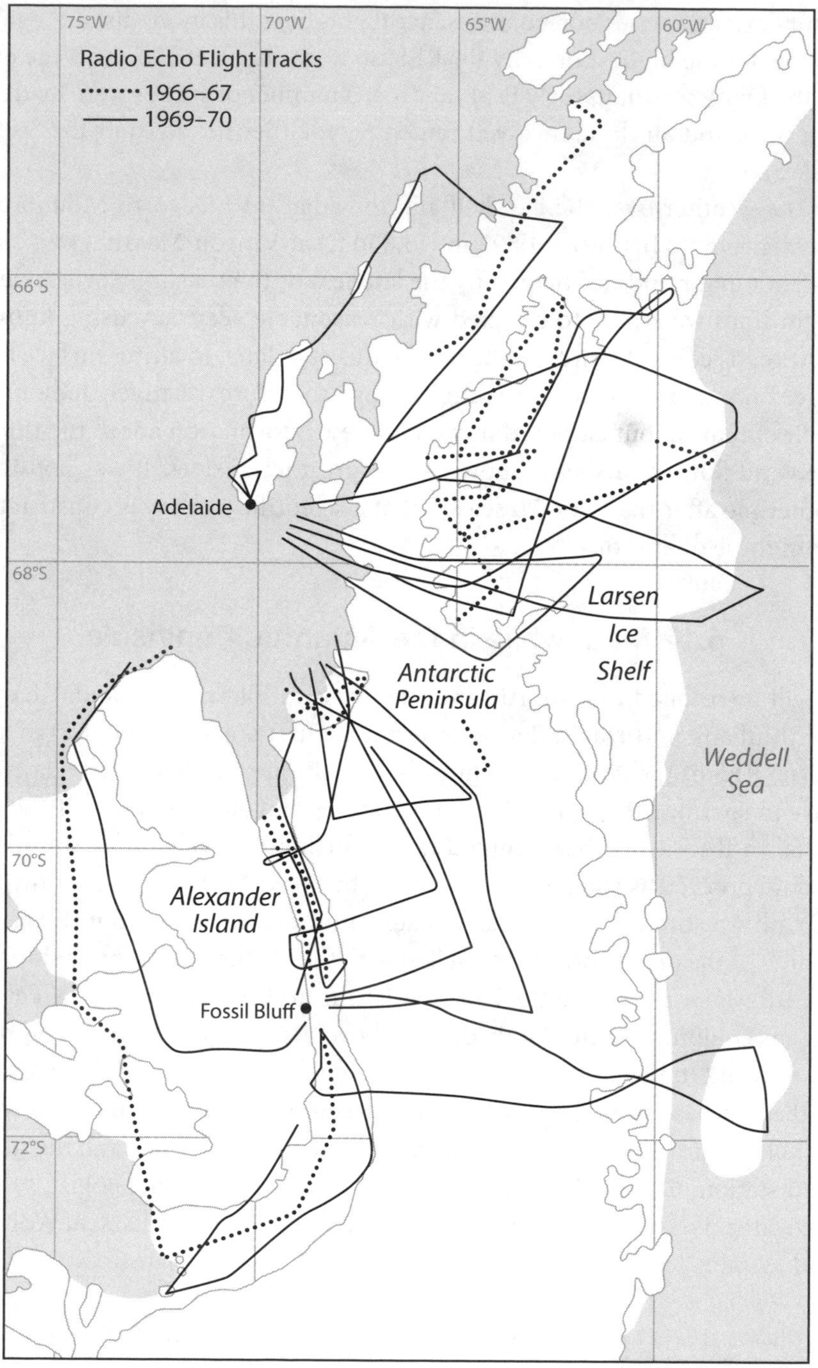

Figure 5.30. RES flight lines conducted in the Antarctic Peninsula. (Courtesy BAS).

arctic Peninsula, where a maximum depth of 1630 m was recorded. Overall, it was a successful season and provided valuable information for later more detailed campaigns.[140]

5.13 The 1969–70 Season in Perspective

The RES flying activities by the SPRI team terminated on 12 February 1970, and all the members returned to New Zealand on 14 February. Figure 5.31 shows the flight lines extending over a wide range of Antarctic terrain. Flight hours attributed to the RES season totalled 330, which included ferry flights to and from Christchurch.

The following is a record of my own experiences:

> For some of us the Antarctic season had been an astonishing time scientifically and experientially. Our survey work had ranged over an enormous expanse of Antarctica from great flat floating tracts of ice shelves to the endless interior . . . ; from the weather-beaten coastal fringes with their sea ice, storms, and incessant cloud to the high, dry and bitter tracts of the interior ice sheet—the coldest place on Earth. We had visited far-off research bases of several nations and met new and interesting colleagues engaged in advancing the frontiers of knowledge. We had seen parts of our planet no person had viewed before and detected with our radars an invisible land beneath the ice. We had experienced the raw and dangerous edge of this elemental continent and yet warmed to its ineffable beauty. Some of us succumbed willingly to the powerful grip of this wild and hostile land, and from which we would never be fully released.

The objective had been to gather data on the thickness of the ice, the form and flow of the ice sheet, and the details of a submerged landscape. The 1967 survey had been extended significantly, and data were generated on an unprecedented scale. It was a turning point, not that this was recognised fully at the time. New avenues of research were opening rapidly from the programme, and as new and inquisitive people joined, the intellectual output swelled. Furthermore, the success of the RES technique began to encourage other groups to develop radar systems in universities and research

[140] Ewen-Smith, B M (1972) (footnote 139).

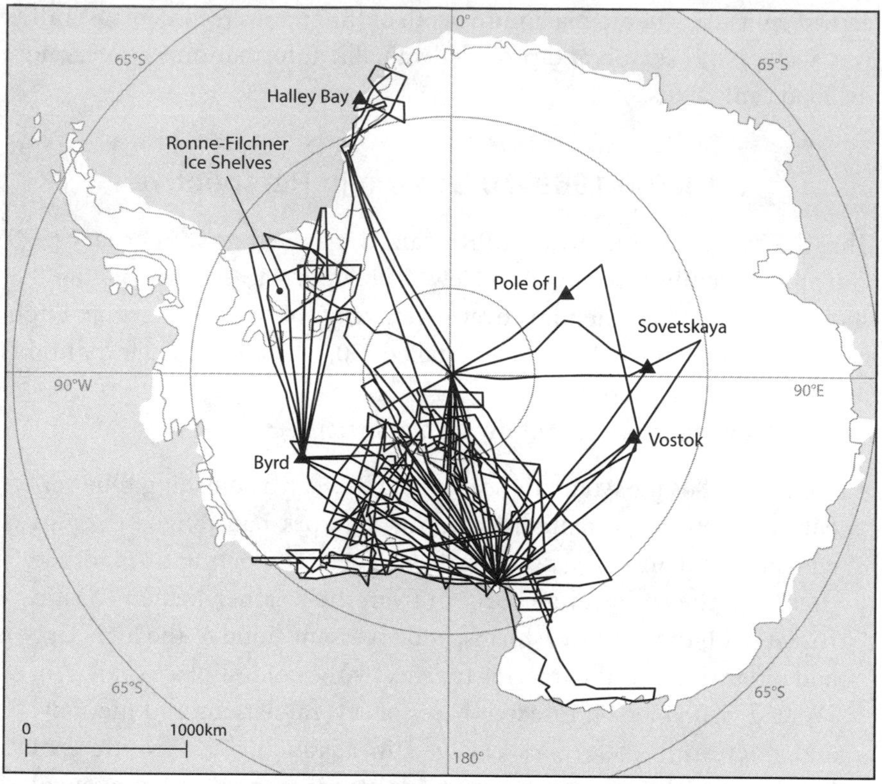

Figure 5.31. RES flight lines conducted in the 1969–70 season. Note the dense radiating lines from McMurdo Station across the Ross Ice Shelf, and the reconnaissance lines over the Ronne-Filchner Ice Shelf. Note the RES flight lines in the Antarctic Peninsula by the BAS are shown at a larger scale in Figure 5.30.

institutes around the world which in the longer term would come together as a new sub-discipline—**radioglaciology**.

In the meantime, a great deal had been learned about SPRI's operational capacity which would augur well for future seasons. The Hercules aircraft had proved a great success. It was a large and stable platform for airborne geophysical work, and there was much space onboard for equipment and crew. Being a military aircraft, it was rugged and capable of modifications to its structure with the installation of aerials and photo ports. The generosity of American colleagues in support of the RES project was acknowledged and seemed assured. An exciting future beckoned.

6

Review and New Plans

The 1969–70 season had ended, and all parties returned to Cambridge; it was time to take stock and plan ahead. There was undoubtedly much optimism and positive thinking in the RES group following the long season. Considerable experience had been gained in operating with the US National Science Foundation and the US Navy in the C-130 aircraft. The RES equipment had performed adequately, but problems had been encountered with the antenna system. Navigation remained a major problem area. Importantly, much data had been gathered that would occupy the team for many months. Momentum had to be maintained. Robin's liaison with the NSF was continuing to be effective, and discussions took place to consider a follow-on season. A critical part of that plan was an emerging international project to investigate the vast interior of East Antarctica. At this juncture, at the commencement of the decade of the 1970s, considerable enthusiasm flourished for widespread collaboration. The Cold War was still holding much of the world in its icy, nuclear grip, but there was a remarkable opportunity to work together in Antarctica through the auspices of its Treaty, now 10 years old and proving a milestone in international cooperation. A central role was being played by the Scientific Committee for Antarctic Research (SCAR) in bringing together and coordinating actions in scientific activity.

6.1 International Antarctic Glaciological Project

Understanding the large-scale dynamics and behaviour of the Antarctic Ice Sheet was stimulating the minds of glaciologists and geophysicists. There was a growing realisation these huge ice masses in the polar regions were not isolated phenomena—a curiosity of those who braved such remote regions of the planet—but ones that played a critical role in the functioning of the world's life-support system. And yet the scientific community was ignorant of many of the important internal aspects as to how ice sheets operate, as well as the cryosphere's external connections to the oceans and the atmosphere. New techniques were being developed that were providing the first and exciting insights. The drilling of deep ice cores had started to reveal a detailed history of ice–climate relations over hundreds of thousands of years, vital for understanding the drivers of long-term ice sheet fluctuations and ice ages. Large-scale modelling of ice flow was emerging as a new sub-discipline made possible by the rapidly expanding capability and computational capacity of digital computers, and theoretical insights into ice sheet dynamics and thermodynamics were suggesting testable hypotheses. Remote sensing was an emerging branch of applied science and although in its infancy, the observations from Earth-orbiting satellites and airborne RES were promising to revolutionise the details, speed, and coverage of data for a number of ice sheet parameters.

In this scientific milieu, it was recognised that individual national programmes were insufficient to tackle the large-scale and diverse questions Antarctica presented. France and the Soviet Union had collaborated on studying the flow line from the Soviet station at Mirnyy on the coast inland towards Vostok Station. This had stimulated discussions with the US, in particular with Bert Crary at the NSF, on possible future and wider cooperative work. If the geographical sector that involved these three nations was to be investigated comprehensively, it would necessitate the participation of the other major country operating there, Australia.

The outcome was the creation of the International Antarctic Glaciological Project (IAGP) at a meeting in Paris in May 1969. It was agreed that the primary effort would be focused on the area bounded by 90°E, the central dome of East Antarctica, and the Ross Ice Shelf, comprising the Wilkes Land and western Ross Ice Shelf drainage systems. A clear set of glaciological aims was formulated requiring national and collaborative programmes with the exchange of scientists between laboratories.[141] This philosophy was seen as necessary to ensure that the project would 'provide complete and comparable results for testing the best theoretical ice sheet models which can be devised at the present time'.[142]

[141] Anonymous (1971) International Antarctic Glaciological Project, *Polar Record* 15 (98): 829–33; Bentley, C R; Budd, W F; Kotlyakov, V M; Lorius, C; and Robin, G de Q (1972) International Antarctic Glaciological Project standardization document, *Polar Record* 6 (101): 349–64.

[142] Bentley, C R (1972) International Glaciological Project, *Antarctic Journal of the United States* 7 (3): 50–52.

From France, Claude Lorius, an insightful and companionable glaciochemist, from the Laboratoire de glaciologie et géophysique de l'environnement (LGGE) in Grenoble was keen to extend French glaciological research. Their plan was to undertake traverses from their coastal station at Dumont D'Urville inland along a line towards Vostok Station, gathering shallow cores for chemical analyses and deeper drilling where appropriate. Bill Budd at the University of Melbourne was the premier glaciologist in Australia and keen to stimulate the Australian Antarctic Division into further work on the dynamics of the drainage basins feeding the East Antarctic coast, and additional drilling on Law Dome, as well as sophisticated computer modelling. Vladimir Kotlyakov, at the Institute for Geography of the Academy of Sciences, Moscow (and later an academic 'survivor' of several Soviet and later regimes), saw considerable advantages of being part of a larger international operation. The Soviet Antarctic Expeditions (SAE) had substantial capability for inland traverses from Mirnyy to undertake surface glaciological measurements of movement and accumulation. At Vostok, SAE was initiating pilot drilling activity to eventually extract deep ice cores for paleoclimate investigations on a massive scale. For the United States the East Antarctic Ice Sheet was at its back door, at McMurdo Station, but it had already committed a high level of financial and logistic support to the Ross Ice Shelf Project and a programme of geological drilling in the McMurdo Ice Free Valleys (Dry Valley Drilling Project—DVDP), so despite US resources, a further large-scale activity would be difficult. The US would, however, contribute logistical support to specific programmes (Figure 6.1[143]).

The first IAGP Council meeting was held in Paris in May 1970, and a wide-ranging plan and data collection specification emerged. It had been decided that a representative of SCAR should also attend the meetings, and Gordon Robin as the new president and a leading glaciologist was invited to participate in that capacity. Underpinning much of the IAGP scientific plan was acquisition of comprehensive data on the surface form and thickness of the ice sheet, and configuration of the sub-ice bedrock surface. The only group capable of tackling such large-scale objectives was Robin's RES programme at the SPRI. This led quite naturally for the US to contribute scientifically through the NSF's support for the RES campaign. It was quickly

[143] *Antarctic Journal of the United States* (1971) 6 (6): 238.

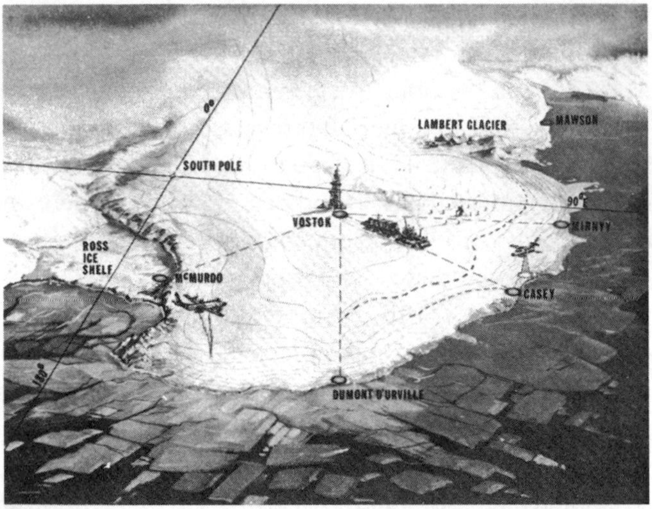

Figure 6.1. Sketch of the IAGP activities. SPRI airborne radio-echo sounding is depicted in the lower left. (Courtesy NSF).

recognised, however, that the UK, through the SPRI, should be a fully integrated member, and an invitation was sent to the Royal Society in London, which immediately nominated Robin.[144] This was a neat and logical solution for the SPRI team, but the pressure was on! The SPRI and its RES results were in the front line with other projected activities needing the ice thickness information and related parameters at an early stage to develop their fieldwork strategies and, later, as regional input for interpretation and modelling. Consequently, the RES field campaigns needed to re-commence as soon as possible, which gave new impetus to solving some of the electronic and navigational problems and deficiencies that had been encountered in 1969–70.

6.2 Aircraft

The shift from the Super Constellation to the C-130 had been of major benefit; the ski-equipped aircraft meant that it could operate Antarctica-wide. Range was satisfactory should there be sufficient opportunities for refuelling at intermediate stations. This procedure was, however, very costly in terms of the inward flights needed to supply fuel at remote locations. A crit-

[144] The UK was formally admitted on 13 July 1973; Bentley, C R (1972) (footnote 142).

ical consideration was whether the RES programme could operate sufficiently early in the Antarctic season to take advantage of the sea-ice runway where take-offs and landings were on wheels not skis. The benefit would be at least one additional hour of sounding per flight. The US programme was typically front-end loaded with a priority to open and resupply their inland bases and deploy field parties; there would be strong competition for aircraft time.

The optimum altitude for RES was within 1000 m of the ice surface when over the ice sheet. This resulted in flying heights over the inland ice (3000–5000 m asl) at which turboprop engines perform much less efficiently and burn more fuel, reducing range. In later seasons considerable attention was paid to achieving the best operating mode to maximise endurance.

Use of the trimetrogon camera array had been very advantageous. Installed by the USGS for its extensive photo-mapping programme, the agency generously supplied complete sets of the photographs taken on the RES missions. Besides their use for fixes, these exceptional high-quality images provided details that augmented glaciological and geographical interpretations.

6.3 Navigation

Perhaps the greatest of all the challenges facing the SPRI programme was accurate flight track positioning, and during the first two seasons this had proved elusive. In retrospect, when global position fixing by satellite has become routine and with accuracies of tens of centimetres, even for fast-moving platforms, the difficulties and high level of uncertainty of location that were wrestled with seem inconceivable. Yet, they were the realities of research in the early 1970s. The 1967 and 1969–70 seasons had used a typical suite of navigation aids available at the time. It was regrettable that several of the instruments which would have assisted and improved track fixing were inoperable—the Doppler and polar path compass.

The parameters recorded on the SFIM chart were helpful, and computer programs were written to reconstruct flight tracks using these data. This was a laborious process, because the traces had to be digitised by hand. Beverley Ewen-Smith spent considerable time investigating the problems of navigation, and this constituted an important part of his thesis.[145] The most

[145] Ewen-Smith, B M (1971) (footnote 80).

useful record was the compass heading, so that significant course changes could be identified readily. In addition to using the SFIM, the aircraft navigator would make regular (half hourly) solar observations to obtain position lines and sun fixes using an aircraft sextant. The accuracy of these observations relied heavily on the proficiency of the navigator, and in both seasons the SPRI team was lucky to have highly experienced individuals who achieved remarkable levels of precision.[146] Within range of mountains or rocky coastal features, vertical or oblique photographs from the trimetrogon camera array could be captured, and fixes to within hundreds of metres could be achieved. The other method was to use the aircraft radar to plot the aircraft's position, and once more this depended on the skill of the navigating officer. Even so, it was estimated that for any one radar fix the error was of the order 2–3 NM (3.7–5.6 km).

The emerging plans for the IAGP presented challenges for navigation of a different order. The area of Antarctica to be surveyed was essentially the quadrant of the continent described by the 90°E–180° meridians. The only way to tackle this enormous region of 4.5 M km^2 (half the size of the US) in a systematic manner was to undertake soundings on a grid with line spacings of 100 km and in some cases of 50 km. The lines, stretching from the Transantarctic Mountains in the east to the East Antarctic coast in the west, were well over 1000 km in length and made a huge demand on consistent, accurate navigation.

Help was, however, at hand! Space exploration, military operations, and commercial air activity also required precision navigation, and those needs had led to the development of a new generation of instruments—inertial navigation systems (INS). First invented for rocketry at the end of WWII and then used in NASA space missions, an INS is a sophisticated automated aid to dead-reckoning navigation. It uses accelerometers to sense aircraft motion, and gyroscopes for measuring rotation, to calculate position and velocity (both direction and speed). By the early 1960s such systems were being used in military aircraft and were increasingly being installed in long-haul jet airliners. INS are particularly suitable for Antarctic flying because they provide position data instantaneously and continuously, are completely

[146] Robert Nyden was a junior polar navigator (Ensign) with VXE-6 during the 1971–72 season and compiled a short article that detailed some of the navigational requirements of the RES programme at that time: 'We called it Ice Sensing', privately circulated PDF document, 29pp.

self-contained, and do not rely on ground stations.[147] In these early days the principal impediment to their installation was the very high cost. By the dawn of the 1970s prices had come down and there was an opportunity to acquire them.

The NSF, in its role as supporting the IAGP through the RES programme, along with the US Navy, undertook to install two IN systems on #320 for the next RES season. This commitment was indeed transformational—it was a step change in the airborne sounding work and shifted the character of data collection from being 'reconnaissance' to fully and effective 'survey'; it allowed very specific research and experimental flights to be conducted where positional accuracy was paramount. Consequently, during 1970 and the early part of 1971 at Cambridge flights were planned to give optimum coverage of the IAGP area within the allocation of flight hours from the NSF.

6.4 Radio-Echo System—Collaboration with the Technical University of Denmark

The SPRI MkIV had worked well and achieved its laboratory performance but had not met expectations because of the problems encountered with the antenna. In Denmark Professor Preben Gudmandsen's group at Danmarks Tekniske Højskole (Technical University of Denmark—TUD) at Lyngby had been developing their system, at 60 MHz, for sounding in Greenland (Figure 6.2). The Electromagnetics Institute possessed a much more sophisticated facility for designing antennas than the one at the SPRI, including a fully equipped anechoic chamber. Gudmandsen's group was already collaborating with the NSF for airborne support in Greenland similar to that being provided to the SPRI in Antarctica and were constructing an aerial system with a high specification and performance that could be mounted effectively and safely on a C-130.

[147] Despite higher levels of accuracy, errors occur from instrumental biases and alignment of the system at the commencement of a flight that lead to 'drift', which can be minimised at the completion of a flight. At the SPRI there was considerable interest in understanding the operation, accuracy, and error problems with the INS. Keith Rose prepared a short report or user guide: Rose, K E (1978) 'Inertial Navigation Systems for Antarctic C-130 Geophysical Operations', SPRI Radio-echo Sounding Programme, 16pp.

Figure 6.2. Preben Gudmandsen. (© ESA–T Schonfelder).

Gudmandsen was a gifted electronics engineer, slight in stature but big in his aims, friendships, and collaborations. The outcome of the discussions between SPRI, TUD, and NSF was an agreement for the TUD-built antennas to be made available for the SPRI operation, and the SPRI would shift the centre frequency of MkIV to 60 MHz. The NSF was to co-fund the Danish development, and the US Navy agreed to the necessary modifications with installation and testing of the C-130 through the US Navy Air Development Center based at Warminster, Pennsylvania. The TUD design moved the antenna from under the tailplane to beneath the main wing, providing a much larger surface area as a reflector. In 1971 the assembly comprised four wooden frames housing butterfly-plan multi-wire dipoles suspended by a series of trapezes made up of steel tubes and fibreglass rods (shown in Figure 6.3).[148] The installation of hard attachment points and cable entry points required opening the wing and undertaking extensive mechanical work. Besides supporting the casings for the main engines and the moveable attitude surfaces, the Hercules wings are 'wet', accommodating several large fuel tanks, and all these elements had to be carefully negotiated during the structural modifications. This work was accomplished once more at the Lockheed plant at Marietta, Georgia. Gudmandsen's group tested the antenna system during two flights in Greenland in September 1971 before it was deployed to New Zealand and thence McMurdo to be fitted for the 1971–72 Antarctic season.

[148] Evans, S; Drewry, D J; and G de Q Robin (1972) Radio-echo sounding in Antarctica, 1971–72, *Polar Record* 16 (101): 207–12.

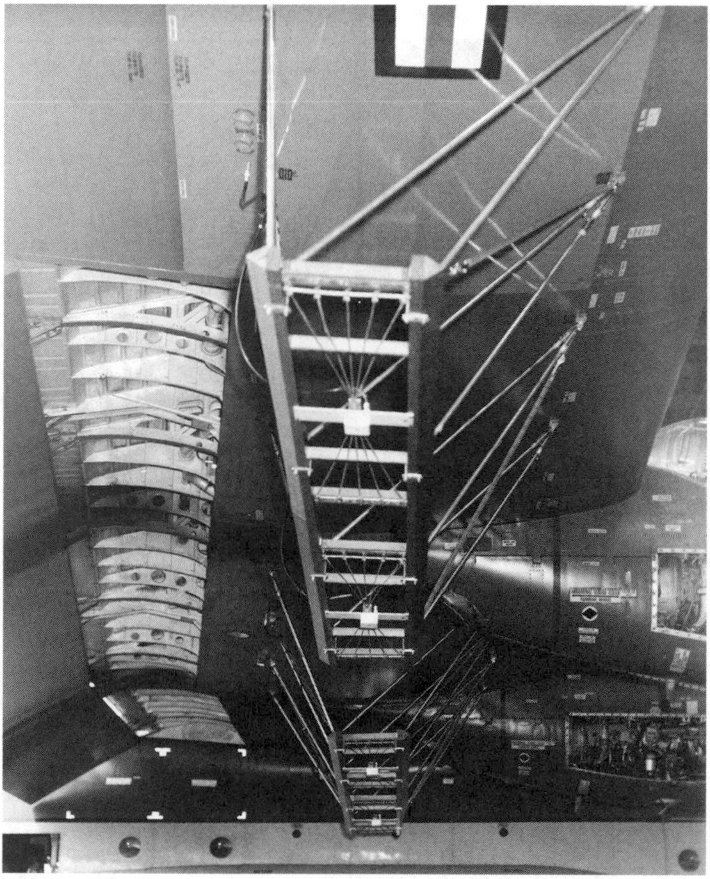

Figure 6.3. The TUD 60 MHz antenna array in 1971 mounted beneath the starboard wing of C-130F #320. In strong air turbulence, even these sturdy wooden frames were seen to flex!

6.5 Deconvolution and Migration

As has already been observed, the relatively wide beam of the RES system illuminates a large area of the land beneath the ice, and the configuration of the antenna results in the shape of the area being an ellipse. The 60 MHz system was typically 110° in the direction of flight and 22° in the transverse direction. Consequently, as the aircraft moves over the surface the radar 'sees' the terrain ahead. The distance or range diminishes as the plane passes vertically over a specific point, and once it has passed over the point, the range increases again as the radar 'sees' the terrain again from behind. For

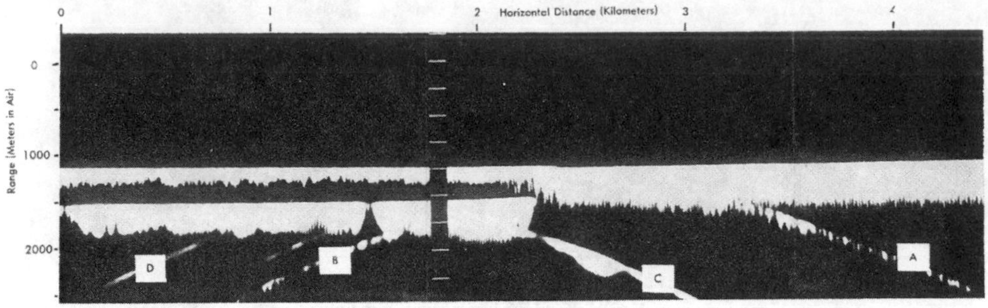

Figure 6.4. RES record from the front of the Ross Ice Shelf (1967). A is the hyperbolic reflection from the exposed ice cliff; B is from the bottom (submerged) corner of the ice shelf front; C and D are reflections from point reflectors on the bottom of the ice (From Harrison (1970); see footnote 149).

example, if the surface is a single point such as a sharp mountain peak, the reflected signal will describe a hyperbola on the RES record. The reflections are always returned at incidence normal to the reflecting surface. This makes interpretation more complicated if there are particularly steep gradients such as deep troughs or very steep mountains. The edges of cliffs such as those encountered at the front of ice shelves and at the submerged base act as corner reflectors and are similarly depicted on the radar records by hyperbolas. An example is shown in Figure 6.4.

Chris Harrison developed a program to 'deconvolve' or 'migrate' the returns from the base of the ice, enabling the reflecting surface to be reconstructed more realistically.[149] Figure 6.5 shows typical ray paths as presented to a radar receiver moving over a wavy surface and the ranges as observed on a radar record, and Figure 6.6 is an example of the output from the Harrison program (see also Figure 4.9).

This was a useful and powerful method of extracting the details of the ice/substrate interface, but the digitisation was very time intensive and realistically could be employed only in small areas where great detail was required, such as cross sections of outlet glaciers. At the larger scale the digitisation process would ignore the hyperbolas and focus on the shortest-range returns.

[149] Harrison, C H (1970) Reconstruction of sub-glacial relief from radio-echo sounding records, *Geophysics* (6): 1099–115.

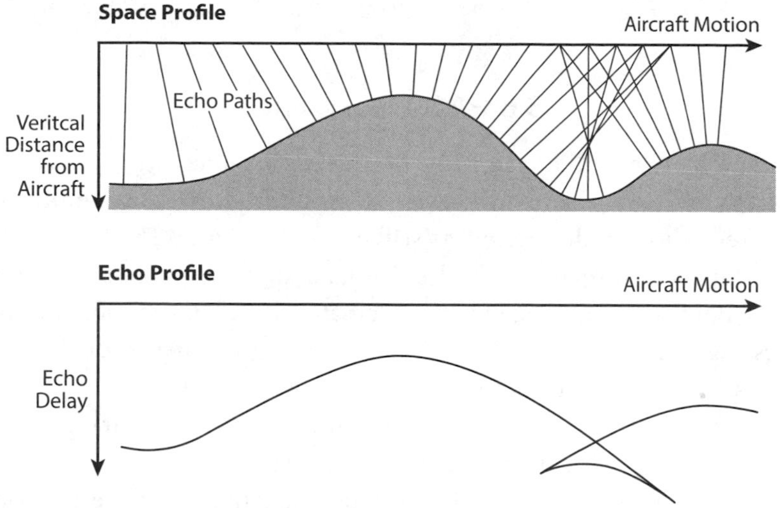

Figure 6.5. Schematic showing the reflections normal to the sub-ice surface (above) and (below) how these combine to trace out a pattern of cusp-shaped echoes (see also Figure 6.6). Note that the steepest slopes cannot be resolved.

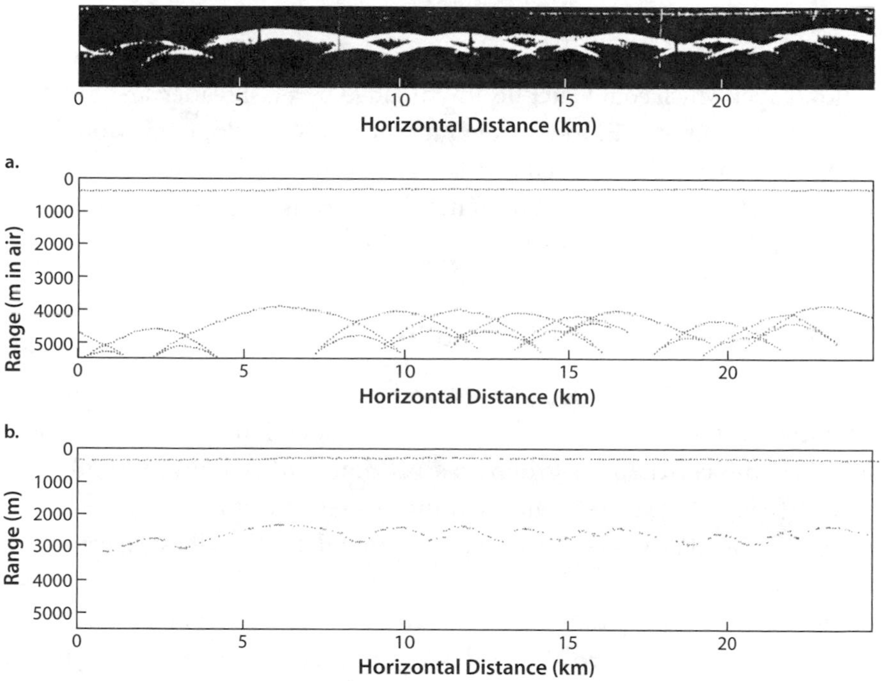

Figure 6.6. Migrated RES record in the vicinity of Vostok Station (1967). Top: RES record of the ice/rock interface. Note many overlapping hyperbolas from undulating terrain. (a) Digitised points from top trace including aircraft height; (b) migrated bed surface (ice surface shown). (From Harrison (1970); see footnote 149).

6.6 RES Recording

Photographic film still appeared to be the best at integrating the RES returns and achieving the highest levels of sensitivity and data compression; digital recording on this scale was still some time away. The RES work in 1969–70 had experimented with the Bimat system for processing the film records, but that had proved to be problematic. More conventional wet-bath processing was agreed to, and the US Navy undertook to provide dedicated facilities, solving the issue of routine post-flight development. The film chosen was RAR (Rapid-access-Recording) at that time a military grade ESTAR-AH based product manufactured by Kodak.

The recording of the navigational data emanating from the INS was a new and challenging problem. The team considered various options; magnetic recording was rejected because the systems available were insufficiently rugged for operations in aircraft in Antarctica. The chosen method was based on sampling the INS output (which was time coordinated with the RES crystal clock), namely latitude and longitude, and transferring the data onto punched paper tape. This process generated a large tape of distinctive yellow paper which could later be downloaded by a fast reader, and the data stored in a computer. In today's digital world this development would appear rudimentary if not antiquated, but a Rubicon had been crossed—there was no further requirement to digitise numerous lines on endless SFIM charts (unless to fill in accidental gaps)! This was progress.

6.7 The Team

The personnel in most university research groups change over time as new people join, and others, particularly research students, depart after completing their theses. The RES group was no different. Bev Ewen-Smith had completed his PhD in 1970 and had published papers on navigation and computer-aided aspects of mapping scattered data, as well as survey work in the Antarctic Peninsula.[150] Chris Harrison was in his third year and busy completing his dissertation, mainly from the data collected in 1967.

David Petrie had returned to BAS for a short period before leaving there, too. He was replaced by a new recruit—Michael Gorman. Michael had been

[150] Ewen-Smith, B M (1971) Algorithm for the production of contour maps from linearized data, *Nature* 234 (5323): 33–34.

an electronics technician with the US Air Force and was something of an Anglophile. He was looking for a new and challenging opportunity. Both his military and technical background proved very valuable, and he remained working with the SPRI, with a few breaks, for 40 years! Julian Paren had also finished his thesis under Stan Evans in 1970, on the electrical properties of ice, which was principally based on laboratory results. Whilst Paren had not been part of the fieldwork team, his experimental studies were highly germane in providing a better understanding of radio wave propagation in ice and snow. Paren had worked for several months at the US Cold Regions Research and Engineering Laboratory (CRREL) in New Hampshire undertaking experiments. Besides confirming earlier measurements of electrical permittivity, his research later assisted in studies of layering in the ice sheet.[151]

Robin had applied for financial support for a new phase of RES research from the Natural Environment Research Council (NERC), to match the generous provision in kind from the US. His application was successful and was accompanied by requests for new PhD studentships, and the group was soon joined by Gordon Oswald (from Bristol), who was to assist Stan Evans in particular with a number of the electronic developments. Evans had another research student, Les Davis (from Canada), who was investigating RES sounding of temperate glaciers. He undertook fieldwork in East Greenland on the Roslin Glacier in the summer of 1970 using a modified version of the higher-frequency SCR-718 radio altimeter and later conducted experiments on the Upper Aletsch Glacier in Switzerland.[152]

Collaboration, it has been seen and demonstrated, is a prominent and widespread feature of research activity in Antarctica, and without the strong support of the US NSF and the common interests of TUD, the SPRI programme would not have flourished. It also became clear that a number of new groups around the world had observed the power and effectiveness of RES and wished to pursue similar work both in Antarctica and other ice-covered areas. In Belgium, Professor Tony van Autenboer and his research colleague Dr Hugo Decleir (at the University of Ghent) were geophysicists who had made a gravity survey in the Sør Rondane in Dronning Maud Land with Expéditions Antarctiques Belges. Importantly, they had undertaken a

[151] Paren, J G (1970) 'Dielectric Properties of Ice', PhD thesis, University of Cambridge, 233pp.

[152] Davis, J L; Halliday, J S; and Miller, K J (1973) Radio-echo sounding on a valley glacier in East Greenland, *Journal of Glaciology* 12 (64): 87–91; Davis, J L (1973) 'The Problem of Depth Sounding Temperate Glaciers', MSc dissertation, University of Cambridge.

detailed airborne radio-echo sounding project in February 1969 using an SPRI MkII sounder purchased through Randall Electronics and with advice and training from Stan Evans in Cambridge. The equipment had been installed in a Cessna 180 aircraft and flown over the Jelbartisen-Trolltunga-Fimbulisen area, out of the South African National Antarctic Expedition (SANAE) Station.[153] Decleir expressed an interest in further developing his expertise in RES, and following discussions with Robin and Evans, it was agreed he would join the team for its next Antarctic operation. He would prove to be a very obliging and effective companion.

6.8 New Plans and Preparations

By the end of 1970 the tempo of interest in East Antarctica from the emerging IAGP was driving and shaping the next field season. With the technological developments that were in the offing and the lead time necessary to accomplish electrical, electronic, and mechanical work on the airplane, the earliest possible season for further sounding would be 1971–72. Even this seemed to be a daunting proposition, given the tasks and the people available to do them. Robin, who had responsibilities as president of SCAR and in Cambridge with Darwin College, as well as running the SPRI, decided not to participate in the season and handed responsibilities to Evans to lead the group. Evans moved ahead with the TUD, and with the assistance of Gorman and Harrison, commenced preparing the SPRI MkIV systems, moving the centre frequency to 60 MHz. This change would also entail linking into the new antennas, and the necessary technical details were discussed with NSF. One important improvement was the design of a logarithmic receiver, which gave a wider dynamic range (–60 dB). The author was to look after the navigation and flight planning and aspects of the glaciological observing. Decleir came over for a brief period of familiarisation and would assist wherever necessary. Oswald joined only in the late summer of 1971 and busied himself with learning as much as he could about the RES system and Antarctica. The group was small, facing a big challenge, but poised and expectant to take the next step.

[153] van Autenboer, T; and Decleir, H (1969) Airborne radio glaciological investigations during the 1969 Belgian Antarctic Expeditions, *Bulletin de la Société belge de géologie, de paléontologie et d'hydrologie B* 78 (2): 87–100.

7

The Continental Survey Begins

A Land Emerges

The summer of 1971 was busy with preparations for the upcoming field season. With a format similar to that of the previous programme, the NSF and US Navy had allocated some 300 flight hours to the RES activity. Little were they or we to know what lay in store in Antarctica that would lead to one of the most logistically vexing and yet scientifically rewarding seasons. It was understood we would be operating in the same C-130, #320. Being one of the older 'F' models, this aircraft did not have the additional wing fuel tanks of later versions; hence, range/endurance was limited. This was frustrating given our plans for the extensive grid over East Antarctica. Nevertheless, the knowledge and experience of operating in this airplane had its advantages.

7.1 To Washington and New Zealand

The SPRI party comprising Evans, Harrison, Drewry, Gorman, and Oswald departed for Washington, DC, on 20 October 1971 and rendezvoused with Hugo Decleir, who had flown in from Belgium. The next two days were spent at the NSF, where we were introduced to Dr Joe Fletcher, the new head of the Office of Polar Programs and a well-known climate scientist with considerable polar experience. We met Phil Smith, who was deputy head and whom we had previously encountered, as well as Bob Dale, who had assisted us in McMurdo during the previous RES season. Dave Bresnahan had recently joined the OPP and became a valued supporter in later years. Ken Moulton, a veteran of the US Antarctic operations, briefed us on the circumstances of the current season and, importantly, informed us that the new admiral (L M McCuddin) commanding the US Navy Task Force in Antarctica was proving a helpful individual. We were shown a film of the flight trials of the new antenna system mounted on #320, which appeared to have been successful. A further visit was made to Bill McDonald at the USGS, who informed us of US programmes for satellite remote sensing; this new era of surveillance was already underway.

We departed from Andrews Air Force Base in the evening of 23 October and flew once again with Military Airlift Command in a C-141 to New Zealand, with stopovers in San Francisco, Hawaii, and Pago Pago in American Samoa. I recall that when we were served our in-flight food, we were given a carton of milk which, on examination, sported the statement 'Guaranteed to contain no animal products'. It was a further reminder we were now operating in a different universe—a US military environment!

The day after our arrival in Christchurch, 27 October, we called on Walt Seelig, the season's USARP representative at the US facility at Harewood Airport. Walt, whom we had met briefly in Washington in 1969, was a seasoned NSF executive who had been working with the Antarctic programme for many years. Tall, tanned, and affable, with a 'mature crew cut', he cheerfully covered a variety of issues regarding the deployment of our aircraft, personnel with whom we would be dealing, equipment, and general logistic issues. We were reunited with our cargo of RES electronic and other gear and liaised with the aircraft maintenance crew at the US Navy hanger at the airport. So far, all systems were normal, and our warm welcome was so different from what we had experienced two years previously.

Our airplane, #320, arrived from Antarctica on 31 October. We met the aircrew who would be flying our missions. Several had brought their families to Christchurch, where they had rented properties for the summer season. We were invited to a couple of parties and gradually began building a rapport.

The complications that emerged over our departure for Antarctica were somewhat symptomatic of the many we were to confront. It was on reflection one of the most difficult periods, when uncertainty, the weather, and interminable problems with aircraft operability plagued not only our project but many other US activities. We were informed that #320 would have to deploy south without all the RES equipment; the remainder would be brought in on a flight shortly thereafter. However, the principal problem was that not all the RES team could be taken onboard, as the flight was completely full of aircrew and other personnel. After negotiation, the squadron agreed that two of our number could fly down to McMurdo on #320—and

Figure 7.1. Tight personnel transport in a crowded C-130. The seats are canvas and webbing. (Courtesy SPRI).

it was decided these should be Evans and Gorman. Evans, as the leader of the team, could liaise directly with the NSF and VXE-6 staff in McMurdo and along with Gorman make some early inroads on equipment installations. The four remaining would have to await another flight, which materialised more rapidly than we had assumed. A C-130 was to leave for McMurdo on 6 November carrying a spare helicopter, and we were told to join that flight, shoehorning ourselves into the hold on the unyielding webbed seats along the side of the plane between the bulk of the helo and the aircraft fuselage (Figure 7.1). We arrived stiff and tired in McMurdo the next day.

7.2 Antarctica—Delays and Frustration

Once in McMurdo we were accommodated in a repellent and foul-smelling set of bunk rooms which went under the apt title 'Vermin Villa'. It was used mainly for short-term transit personnel deploying to other bases or into the field. Since we were to be working out of McMurdo for the next two months and operating at very unsocial hours, we considered we needed somewhat better facilities. The new 'USARP Hotel' had available accommodation, and following discussion with Chris Shepherd, the NSF representative in

Figure 7.2. Buildings in the snow. Left: typical Jamesway at Siple Station. (Courtesy NSF, USAP Photo Library). Right: Thiel Earth Sciences Lab with Observation Hill behind.

McMurdo, we were soon all ensconced in much more agreeable shared twin rooms. Evans had negotiated facilities in the Field Center which would serve very adequately for any laboratory work on the RES systems. Additional space was made available—three rooms in the Earth Sciences Laboratory, where we had been comfortable two years previously (Figure 7.2). Indeed, some of the filing cabinets still had our SPRI labels stuck on them from 1969. We had two rooms for flight planning and general work, and one, equipped with a light table, for sorting and inspecting the RES films.

The first flight was a dry run to the South Pole on 9 November without the RES operating; indeed, the antennas were not yet fitted. This was an opportunity for Hugo Decleir to familiarise himself with the configuration of the C-130 and the US Navy operating procedures, as well as to obtain a feel for the terrain—the ice shelf, mountains, and plateau. On 10 November a formal planning meeting was held with the NSF, VXE-6 Squadron, and #320 airplane commanders to discuss details of the RES missions and necessary procedures. There were 14 participants, including Cdr Claude 'Lefty' Nordhill, the commander of VXE-6; pilots Elliott, Gunning, and Couch; as well as the NSF representative. This proved a positive introduction to our operations. Two days later the four 60 MHz antennas with their mass of bracing struts were assembled and mounted beneath the wings of #320.

On 13 November the first full test flight was undertaken with equipment and aerials operational. The seven-hour mission left McMurdo with the intention of evaluating the performance of the radar ensemble over different ice environments. We commenced by flying over the Ross Ice Shelf to its ice front (see Figure 5.20) and then continued across the open Ross Sea to undertake a series of manoeuvres to evaluate the directional properties of the antenna. In a mode similar to that used in previous seasons, this entailed flying at fixed bank angles and recording the power of the returned signal as described in section 5.4. A small bias was discovered with the main radar beam skewed some 5° to starboard. However, there was some good news: first, the system sensitivity was significantly higher than in 1969, and, second, there was much better matching between the antenna array and the transmit/receive switch port. The latter lessened any spurious ringing in the cables and resulted in a cleaner record of the returned signals. The flight continued to the Nimrod Glacier, passing along its length to the high plateau. Admiring the spectacular peaks and glaciers of the Transantarctic Mountains as we returned to McMurdo reminded us how great a privilege it was to fly amongst these remote and desolate ranges and for some of us how much we had missed the mental and physical stimulation of this harsh but beautiful land.

Back at base we assessed the test flight and were well satisfied with the performance of the SPRI system and the new TUD antennas; they met our expectations. It became apparent that the new configuration was going to open up exciting opportunities for us in the deep ice of East Antarctica. The team was enthusiastic and fully ready to commence operations. The first science flights were planned to commence the next day, 14 November, with two missions scheduled, including a segment devoted to an aerial photographic task for the USGS.

With everything poised for starting the programme our hopes were shattered when we learned that early in the morning a trainee pilot had overtorqued the outboard port engine, which had incurred some structural damage. The strict safety code followed by the US Navy necessitated that the aircraft be returned to Christchurch for crack tests to be conducted on the engine mounting. We were told this would take three days. Little did we know that three days would turn into more than 30 before we could recommence our programme; the frustration and chaos of the intervening period was exceptionally sapping.

Several minor diversions occupied members of the team during this uncertain stretch of 'down time'. The new inertial navigation units required accurate geographic coordinates to initialise the system and determine closure errors upon return. We considered the set of coordinates the Navy

Figure 7.3. Decleir (left) and the author surveying Williams Field air facility; White Island is in the background.

used for Williams Field to be only approximate—good enough to get the aircraft back to McMurdo but not sufficiently exact for scientific missions. Consequently, Hugo Decleir and I planned and then conducted over a few days a small geodetic survey to accurately position several features around the air facility where our aircraft could set the INS with precision (Figure 7.3).

The second activity was a mini-expedition into the ice-free Dry Valleys of Southern Victoria Land, where Decleir and I were deployed by helicopter and camped for a week to study the periglacial features of the dramatic inner region of Wright Valley (Figure 7.4). The foray relieved the monotony of unproductive life in McMurdo, and organising an expedition-within-an-expedition was to prove useful experience and a template for future occasions when an additional research project was undertaken in the Dry Valleys.

In the meantime, Mike Gorman had arranged to undertake a survey of the lower part of the Ferrar Glacier to measure its speed and was flown out by helicopter with a set of survey equipment.

On 17 November the plane was still in McMurdo and had not yet flown back to Christchurch; we could see that further delays would place mounting pressure on our programme. What followed was, in retrospect, a com-

Figure 7.4. Upper section of Wright Valley with massive screes on the valley flanks and, in the valley bottom, two small frozen lakes. Note thick dark layers of dolerite sills in the upper middle distance.

bination of serious errors in management of the US programme, a period of prolonged bad weather, and a succession of technical problems with the C-130 fleet.

The USARP operational plan in the 1971–72 season was highly committed—some might say over-programmed. A major project was to establish a new US station at the base of the Antarctic Peninsula—Siple. This necessitated the transport of very large quantities of cargo by C-130 and most of the flights required refuelling at Byrd Station. The fuel had to be flown to Byrd from McMurdo, and the resulting logistics cost in terms of aircraft hours was immense. The maintenance of South Pole Station and its scientific programme was always paramount, and new field-base projects had been agreed to with a high demand on C-130 time.

Two and a half weeks into the field season the US Navy had two aircraft grounded with serious problems in addition to #320. Discussions were being held regarding the viability of our RES programme. One scenario was that it might shrink to one week of flying at the end of November followed by another at the end of December. This would severely curtail the overall flying hours we had planned. During the next two weeks no clear strategy emerged on handling the competing demands and priorities for air support. At almost every enquiry there was prevarication and procrastination.

On 4 December came the worst news. A C-130 crashed on take-off at D-59, a traverse station about 250 km inland of the French base Dumont D'Urville on the East Antarctic Plateau. Fortunately, nobody on the plane was injured, but the aircraft suffered severe damage when two jet-assisted take-off (JATO) rocket bottles broke loose and destroyed no. 2 inboard (port) engine, and fragments damaged no.1 engine. Furthermore, the nose ski buckled on landing. The Hercules had been supporting a party on the French traverse from Dumont D'Urville as part of the IAGP activities. An inspection of the airplane concluded it could not be repaired, and it was abandoned on the ice sheet.[154] The pressure on the remaining three aircraft and the flying schedule had reached the breaking point.

Of course, altruism is not the most readily observable hallmark of hungry, enthusiastic, and self-motivated research scientists. For the RES team already frustrated by endless delays, these developments were hugely disappointing, but the team was not going to give up its fight to rescue some sort of credible programme from the season of disasters. We had to recall that the IAGP was depending on the timely delivery of SPRI ice thickness data. After careful consideration it was decided to give the Navy an ultimatum—that the RES programme get 100 hours (at a minimum) of flying *immediately*, or the whole team would go home!

This blunt but necessary approach caused a minor volcanic eruption amongst the USARP and Navy staff, especially as it was made known that Evans had booked a phone call to Gordon Robin in Cambridge. Robin was president of the SCAR and in a powerful position to assist behind the scenes and had very good connections with the senior staff at OPP in the NSF in Washington. Undoubtedly, this psychology exerted considerable pressure in McMurdo, and the situation changed dramatically! Evans was called to a meeting with Admiral McCuddin during one of his infrequent visits to the 'Ice', the outcome of which was that following further consultation, flying for the RES programme could re-commence on 14 December and continue until the end of the calendar year.

[154] Nordhill, C H (1972) Air operations Deep Freeze 72, *Antarctic Journal of the United States* 7 (5): 215–17.

7.3 The Science Begins—Eventually!

On 19 December 1971, later than expected owing to bad weather during the previous few days, and more than a month since we had been buoyed with high expectations from our test flight, the first of the fully science-focused flights took off at 0200. But during this mission further problems were encountered—one of the struts holding an antenna array broke, and shortly thereafter another snapped off completely and necessitated cutting short the flight. Nevertheless, we had achieved 7½ hours of high-quality sounding, and we began to perceive the power of the new system. Out on the plateau the team had measured ice depths of 4.2 km. Furthermore, the new punched paper tape system for recording the navigational data had worked well. If we could fix the antenna hitch, it was time to speed up the operation and commence flying around the clock!

Back at Williams Field the snag over the breaking struts became the headache of Bob Ball, a hefty, unflappable Navy chief warrant officer cast in the mould of the can-do US military men of Hollywood films! With energy and amazing resilience to the piercing cold conditions on the exposed ice shelf, he worked doggedly to solve the mechanical problems and throughout the next three weeks kept the antenna structures operating with hardly any interruption.

Over the following days the team steadily built up the hours and the grid of flight lines over East Antarctica. There was, of course, a short break for Christmas, but we were back flying on 26 December, by which time eight missions had been flown in a total of 80 hours. The results were excellent, and a new maximum ice depth of 4540 m had been recorded. By 30 December, 130 hours had been clocked. Bad weather at McMurdo on that occasion required we spend a night at South Pole Station, which was an interesting experience.

During some of the spare time I was shown the 'frozen' food store, which may seem a contradiction in terms given the average temperature at this place is about –50°C. The bulk frozen food, flown in from the US via New Zealand and McMurdo, cannot be kept in the heated base buildings, so a large shallow-angle shaft had been driven into the ice into which a truck can be guided. Cases and boxes can then be stacked here in a very effective refrigeration system. I was shown down the shaft, and as we progressed, by tens of metres, past the stacked stores, I remarked that there seemed to be a great deal of foodstuffs, even given the need to have supplies for an emergency winter. My companion from the base explained that the operation always ordered more food than could be consumed, and so, over time the line of boxes moved steadily outwards. He remarked that at the very end there was even food from the building of the base in the IGY!

The last flight was completed on 4 January, bringing the season total to 16 missions, 162 hours of flying, and some 60,000 km of track flown. It was a highly creditable, high-pressure performance by the team, given that at one point we were on the cusp of aborting the operation. Even so, it should be recalled that this was only half the total hours that had been planned back in the summer of 1971.

By and large, the majority of the fights were uneventful as the aircraft cruised at low level above the seemingly endless ice sheet. Gordon Oswald has clear memories of some of those missions:

> Those operating the radar had the racks of equipment in front of them, which needed attention. Transmitter power, clean waveforms, switch positions, INS monitoring, film replacement etc. needed checking (modern RES equipment includes digital data recorders. In 1971 . . . we had multiple 35mm film cameras recording oscilloscope traces and needing new film every few hours).
>
> Flying steadily at hundreds of metres over the ice, hour after hour, excitement was hard to find. However, there was a drama unfolding on the oscilloscope traces, which showed, for those privileged to see it, what the radar saw. Looking out of the C130 porthole, was the smooth white surface of snow covering the ice. Looking at the traces, a patient eye could see the radar reflection from perhaps 3000 metres beneath. . . . Just sometimes, movements in the trace could tell the operator that under the smooth surface, 1,000-metre mountains and steep escarpments were buried, invisible.[155]

In 1969–70 several flights had taken the opportunity to sound the great outlet glaciers of the Transantarctic Mountains on either the outward or return legs. There were still several that had not been flown, particularly in Northern Victoria Land. Flying through these massive defiles—more than 100 km long and unsurpassed in scale anywhere else in the world—was exciting stuff! Hugging the ice surface, we would be dwarfed by the sheer magnitude of the scenery. But there was a sting in the tail. The glaciers acted

[155] Oswald, GKA, personal communication, December 2020.

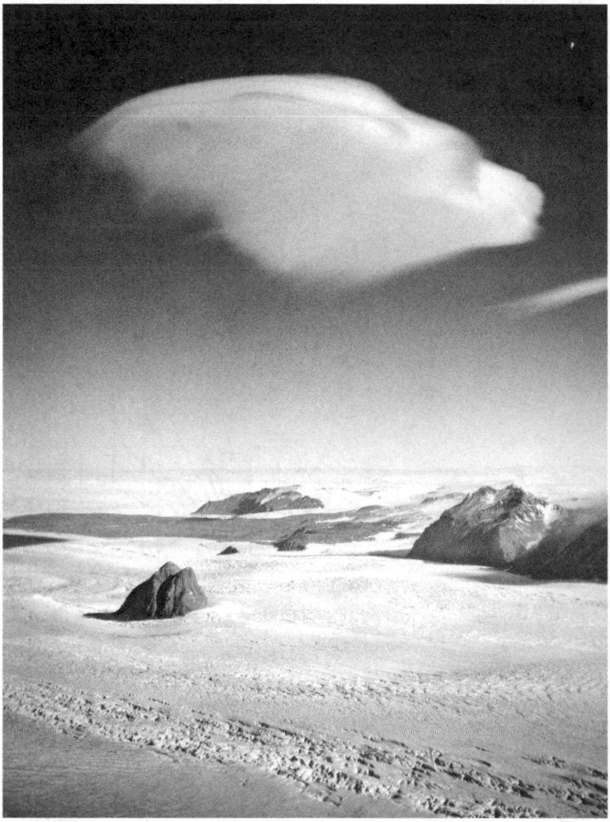

Figure 7.5. Isolated Hansen Nunatak on the Reeves Glacier, northern Victoria Land, with its standing lee-wave cloud indicating the presence of possible strong vertical uplift and turbulence!

as natural pathways for the down-slope drainage of cold air from the inland plateau, that is, katabatic winds. The air, tumbling down and channelled by the mountain walls, was turbulent and fierce. Flying close to the surface of the ice, even our bulky Hercules was tossed around like a cork, and the sturdy wooden mountings for the RES antennas would be seen to flex wildly.

On one of these occasions, descending the Reeves Glacier, we noted a large isolated nunatak in the centre, around which the ice was flowing (Figure 7.5). It was an ideal place to obtain a very precise navigational fix if we were to fly over it. From the flight deck I noticed a lee-wave cloud sitting over the nunatak and above our much lower flight level. Located behind the pilot, I was standing with a button to press as we passed over the top of the nunatak to mark the RES film. On reaching the nunatak, the aircraft was

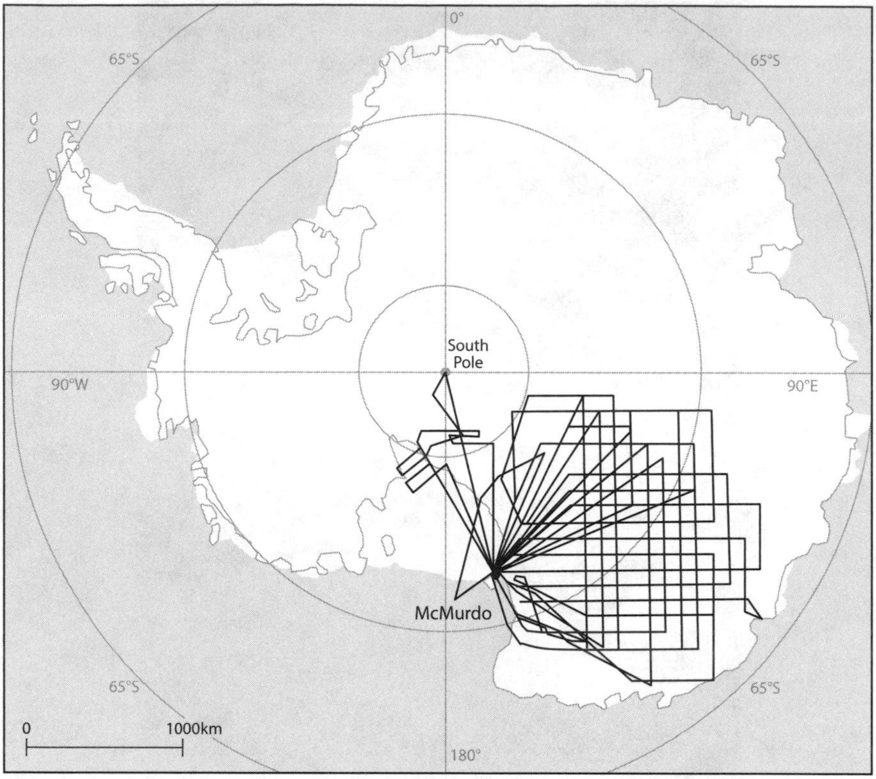

Figure 7.6. RES flight lines conducted in the 1971–72 season.

forced violently upwards by the airflow over the peak. The g-force was so great my legs collapsed, and I found myself sprawled on the floor of the flight deck. The turbulent event was short-lived, and I rose to my feet again. I was asked if I was injured, since I was not strapped in. All I could answer at the time was, 'I pressed the button; I got the fix'.

In most cases the bumpiness would continue for a long period, but eventually we would pitch out of this torment and onto the vast plain of the Ross Ice Shelf, with its calmer, smoother air. Over the course of our RES programme of several years I had the opportunity to fly down every one of the major outlet glaciers, and in many ways each was an unforgettable experience.

The shortage of aircraft had an impact on our return to Christchurch and thence to the UK following the completion of the RES programme (Figure 7.6). Freight was to be shipped north on a US Navy cargo vessel and would be delayed getting back to Cambridge. Two hundred personnel were also scheduled to be transferred to New Zealand by ship. Fortunately, the SPRI team was not amongst them, and our flights north were to follow quickly. We left McMurdo on 7 January.

7.4 A Land Emerges

The 1971–72 Antarctic season had delivered scientifically despite its substantial abbreviation and the frustrations that delays and aircraft mechanical problems had posed. The continental-scale survey in East Antarctica confirmed the efficacy of the improved SPRI MkIV system, the outstanding performance of the Danish aerials, and the step change in navigation to the inertial systems. The quality of the data overall was excellent. Now, the task was to commence reducing the records and to begin piecing together in a systematic manner the patterns we had seen emerging during the operations.

The several members of the team who had participated in the season returned with enthusiasm and a host of ideas and concepts to be worked upon. Chris Harrison had a clear brief—to finish his thesis. Gordon Oswald was to examine the physical nature of reflections from the base of the ice sheet, where he had observed considerable differences in returned power; he suspected a number of these with very high reflection coefficients indicated water at great depth. His work would lead to some startling results. Gordon Robin, although not part of the season's team, was anxious to incorporate the accurately navigated flights over the Ross Ice Shelf into his study of its dynamic behaviour.

I was to use the full geographical coverage of radar data to map out and investigate the large-scale land surface beneath the ice in East Antarctica; chart its mountains and basins; and commence a study of its likely geological evolution, connections with other areas of the former super-continent of Gondwana, and the role of these interior highlands in the early evolution of the ice sheet.

There were other tantalising topics for research, such as investigating the internal reflecting horizons. We now had much more detail of the layering that extends laterally on a scale of hundreds of kilometres and to great depth. The hypotheses and speculations that had arisen from the oversnow studies along the Camp Century trail in Greenland could be tested. What would these layers disclose about physical and chemical processes at the surface and within the ice sheet, and how old were the deepest observed? How far could they be traced, and what would they reveal at continental scale about the flow and dynamics of the ice sheet, perhaps hinting at climatic changes over several millennia? There was much to occupy our small band of glaciologists for months, indeed years to come. We follow the story of the layers more fully in chapter 14.

7.5 Mapping of East Antarctica

The 60,000 track-kilometres of radio-echo soundings required immediate attention to turn them into data that could, in the first instance, be mapped to produce charts of the morphology of the surface of the ice sheet, ice thickness, and the sub-ice topography. The prospect was enticing. What lay below this vast area—great plains, more mountains chains, scattered highlands? We had already identified and had glimpses of the Gamburtsev Mountains, for example, from the earlier RES work, but now we had a systematic grid to assist in this survey.

Gordon Robin recognised the challenge and was successful in obtaining further funding from the Natural Environment Research Council and proceeded to recruit additional staff who would assist in the task of data reduction. The many hundreds of metres of RES 35mm film needed to be digitised to extract the ice thickness. In addition, the information had to be compiled into a database. A bright graduate, Joyce Whittington, who had joined the team as a research assistant the year before to work on earlier records, embraced the work with alacrity. Soon, substantial quantities of material were being acquired and combined with that produced by others in the team who took their share of the somewhat monotonous process of digitising.

The reduction of the navigational data was revolutionised by the IN systems. It was estimated that of the time taken to process the navigation and ice thickness data, the former occupied no more than 10% compared with 70%–80% for the 1967 season! The drift of the INS units was determined from numerous missions and ranged from 0.5 m s^{-1} to as little as 0.07 m s^{-1}, the average being about 0.2 m s^{-1}. Aerial photographs continued to be used for making precise fixes, and closure errors could be redistributed along the flight track so that the position error should not exceed 5 km anywhere and in many areas would be considerably less.

As discussions about the reduction of the data proceeded it became apparent a decision was needed as to how to present the principal parameters from the gridded data at a continental scale. Various sketch maps depicting results from specific areas under investigation had already been published in several papers. Draft maps had been prepared of the Ross Ice Shelf and in manuscript form for East Antarctica and had been distributed to the IAGP partners attending the 12th Meeting of the Scientific Committee on Antarctic Research (SCAR) in Canberra in August 1972. It was agreed that the results of the first three seasons of RES should be prepared for formal publication, and NERC agreed to fund production.

It was clear that the continental mapping process required the skills and techniques of a professional cartographer. The BAS had for some years been undertaking detailed geodetic mapping of the Antarctic Peninsula and working closely with, at that time, the Directorate of Overseas Surveys (DOS). Alan Clayton, who had worked for DOS but also spent two years at Halley Bay, came to Robin's attention and joined the group, bringing a fresh approach to the work, along with an easy-going and friendly manner.

There were significant differences among the data collected in each of the first three Antarctic RES seasons, principally in terms of navigational accuracy. If a definitive map was to be constructed, the flight-line positioning had to be comparable, and choice of data demanded careful consideration. The base control would comprise the accurately positioned grid of flight lines from the 1971–72 season. The dense network of flight lines around the inland flank of the Transantarctic Mountains from 1969–70, for example, could be included to extend the coverage, as they were well controlled. The final selection incorporated 16 flights from 1971–72, seven from 1969–70, and four from 1967. These totalled 280 hours of flight time and almost 100,000 km of continuous sounding. Surface elevations were available along all the lines, and bedrock data along approximately 70% of them. It was decided that the maps would depict the flight lines to show the distribution of the data and give the reader a clear indication of the degree of interpolation. When the flight lines were plotted on a map at a scale of 1:5 million, the thickness of the flight lines (0.5 mm) was equivalent to 2.5 km on the ground—corresponding to the modal value of the navigation errors.

7.6 The Ice Sheet Surface

The RES had been designed primarily to measure ice thickness—the great unknown of the ice sheet. However, the surface shape of Antarctica, whilst a much more slowly varying quantity and more readily accessible to "ground" surveying, was also poorly mapped. The elevation of the ice sheet is an important glaciological parameter, as at regional scales the slope of the ice surface creates a pressure gradient that drives the ice flow, at approximately 90° to the regional trend of the surface contours. The RES system measures the time interval between the transmission of the signal from the aircraft and its return from the ice surface. Combining these data with the aircraft altitude allows determination of the surface elevation.

The network of flight lines over East Antarctica made it possible to map the surface height with a greater degree of accuracy and consistency than had hitherto been obtained. Indeed, it was essential to all studies of the bedrock beneath the ice sheet to have accurate surface elevations when subtracting the ice thickness. It should again be recalled that this was an era of geophysical exploration just before the dawn of satellite radar and laser altimetry, which would, within the following decade, transform data gathering and provide unimaginable detail of the surface of the great ice sheets and their temporal changes (see chapter 17 for some comparisons with the latest ice surface mapping).

The principal source of error in ice sheet surface altitudes from the RES survey was the height of the aircraft. This information was coarse owing to instrumental inaccuracies, the large-scale changes in the atmospheric pressure surfaces over Antarctica, and complications from the aircraft's changing altitude. It was considered that the errors could be up to 100 m and unacceptable! To reduce the uncertainty, the elevations on all the flight lines were 'fixed' to a few control heights. These comprised the known altitude for the start and the termination of a flight at McMurdo, sea level should the flight transit over open water, and some traverse stations determined by geodetic levelling, which unfortunately were very few. Differences between fixes were linearly interpolated.

Even after these corrections were applied, RES ice surface elevations were found to differ at the intersection of two independent flights. This is where we had a breakthrough. Alan Clayton with his experience at DOS suggested that with an extensive grid we could adjust the differences at the crossover points to minimise the errors. The DOS had a least-squares computer program that would make the adjustments—and so the data were prepared, and the exercise was run. The results were encouraging, but we had another opportunity to improve the heights further.

On sabbatical at the SPRI for a year was Dick Jensen from the Department of Meteorology at the University of Melbourne. Jensen was a friendly

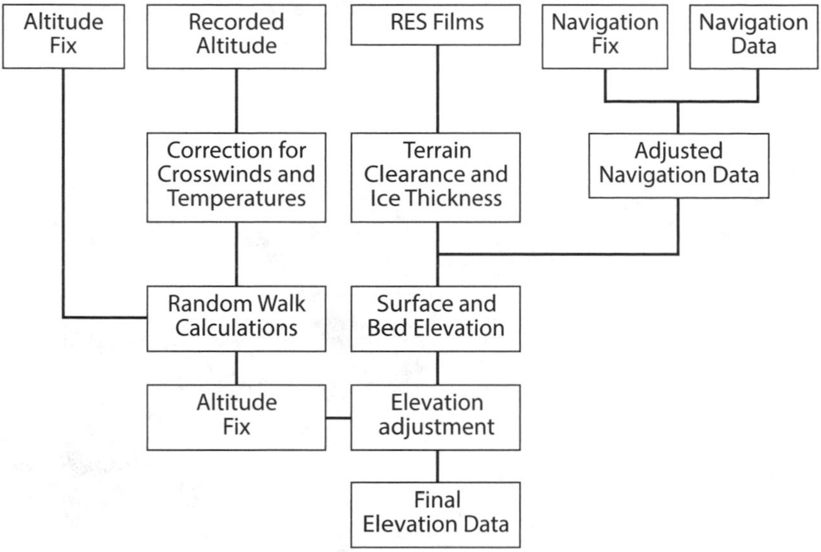

Figure 7.7. Flowchart for the reduction of the RES data.

computer geek working closely with two glaciological colleagues in Melbourne—Bill Budd and Uwe Radok. Together they were preparing a compendium of physical characteristics of the ice sheet derived by calculation from extant measurements and then interpolated for the whole continent, albeit at a small scale. Jensen offered to write a program to minimise and re-distribute these nodal errors using a random-walk technique. This took some weeks, but when we saw the results for the same data as those used for the least-squares adjustment, we observed they were very close indeed and gave us enormous confidence that we had done our best to reduce the uncertainties in the surface elevations. We deduced these to be no more than 30 m anywhere over the area we mapped.[156] These techniques became the standard by which we would minimise errors in areal mapping and played a significant part in the compilation of the Antarctic Folio, described in chapter 14. Figure 7.7 shows the generalised flowchart for making these various adjustments.

Clayton undertook the contouring from the adjusted data using colour-coded flight-line plots of elevation at a scale of 1:2 188 800 (the scale of the US Navy Oceanographic Office Air Navigation Charts of Antarctica) and then reduced them to 1:5 million (an often-used scale for the continent).

[156] Drewry, D J (1975) (footnote 134).

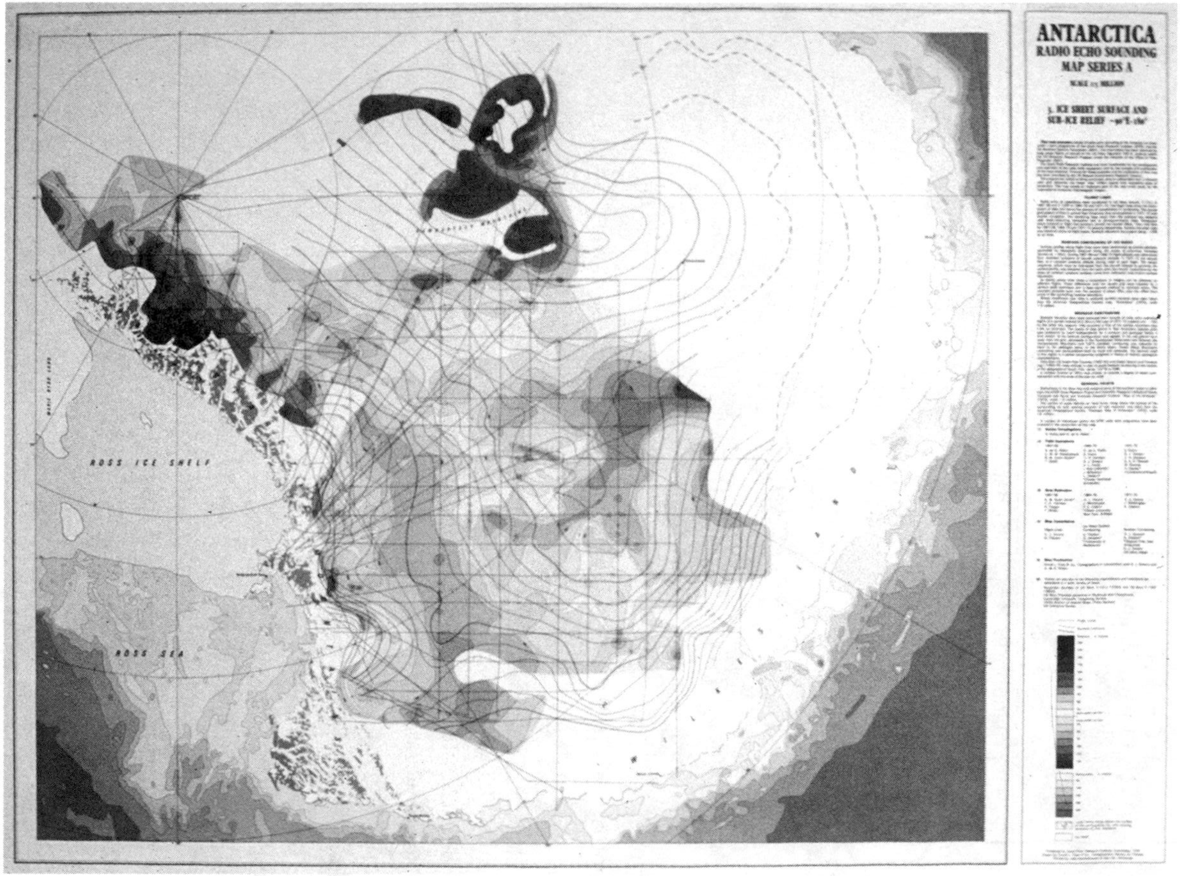

Figure 7.8. SPRI map (Antarctica: Radio-Echo Sounding Map Series A 3. Ice sheet surface and sub-ice relief: ~90°E–180°) depicting the ice sheet surface contours in red and the sub-ice land surface contours in various shades of yellow and green. Contour interval 250 m. Original scale 1:5 million.

Figure 7.9. Drygalski Ice Tongue, a 50 km–long floating extension of David Glacier, northern Victoria Land.

The features that emerged from the analysis (Figure 7.8[157]) provided great detail. In particular, it was exciting to see three high regions within the interior—we called them by the unexciting names Domes A, B and C! These are centres of ice outflow from the interior to the ice sheet margins. Dome A, in the very centre of East Antarctica, is the highest part of the whole of the Antarctic Ice Sheet, at about 4100 m asl. Dome B turned out later to be more of a long, flat ridge many hundreds of kilometres long at 3800 m descending from Dome A. Both Domes A and B are located over the irregular highland region of the subglacial Gamburtsev Mountains. Dome C lies further north along the 130°E meridian at about 75°S with an elevation of 3200 m. Another important feature depicted by the surface map was the shape and extent of the drainage basins of some of the large outlet glaciers flowing through the Transantarctic Mountains, namely, the David Glacier in northern Victoria Land and the Byrd Glacier south of McMurdo, at 80°S. Their catchments were gathering ice from deep within the continental ice sheet and, in the case of the former, feeding icebergs into the Ross Sea (Figure 7.9), and in the latter as a very distinctive lobe of thick ice into the Ross Ice Shelf.

These outcomes were very relevant to Claude Lorius and the personnel of the French Antarctic programme, who were planning to establish a

[157] Drewry, D J (1975) (footnote. 134).

summer station within the interior where they would drill through the ice sheet for climate studies. Dome C looked to be the ideal location, as it provided a centre of outflow with thick ice (of the order of 3200 m), enabling a long historical record of environmental change to be extracted and with minimal complications introduced by the lateral ice flow. It could also be accessed by an oversnow traverse from the French coastal station at Dumont D'Urville. Bill Budd's Australian glaciologists were also keenly interested in the emerging data, as they, too, had planned a series of traverses to study ice sheet dynamics inland from the Law Dome as part of the IAGP. The area covered by the RES survey did not extend greatly into the region being investigated by the Soviet Antarctic Expedition, especially towards the coast and their station at Mirnyy. Nevertheless, the new maps were of considerable interest to them for the interior region around Vostok and the Gamburtsev Mountains.

It was clear that the grid we had completed had revealed a giant sector of East Antarctica in unparalleled detail, but to answer further questions on the form and flow of this great ice mass we needed to extend the coverage and density in future seasons.

7.7 The Sub-Ice Morphology

We were commencing a new period in the exploration of Antarctica, peeling back the ice to reveal a landscape on a continental scale never seen by humans and isolated for several millions of years from the rest of the world by its frozen shell. This was the last great undiscovered terrestrial realm of our planet. Of course, earlier seismic and gravity observations had provided an exciting hint at what lay beneath the ice, but now with exceptional detail and new discoveries we were beginning to discern the patterning of the terrain. As we combined the marked-up flight lines with the bedrock elevations and commenced the contouring we were amazed at the picture that emerged. Along the whole length of the Transantarctic Mountains, as we had glimpsed in the Queen Maud Mountains region, the structurally controlled topography dipped away inland beneath the ice sheet to pass below present sea level into an extensive basin but one still presenting a diverse relief (section 8.4 and Figure 8.7). This is the Wilkes Basin, extending from close to the coast at 69°S to then narrow and eventually die out against upland blocks at 81°S—a major structural feature of Antarctica

At Dome C a substantial massif was discovered covering an area in excess of 75,000 km^2, with isolated peaks at elevations of 1000 m asl and considerable tracts more than 500 m asl. Because the weight of the overlying ice depresses the crust, it was possible to use the ice-thickness data to calculate the height of the landscape after removal of the ice and full isostatic recovery. It was found to be on average about 800–1000 m. We shall be discussing this 'rebound' effect in more detail in section 14.8.7.

To the west and south, steep escarpments were evident, plunging 1700 m into a narrow and very deep trough 750 m below sea level and 150 km in length. We identified a series of wider depressions bordering the other sides of the massif, extending also to 750 m below sea level and up to 100 km in extent. To the southwest a broad, deep trough separates this massif from the eastern extension of the mountain ranges around Vostok Station and appears to be quite geologically distinct. All this information was highly pertinent in understanding the glaciological setting for the planned deep drilling at Dome C.

To present the results in the most professional manner possible we searched for commercial cartographers to produce three separate maps covering the quadrant of the continent between 90°E and 180°: the ice sheet surface, the sub-ice relief, and a map combining these two parameters. We decided on a scale of 1:5 million, compatible with a number of other published maps of Antarctica at that time.

The choice of cartographic company had a curious story behind it. Robin was acquainted with John Bartholomew, who owned the renowned cartographic and publishing company John Bartholomew and Sons, based in Edinburgh, Scotland, and had contacted him regarding the possible production of these Antarctic sheets. A little later, at a British Cartographic Society meeting in Edinburgh, Bartholomew met up with David Fryer whose own cartographic company, David L Fryer and Co., was operating out of Henley-on-Thames. Bartholomew suggested to Fryer that this Antarctic project might be one he would like to take on in respect of the cartography, with the maps being printed in Edinburgh.

Bartholomew invited Fryer to accompany him the following day to Cambridge, where he already had an appointment with Robin regarding this map project. They travelled together by train, and Fryer was introduced at the SPRI to Robin, who took both to dinner at Darwin College. The project was offered to Fryer, whose cartographic manager, Bob Hawkins, was immediately enthused upon Fryer's return to Henley and commenced working with Robin and the author on preparing fine-line drawings that would later be turned into four films for the colour printing process.

The relationship worked extremely well, and there were frequent visits between Cambridge and Henley. Additional data were added, including the bathymetry of the Ross Sea, continental shelf, and ocean areas. These data and the coastline were taken from the 1973 Soviet map of Antarctica produced at a scale of 1:5 million[158] It was also important to depict the areas of rock exposed above or beyond the ice sheet, such as the Transantarctic Mountains, comprising a significant arc of terrain on the map. This information was abstracted from the American Geographical Society's geological map of Antarctica.[159]

The completed map films were despatched in September 1974 to John Bartholomew for printing in Edinburgh and officially published by the SPRI later that month (Figure 7.8). An article describing the preparation and production of the maps and discussing the principal features they depicted appeared in *Polar Record* the following year, accompanied by a folded copy.[160] This whole process was a considerable achievement and laid the foundations for a further productive relationship with David Fryer and Bob Hawkins on a much larger scale project some years later.

7.8 Lakes beneath the Ice

One of the recurring truisms of science is that you should always be prepared for the unexpected, and nature rarely disappoints! During the many years of radio-echo sounding of the Antarctic Ice Sheet there were remarkable discoveries, technical innovations, emergence of new fields of investigation, and fresh insights into this vast icy region. One of the most astonishing was the identification of numerous bodies of free water—some of immense size—which were called 'lakes', beneath the great expanse of ice sheet.

Looking back to 1967, the longest flight that season stretching out deep into East Antarctica extended over the abandoned Soviet camp at Sovetskaya. It will be recalled that just beyond the station the radar picked up very strong signals from the base of the ice at a remarkable depth of about 4200 m

[158] USSR State Research Project and Scientific Research Institute of Ocean Transport and Arctic and Antarctic Research Institute (1973) 'Map of the Antarctic', scale 1:5M.

[159] American Geographical Society (1972) 'Geologic Map of Antarctica'.

[160] Drewry, D J (1975) (footnote 134).

that persisted for just under 10 km (see Figure 4.10). The radar with its system sensitivity should not have been capable of detecting returns at this depth.

Working backwards, Robin and Evans speculated on the boundary conditions required to generate a detectable return. In making such an assessment, they needed to consider three principal factors. The first was absorption of the radio waves travelling through the ice (dielectric absorption), which is dependent on the ice temperature and can be calculated. The second factor was the loss due to spreading of the radio waves from the transmitting point and the characteristics of the antennas and can be derived empirically. The final and crucial element was the loss of signal strength at the base of the ice as signals are reflected.

For ice resting on a rock surface the reflection loss can be about –12 to –24 dB, depending on the composition of the strata. If the aircraft was flying over an ice shelf floating on sea water, the reflection coefficient would be very much higher, about 0 to –3dB, with almost all the incoming energy being reflected. Only by factoring in a reflection coefficient from an ice/water interface was it possible to account for the strong signal returned from the area beneath Sovetskaya. The conclusion was that 'at Sovetskaya the ice is floating on a water layer of thickness greater than 1 m'.[161] Furthermore, the bending of ice layers detected in the radar record above the water reflection suggested the ice was dipping into a bedrock basin where, presumably, the water was collecting. This was the first time that geophysical evidence had directly indicated the presence of extensive water at the base of the interior ice sheet and that the ice at the interface was at the pressure melting point.

In the late 1960s and early '70s it was well understood that the base of many temperate glaciers such as those found in the European Alps or in the North American Rockies was frequently at the pressure melting point. This condition both assisted the sliding motion of the ice as well as resulted in the production of films and small channels of water which would eventually coalesce and discharge from the front of the glacier as a meltwater stream.

However, in Antarctica, because of the intense cold, it was generally considered that the bulk of the continental ice sheet was frozen to its bed, or cold-based; there was little evidence to verify this supposition besides

[161] Robin, G de Q; Swithinbank, CWM; and Smith, BME (1970) (footnote 86).

theoretical considerations. Virtually no water was observed issuing from the frozen fronts of those ice walls that could be investigated at the margins of the ice sheet, and nothing was known of the ice at great depth until 1969, when a hole was drilled through the West Antarctic Ice Sheet at Byrd Station (Figure P.2). At a depth of 2164 m the ice in contact with the bed was found to be at the pressure melting point (–1.6°C), and the water produced by melting, which rose up the drill hole to a height of 50–60 m, was estimated to be in a thin layer a fraction of a millimetre thick.[162]

This situation was not considered to be typical of the majority of the East Antarctic Ice Sheet, which was at a higher elevation (up to 4000 m compared with 2000 m in West Antarctica) and therefore considerably colder, with average surface temperatures of –60°C to –80°C. Nevertheless, theoretical considerations of the temperature within ice sheets by Robin had shown that under thick ice, temperatures at the base could reach the pressure melting point.[163] The calculation depended upon the average surface temperature of the ice, the downward transport of some cold due to snow accumulation, dissipation of heat into the ice, and, finally, the amount of heat conducted upwards from the earth (geothermal heat). Igor Zotikov, a glaciologist at Moscow State University, following Robin's studies, calculated that the temperature at the base of the ice sheet was at the pressure melting point over extensive areas and could, therefore, be resting on layers of water.[164]

Sovetskaya had been a single spot where there was a strong supposition of basal water (Figure P.2). At the SPRI it was considered that during the 1971–72 season similar areas of melting might be encountered at the base of the East Antarctic Ice Sheet. This aspect was not seen as a major priority at the time—noteworthy but of ancillary interest. The radio-echo film records were scrutinised during the operations in Antarctica, but the punishing schedule of flying did not allow anything but a check on quality. It was not until the team was ensconced back in Cambridge that Gordon Oswald

[162] Gow, A J (1970) Preliminary results of studies of ice cores from the 2163-m deep drill hole, Byrd Station, Antarctica in Gow et al., International Symposium on Antarctic Glaciological Exploration (ISAGE) IASH Publication 86, 78–90.

[163] Robin, G de Q (1955) Ice movement and temperature distribution in glaciers and ice sheets. *Journal of Glaciology* 2 (18): 523–32.

[164] Zotikov, I (1963) Bottom melting in the central zone of the ice shield of the Antarctic continent and its influence upon the present balance of the ice mass. *Bulletin of the International Association of Scientific. Hydrology* 8:36; Zotikov, I A (2006) *The Antarctic Subglacial Lake Vostok: Glaciology, Biology and Planetology*, Berlin: Springer-Praxis, 139pp.

began examining the character of reflections in central East Antarctica. He describes the circumstances: 'Significant and persistent increases had been observed in the reflection amplitude, followed by retreats to a subjectively "normal" level. These were found in the records, in each case in association with changes to a very smooth bed reflection (see Figure 7.10), strongly suggesting a stable fluid interface'.[165] Oswald went on to use the ratio of the surface to the bottom slopes to estimate the density of the fluid beneath the ice, which was between 0.99 and 1.01 and within 1% of the density of water. The question of the composition of the fluid was solved.

From Oswald's detailed scanning of the records 17 locations were identified where the returned radar signal showed the characteristics of being from an extended smooth, almost horizontal surface exhibiting specular reflection and entirely commensurate with an ice/water interface (Figure 7.10[166]). As with the earlier case at Sovetskaya, the RES profiles from the 1971–72 season suggested the presence of rock basins, with the bed sloping down to pass beneath the water surface (below which nothing can be detected, as there is virtually no radar penetration into water). The areas had widths along the line of flight of 1–15 km and were termed 'lakes'.[167]

These were remarkable scientific findings owing to their distinctive and widespread occurrence, and they began to transform our understanding of the dynamics of such large ice sheets, both present and past. The story of the lakes received considerable publicity in the scientific press; their presence seeming counter-intuitive in such a frigid continent. Aside from Zotikov, modellers of ice sheet behaviour had not predicted the melting condition at the base in the Dome C region but were quick to incorporate the evidence into their work. We shall see that even more lakes were identified in later seasons of RES (Figure 7.11). Some lakes displayed dimensions that startled us all.

One of the issues that attracted the most attention was the possible composition of the water within these sub-ice lakes. Questions arose as to how long the lakes had been in existence at the base of the ice and isolated from the rest of the Antarctic and global environment (which could be many tens

[165] Oswald, GKA, personal communication, January 2021

[166] Oswald, GKA (1975) 'Radio-echo Studies of Polar Glacier Beds', PhD thesis, University of Cambridge, https://doi.org/10.17863/CAM.40101; Oswald, GKA; and Robin, G de Q (1973) (footnote 167); Oswald, GKA (1975) Investigation of sub-ice bedrock characteristics by radio-echo sounding, *Journal of Glaciology* 15 (73): 75–87.

[167] Oswald, GKA; and Robin, G de Q (1973) Lakes beneath the Antarctic Ice Sheet, *Nature* 245 (5423): 251–54.

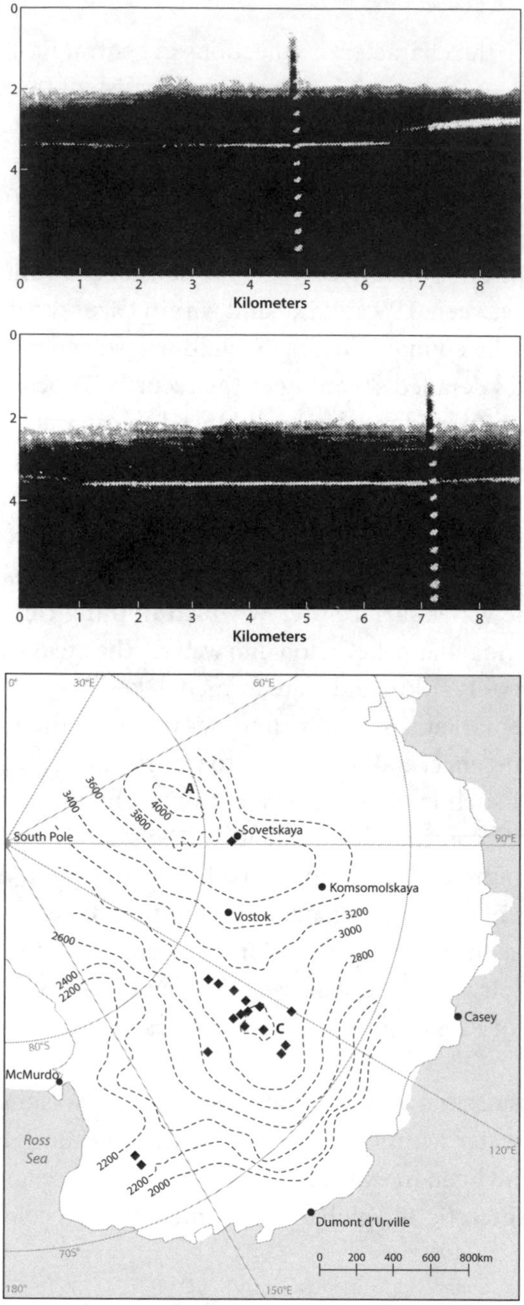

Figure 7.10. Subglacial lakes. Above: two RES profiles (recorded in 1972–72) depicting strong returns from the base of the ice interpreted as lakes, SPRI MkIV 60 MHz sounder (both ice depth scales in km). Below: the location of sub-ice lakes in East Antarctica from the 1971–72 season. A and C indicate the main domes of the East Antarctic Ice Sheet. (Courtesy International Glaciological Society).

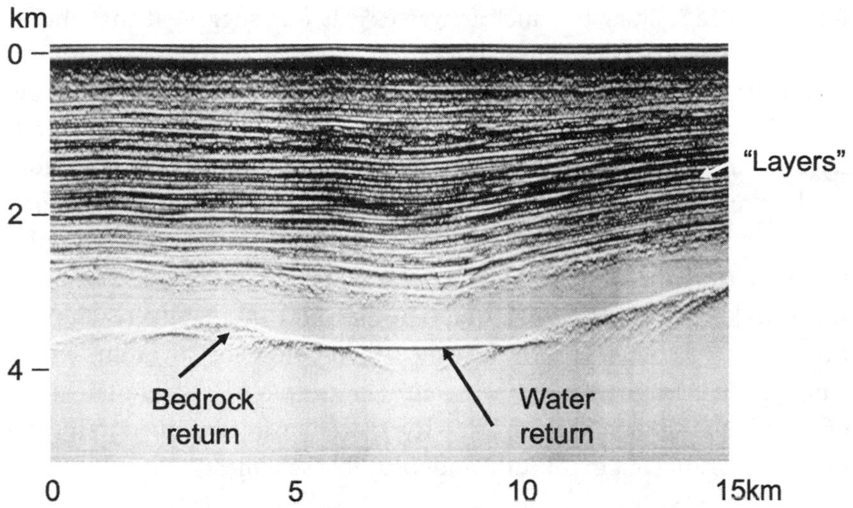

Figure 7.11. A 60 MHz record showing returns from a small sub-ice lake (approximately 5 km in section). Well-developed internal layers are also shown.

if not hundreds of thousands of years) and what might be the chemistry of the water. A whole range of speculations considered whether these lakes might contain organisms or life forms—extremophiles. Although these lakes were mostly located in discrete basins, it was conjectured that some of them might be linked by sub-ice films or channels of water. We shall return to this topic in chapter 17.

The discovery of lakes had other unexpected and environmentally important ramifications. In the 1970s, particularly in the US, nuclear power plants were rapidly being constructed for energy production. But as their numbers grew so did the problem of disposing of high-level nuclear waste, much of which was deposited in large above-ground tanks that were vulnerable to failure, natural disasters such as earthquakes, or even sabotage. Investigations to find long-term stable repositories for nuclear waste considered locations such as mine shafts and holes drilled deep into stable, ancient rock units.

Brothers Bernhard and Karl Philberth, and Ed Zeller,[168] had proposed that the Antarctic Ice Sheet might offer an attractive, remote, deep, and

[168] One of Zeller's co-workers, Ernest E Angino, was a short-term visiting scholar at the SPRI, during which time it was possible to discuss with him these ice sheet burial concepts for nuclear waste.

long-term repository for nuclear waste.[169] It was suggested that the heat generated by the high-level waste would enable the containers to gradually melt their way to the base of the ice sheet, where they would be locked away safely for many tens of thousands of years. Besides the logistical vulnerability, cost, and political issues related to the Antarctic Treaty's prohibition of nuclear materials on the continent, a clinching scientific argument for rejecting such propositions was the identification of water at the base of the ice sheet from the SPRI radio-echo soundings. The possibility of nuclear waste leaking into subglacial water bodies and eventually reaching the ocean as their containers were crushed by the immense overburden pressure argued strongly that such a scheme for deep burial would fail any test of future environmental security.[170] The concept was consigned to the dustbin of interesting but thankfully undeployable schemes.

7.9 Quo Vadis?

Much new science was emerging from the RES research, and a new era of survey work had commenced. Despite the frustration caused by the serious curtailment of the past season and resulting disappointment that the full potential of the RES operation had not been realised, there was nevertheless great energy and optimism for what the future might hold.

[169] Philberth, B (1959) Stockage des déchets atomiques dans les calottes glaciares de la Terre, *Comptes Rendus Hebdomadaires des Séances de l'Académie des Sciences* (Paris) 248 (14): 2090–92; Philberth, K (1977) The disposal of radioactive waste in ice sheets, *Journal of Glaciology* 19 (81): 607–17; Zeller, E J; Saunders, D F; and Angino, E E (1973) Putting radioactive wastes on ice. A proposal for an international radionuclide depository in Antarctica, *Bulletin of the Atomic Scientists* 29 (no. 1): 4–52.

[170] I am obliged to Gordon Oswald for reminding me of this issue of Antarctic nuclear waste disposal.

8

New Impetus

Returning from Antarctica in early 1972 the SPRI team, as recounted, were elated with the achievements of the airborne programme, the quality of the data collected, and the prospect of new and exciting scientific revelations that would ensue from their analysis. Those caught up in the programme had enthusiasm and energy for future field seasons to see the task completed—the mapping of the whole of Antarctica! Even, however, in the most expansive and fervent moments none of us felt that we could fully achieve that goal. For the last three seasons the generosity of the NSF and US Navy had been outstanding, but neither their aircraft logistics nor their financial assistance would be able to support such a fanciful programme. It has always been a humbling parable of Antarctic science that no one country has the resources to undertake substantial projects on a continental scale but requires cooperation with other nations and pooling of assets to achieve its goals and vision, and so it was at the SPRI.

We had already embarked on that journey of cooperation by bringing together our own expertise in RES, glaciology, and geophysics, and the electronics skills of the TUD in Denmark, with US logistics capabilities to bring off a great scientific feat. There was much more we could undertake within the capacity of the existing programme, but it would be important to encourage other nations with operational facilities at bases and research stations located around the continent to undertake RES to fill in the substantial areas which our work could not reach.

Such RES competence was indeed being acquired by other national programmes and research groups. The US and USSR had been involved in early developments of the technology. The Arctic and Antarctic Research Institute in Leningrad (St. Petersburg) had developed several radar systems of their own; the RLS-60–67 with a centre frequency of 60 MHz had been installed in aircraft, with Soviet scientists keen to operate in large tracts of East Antarctica, beyond the range of the US C-130 aircraft, flying radar lines out of their stations at Mirnyy, Molodozhanaya, and Novolazarevskaya.[171] The US interests were more specialised and focused principally on activity at the University of Wisconsin, where Charles Bentley and a team of graduate students and post-doctoral researchers, including John Clough and Ken Jezek, were working with several echo sounders, among which were the SPRI MkII and USAEL systems.

Meanwhile, Trevor Schaefer of the South African National Antarctic Expedition had spent time in Cambridge with Stan Evans to gain experience and with colleagues had operated an SPRI MkII sounder on inland traverses out of the SANAE Station in 1971 and 1972.[172] The success of Belgian scientists in an adjacent region of Dronning Maud Land has already been reported. Vince Morgan with the Australian National Antarctic Research Expeditions (ANARE) had made airborne RES flights over the Lambert Glacier region using a 100 MHz sounder with a system sensitivity of 175 dB. Their radar was constructed at the Australian Antarctic Division by Ian Bird, who, like Schaefer, had worked with Stan Evans and had participated in the SPRI 1969–70 Antarctic RES season.[173]

8.1 A Tripartite Agreement

To consolidate the modus operandi of SPRI, TUD, and the NSF a meeting was called in Cambridge for 19 November 1973 to formulate an agreement to rationalise the research, fieldwork, logistics, and ancillary activity among these primary players. Robin hosted on behalf of SPRI, and Gudmandsen attended from TUD. An understanding was secured setting out the principal points that had been agreed. TUD would continue development and become primarily responsible for radio-echo sounding equipment and antenna design. SPRI was to concentrate on the Antarctic plans, reduction and interpretation of records, development of new techniques for processing data, and collaboration with TUD on technology as appropriate. NSF undertook to provide significant funding for the manufacture of the radars and aerials, to manage logistics requirements, and to investigate the installation of airborne magnetometry, as well as aimed to procure an automatic

[171] Bogorodsky, V V; et al. (1985) (footnote 43).

[172] Schaefer, T G (1973) Radio-echo sounding in western Dronning Maud Land, 1971, *South African Journal of Antarctic Research* no. 3, 45–59.

[173] Morgan, V I; and Budd, W F (1975) Radio-echo sounding of the Lambert Glacier Basin, *Journal of Glaciology* 15 (73): 103–11.

recording system for navigational and other data. The latter installation, it was argued, would be available for use by airborne projects other than RES. TUD and SPRI agreed to make RES records available to groups or individuals sponsored by the three bodies following a set period during which the two primary organisations would have exclusive-use access. We shall learn later that this subject of data sharing was to prove an Achilles heel in the partnership arrangements. Subsequent to the agreement TUD commenced development of new radio-echo apparatus for the 1974 Greenland and a 1974–75 Antarctic season. SPRI invested a preliminary sum of £30,000 in initiating a development of automated film analysis.

8.2 Cambridge Activities

At the SPRI work moved on apace. Chris Harrison completed his thesis on radio propagation effects in the latter part of 1972[174] and the author was caught up in both the reduction and interpretation of the last field season to complete his own thesis.[175] Investigating basal reflections across the network of East Antarctic flights for evidence of sub-ice water occupied Gordon Oswald. Creating sadness for us all was the departure of Stan Evans—the guru of RES. He had decided that his position at the SPRI did not provide sufficient security and was seeking a more permanent position within the university. He left to travel the short distance to the Department of Engineering on Trumpington Street (all 300 m) and took on a teaching and supervision role at Jesus College but remained interested in the SPRI work and was always available for consultation.

Gordon Robin recognised the need to replace Evans with suitable technical expertise. By chance he received a letter from David Meldrum, at that time working on geophysics projects at Manchester University, enquiring as to whether there were any positions available at the SPRI. Meldrum was a talented physicist, originally from St. Andrews University, who had spent time at the Cavendish Laboratory in Cambridge. He was also a mountaineer and keenly interested in the polar regions, having participated in expeditions to West Greenland from St. Andrews. Meldrum also had joined the final field season in East Greenland organised by geologist Peter Friend, who

[174] Harrison, C H (1972) 'Radio Propagation Effects in Glaciers', PhD thesis, University of Cambridge.

[175] It would not be submitted until the summer of the following year. Drewry, D J (1973) (footnote 124).

was based in the SPRI. To Robin and Evans, he seemed the ideal person to recruit. At the end of an interview with Robin, Evans, and Oswald, Robin said, 'He seems to be asking all the right sort of questions', to which Evans replied, 'We need him to be coming up with the right sort of answers!'[176] Meldrum joined the group in 1974.

8.3 Devon Island—An Arctic Foray

The success of the Antarctic radio-echo sounding had stimulated other scientists to seek cooperation with the SPRI to bring its expertise with the technique to bear on their glaciological problems. In 1972 an invitation came from Dr Stan Paterson in Canada to join a field programme on Devon Island in the Northwest Territories (now Nunavut) (Figure P.3) in June 1973. Paterson headed an impressive team with a world-class glacio-chemistry laboratory for ice core studies within the Department of Energy, Mines and Resources. He had also written the first comprehensive textbook in English on glacier physics a few years previously. Gordon Robin was keen to cooperate and perceived the possibility of some interesting research.

The Canadians had drilled a hole through the Devon Island Ice Cap to a depth of 299 m and extracted a core for climate and wider environmental studies. The borehole would provide an opportunity for a series of experiments in radio wave propagation. It was also known that the ice cap was situated over a major geological boundary which could yield interesting studies of the changes in basal reflectance and micro-relief. Detailed thickness profiles along flowlines would assist Canadian colleagues in modelling the behaviour of the ice. There was also the possibility of undertaking tests to ascertain whether ice movement could be determined from an examination of the fading patterns of the returned signal from the bed over time. Plans for these several projects were assembled.

The fieldwork on Devon Island proved very productive. It involved Gordon Robin, Gordon Oswald, Michael Gorman, Alan Clayton, and Chris Doake, who joined from the BAS. Some of the results of this work are reported in later chapters.

The transport logistics proved lengthy and tedious. After the group arrived in Toronto from the UK, the first stages of the journey to Devon Is-

[176] David Meldrum, personal communication, October 2021.

land were provided by the Defence Research Board of Canada, courtesy of Geoffrey Hattersley-Smith. The team boarded an RCAF Hercules in Toronto, flying the first leg to overnight at the USAF base at Thule, in northwestern Greenland. The second leg proceeded via Alert, a 'watch station' on the northernmost tip of Ellesmere Island, and ended at Resolute Bay, Cornwallis Island (Figure P.3).

Resolute Bay is the major transport hub for the Canadian High Arctic. Here the group came under the auspices of the Polar Continental Shelf Project (PCSP) for accommodation as well as logistics. Transport to Paterson's substantial base camp on the ice cap was by a ski-equipped Twin Otter flown by Bradley Air Services. A separate camp for the SPRI group was set up a few hundred metres away, with storage tents and a large prefabricated Nissen-style hut for living quarters. The team was able to settle in straightaway.

The radar equipment, comprising the SPRI MkIV 60 MHz sounder, was quickly unpacked. Michael Gorman recounts: '[T]his equipment had been designed for a comparatively cosy airborne environment, not the physically harsh conditions on the surface of a polar ice cap. After some thought, a plywood cabin was created, and based upon a metal sledge. This was large enough for the three racks of the MkIV equipment, with space for an operator to sit above two large batteries providing DC power (Figure 8.1). The operator in the cabin watched and controlled the radar'.[177]

Trials soon demonstrated the heavy cabin and equipment required two snowmobiles towing in tandem. The RES antenna was trailed behind the sledge (Figure 8.2). A total distance of about 200 km was surveyed and ice depths of 400 to 800 m. Data were recorded in the form of continuous ice profiles, as in Antarctica, and also as A-scopes at defined intervals. Surveys using a new SPRI 440 MHz echo sounder were interleaved with the general RES survey of the ice cap, as it was not possible to deploy both 60 MHz and 440 MHz systems simultaneously.

Oswald's experiments to identify the geological boundary from the strength and character of the reflections from the ice/rock interface were successful. In Figure 8.3, sites to the west (T2 and T3) yielded estimated reflection coefficients lower by 6 to 10 dB than those of the spots to the east (-11 ± 4 dB).[178]

[177] Michael Gorman, personal communication, March 2021.

[178] Oswald, GKA (1975) (footnote 167).

Figure 8.1. Devon Island project. SPRI 60 MHz equipment installed in a small caboose. Top: oscilloscope-mounted Shackman camera (05); centre: oscilloscope and camera for A-frame recording; bottom: SPRI MkIV system. (Courtesy M R Gorman).

Figure 8.2. RES caboose and antenna (note the bicycle-wheel odometer) with camp buildings in the background. (Courtesy M R Gorman).

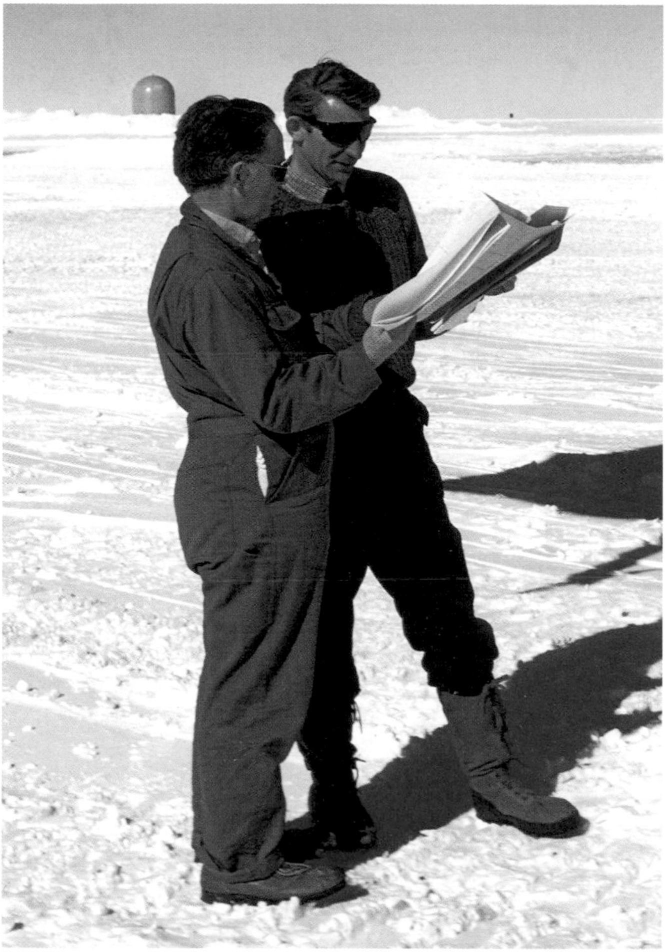

Plate 1 (dedication). Gordon de Quetteville Robin (left) and Stanley Evans (McMurdo Station 1969).

Plate 2 (fig. 2.1). Charles Bentley (left) and the author at a conference of the Scientific Committee on Antarctic Research (SCAR) in São Paulo, Brazil, in July 1990.

Plate 3 (fig. 3.3). Early radio-echo sounding on the Brunt Ice Shelf by Mike Walford (left) in January 1963 from the UK base at Halley Bay. A sledge of fuel and a live-in caboose (attended by Douglas Finlayson) are pulled by a Muskeg tractor. (Courtesy David Petrie).

Plate 4 (fig. 4.12). C-121J "Phoenix" and air crew at McMurdo (Robin (left) and Swithinbank on the extreme right of back row; Ewen-Smith on the extreme right in front row). (Courtesy SPRI).

Plate 5 (fig. 5.7). McMurdo Station from Observation Hill, circa 1977. View across sea ice in McMurdo Sound to the Royal Society Range.

Plate 6 (fig. 5.12). The aircrew greeted us one morning with this version of the use of the RES aerials.

Plate 7 (fig. 5.17). Mount Erebus (3794 m) the ever-present volcanic mountain overlooking McMurdo Sound, with a gentle plume of gas emanating from its summit crater.

Plate 8 (fig. 5.20). The Ross Ice Shelf, looking north. In the distance (left) is Ross Island. The ice cliffs are 30–35m high.

Plate 9 (fig. 5.23). Refuelling at Halley Bay from 50-gallon barrels! No.3 engine propeller is feathered, that is, shut down owing to an oil leak.

Plate 10 (fig. 7.1). Tight personnel transport in a crowded C-130. The seats are canvas and webbing. (Courtesy SPRI).

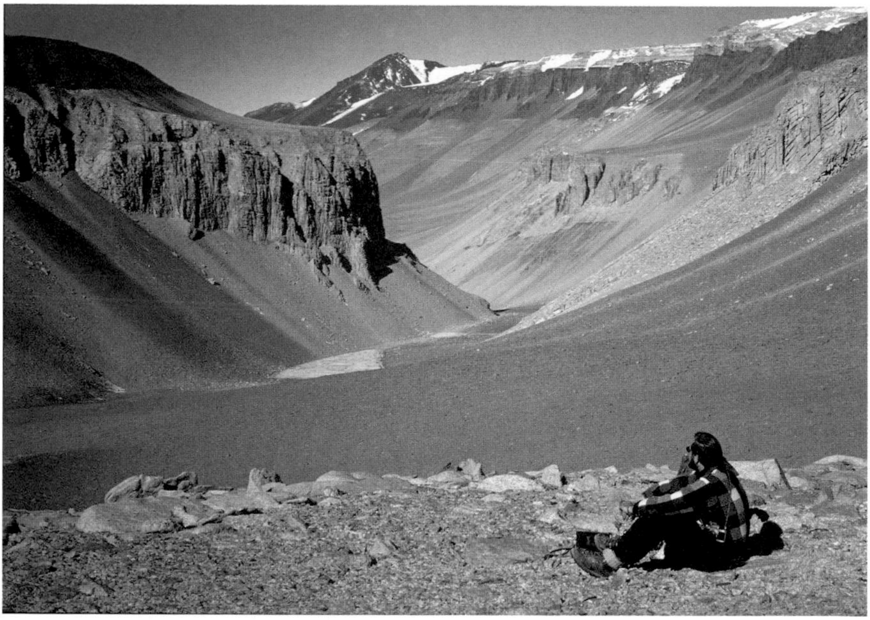

Plate 11 (fig. 7.4). Upper section of Wright Valley with massive screes on the valley flanks and, in the valley bottom, two small frozen lakes. Note thick dark layers of dolerite sills in the upper middle distance.

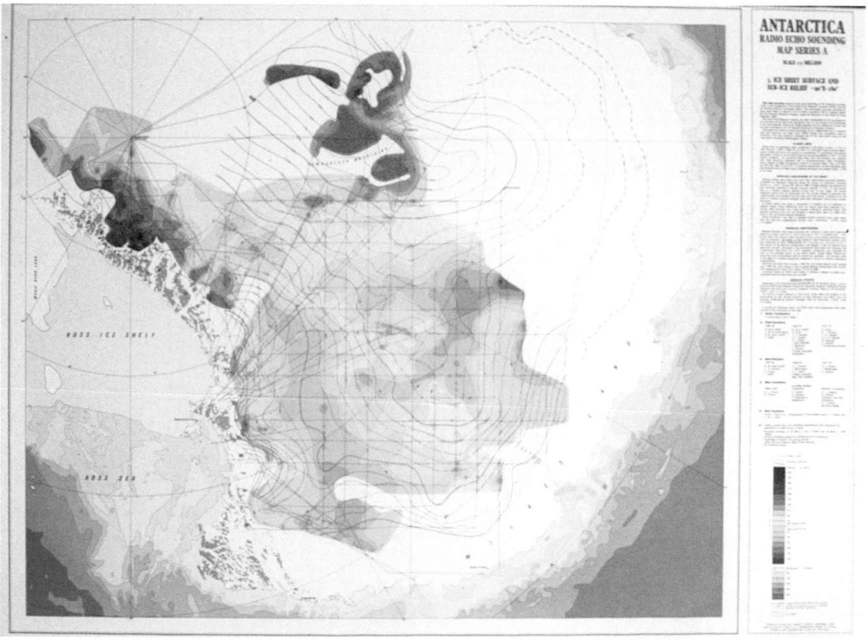

Plate 12 (fig. 7.8). SPRI map (Antarctica: Radio-Echo Sounding Map Series A 3. Ice sheet surface and sub-ice relief: ~90°E–180°) depicting the ice sheet surface contours in red and the sub-ice land surface contours in various shades of yellow and green. Contour interval 250 m. Original scale 1:5 million.

Plate 13 (fig. 7.9). Drygalski Ice Tongue, a 50 km–long floating extension of David Glacier, northern Victoria Land.

Plate 14 (fig. 8.4). Doake (left) and Robin conducting RES ice movement experiments on Devon Island, Canada. The antenna was placed a few centimetres above the snow surface on fibreglass tracks in the direction of ice flow. It was then moved steadily to record the fading pattern. The traverses were repeated with the tracks shifted laterally by 20 cm to map an area of 8 m×2 m. (Courtesy M R Gorman).

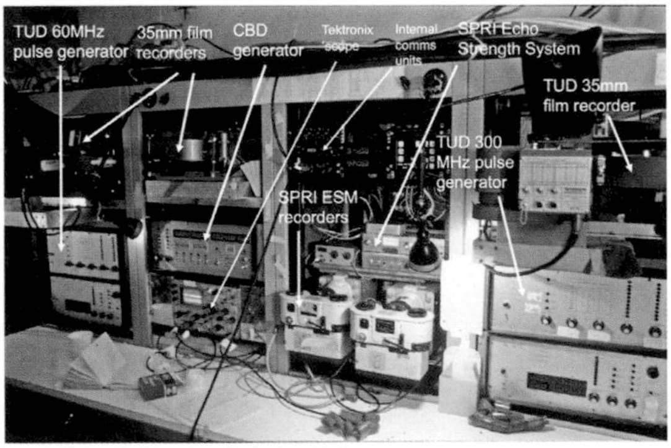

Plate 15 (fig. 8.12). TUD RES systems with 60 MHz and 300 MHz radars, various photographic recording cassettes and cameras, and SPRI echo-strength device. All pallet mounted inside C-130 cargo bay (1978–79 season).

Plate 16 (fig. 8.8). Royal Society Range. Gently westerly (inland) dipping sequences of the Beacon Supergroup overlying crystalline basement rocks (foreground). Mount Lister is the highest peak (4025m). Peak on left is Mount Rucker (3816m) with Walcott Glacier lobe descending. An arm of Koettlitz Glacier is in the immediate foreground.

Plate 17 (fig. 8.9). Sub-horizontal Karoo Supergroup sequence with its associated dolerite sills, at Mont aux Sources, Drakensberg Mountains, South Africa.

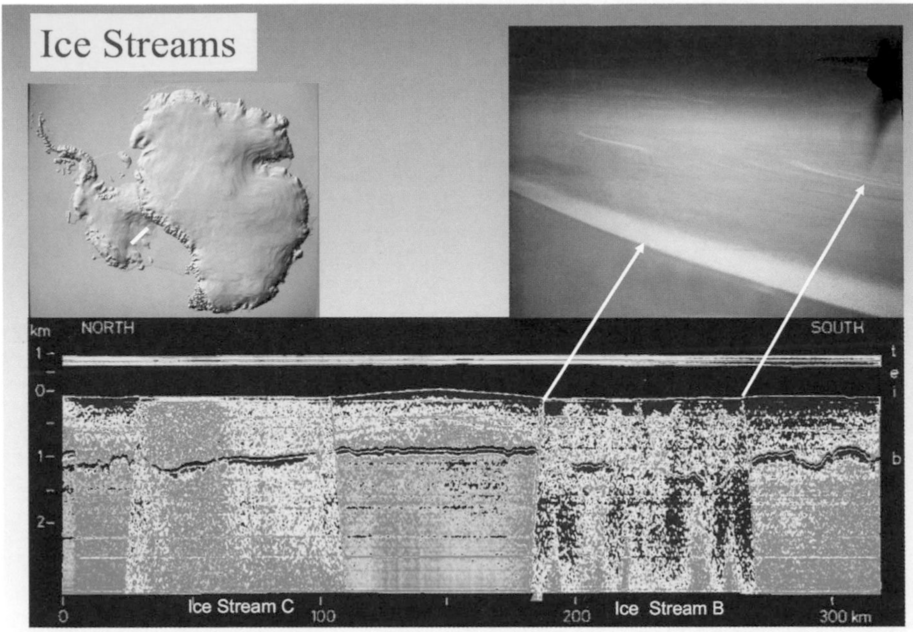

Plate 18 (fig. 9.5). Ice streams in Marie Byrd Land (Siple Coast). Bottom: RES record, in false colour (located on the map). Margins of B and C delimited by the deep clutter echoes from crevassed shear bands depicted in the photograph of Ice Stream B. The ice stream flow is left to right.

Plate 19 (fig. 9.11). JATO take-off from the Shackleton Glacier, central Transantarctic Mountains. (Courtesy Charles Kaminski and NSF, USAP Photo Library).

Plate 20 (fig. 11.3). On occasion, transfers between McMurdo (The Hill) and Williams Field were by helicopter (a 5- to10-minute ride!).

Plate 21 (fig. 11.7). Checking the 60 MHz aerials.

Plate 22 (fig. 11.8). Honeywell chart recorder with a strip profile emerging. To the right is the high-speed magnetic tape recorder for the ARDS (navigational and magnetics data). (Courtesy SPRI).

Plate 23 (fig. 11.14). The Ellsworth Mountains looking east. The highest peak in Antarctica, Vinson Massif at 16,080 feet (4902 m), is distinguishable as the first block of mountains on the left. Note the stacked lee-wave clouds over the mountains—bumpy flying if you went there!

Plate 24 (fig. 12.2). Walker Peak in the Dufek Massif. It is possible to distinguish some of the igneous layering in the upper part of the spires.

Plate 25 (fig. 12.7). Part of the SPRI team at McMurdo. Left to right: Gisela Dreschhoff (NSF Aircraft projects manager); Charles Swithinbank (BAS with Hughes's Byrd Glacier Project); Finn Søndergaard (TUD); Ed Jankowski, Ben Millar, David Meldrum, John Behrendt (USGS).

Plate 26 (fig. 12.16). Snout of Taylor Glacier and frozen Lake Bonney in the foreground.

Plate 27 (fig. 12.20). David Meldrum and the author taking a snack beneath the front of Hobbs Glacier. All the baggage is emergency survival equipment should there be delays with the helicopter pickup.

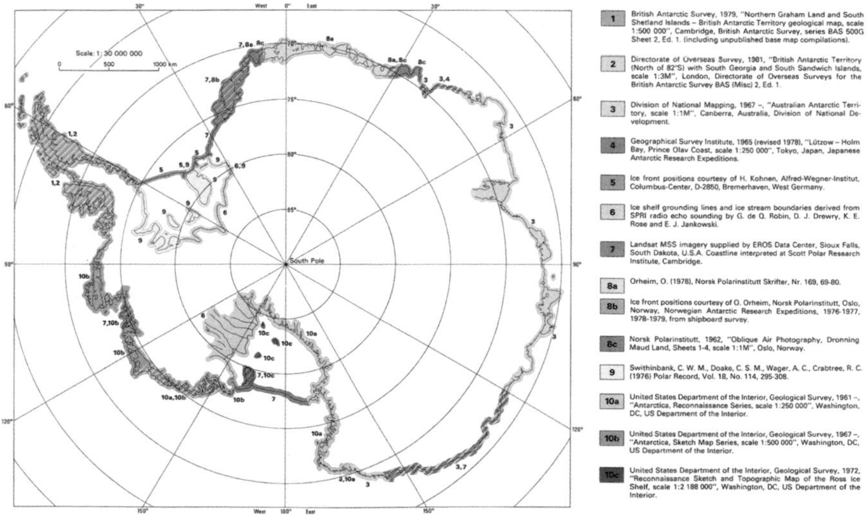

Plate 28 (fig. 14.3). Antarctic coastline showing the various sources used in its compilation. (Fom Sheet iv of the SPRI Folio).

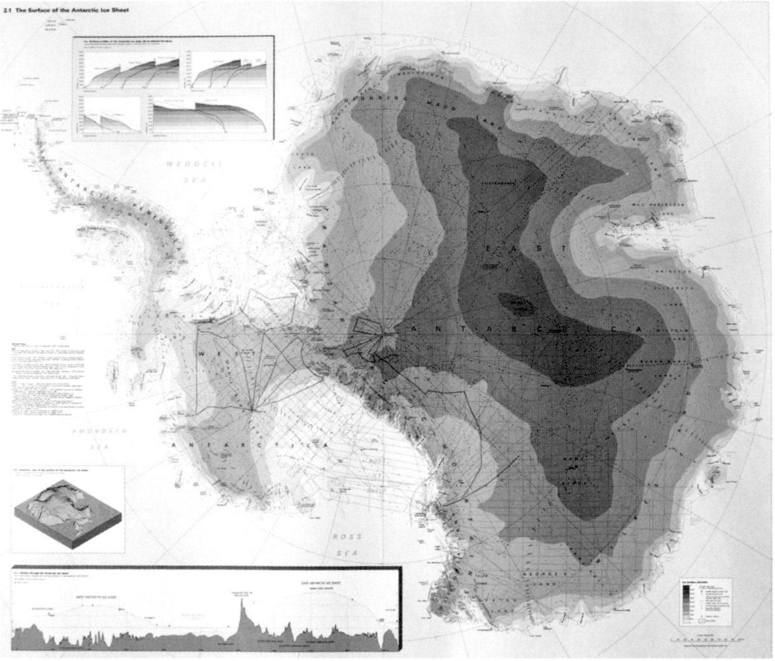

Plate 29 (fig. 14.5). Surface of the Antarctic Ice Sheet. Original scale 1:6 million. Contour interval 100 m. Also depicted are several representative profiles of ice sheet flowlines (top left), an isometric 3-D ice surface plot, and at the bottom left a detailed cross section of Antarctica from west to east. (From Sheet, 2 SPRI Folio).

Plate 30 (fig. 14.7). The bedrock surface of Antarctica. Original scale 1:6 million. Contour interval 250 m. Note the continuous contouring from the land onto the continental shelf and then into the deep ocean. (From Sheet 3, SPRI Folio).

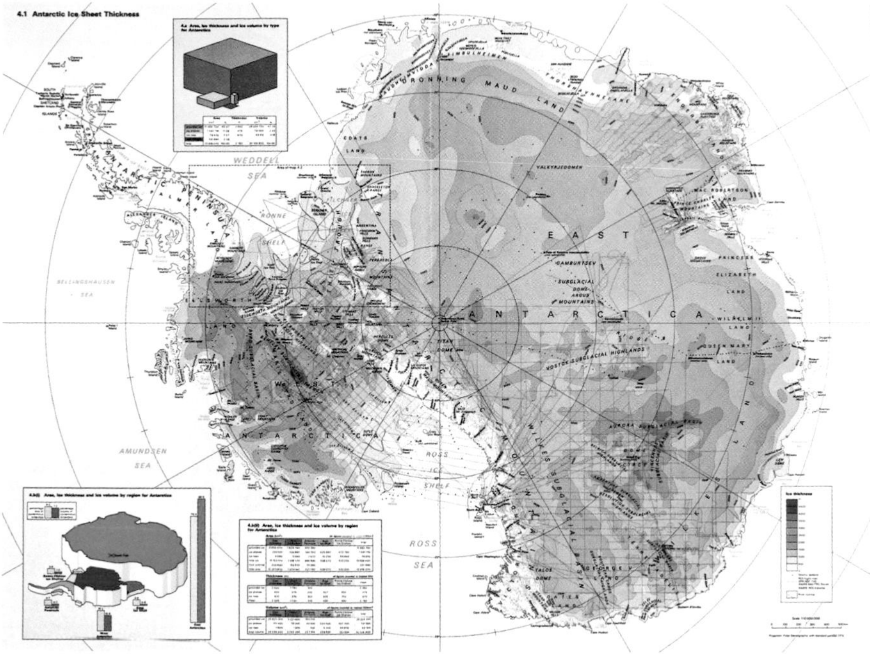

Plate 31 (fig. 14.8). Ice thickness with information on areas and ice volumes. Contour interval for the continental ice sheet 500 m, for the Ronne-Filchner Ice Shelf 100 m. Original scale 1:10 million. (From Sheet 4, SPRI Folio).

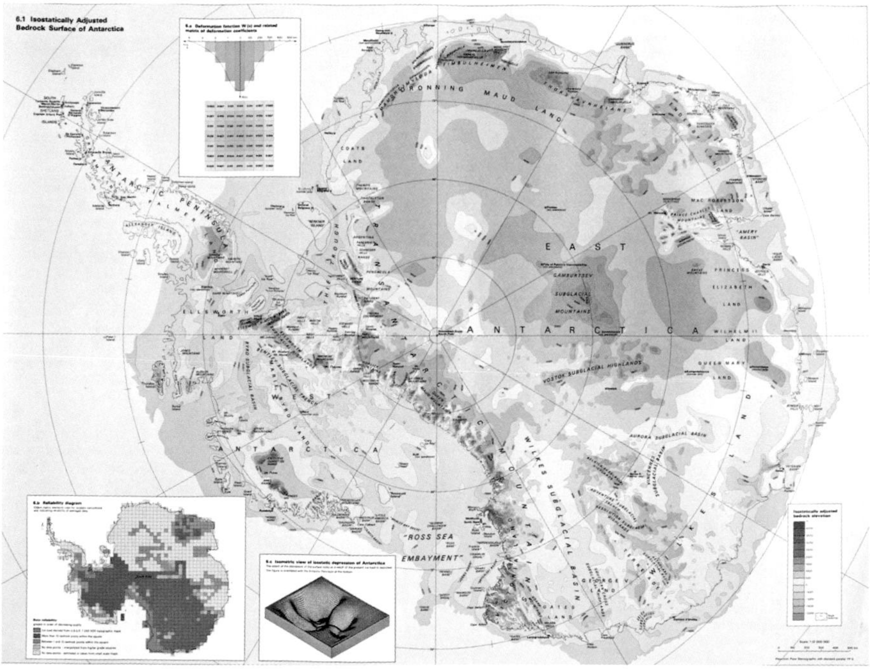

Plate 32 (fig. 14.9). The isostatic rebound map of Antarctica. Contour interval 500 m. Original scale 1:10 million. (From Sheet 6, SPRI Folio).

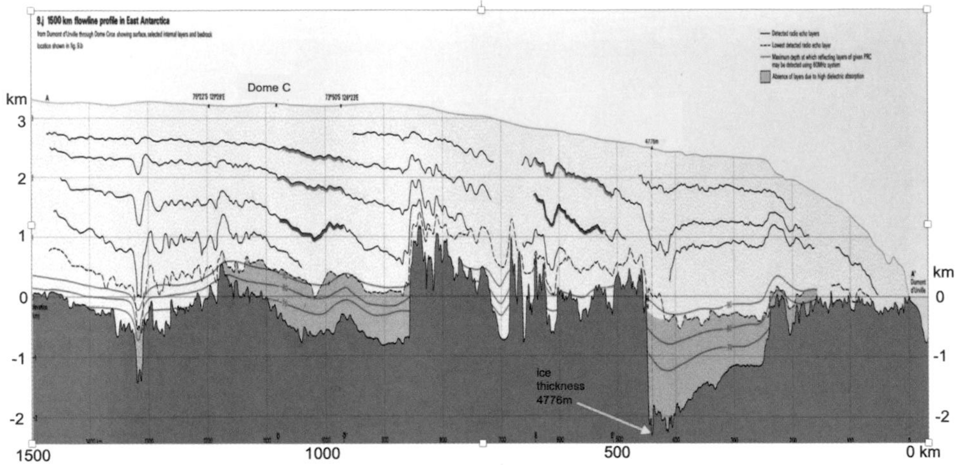

Plate 33 (fig. 14.13). Profile through the East Antarctic Ice Sheet from Dumont D'Urville (right) through Dome C towards Vostok Station showing long-distance continuity of selected layers from RES. These are the soundings along the line of the French Traverse to Dome C described in sections 6.1, 7.6, and 8.8.1. Vertical displacement of layers due to flow over irregular topography is also shown. The deep trough inland of the coast was the location of the deepest ice measured during the NSF-SPRI-TUD programme (4776 m). (From Sheet 9, SPRI Folio).

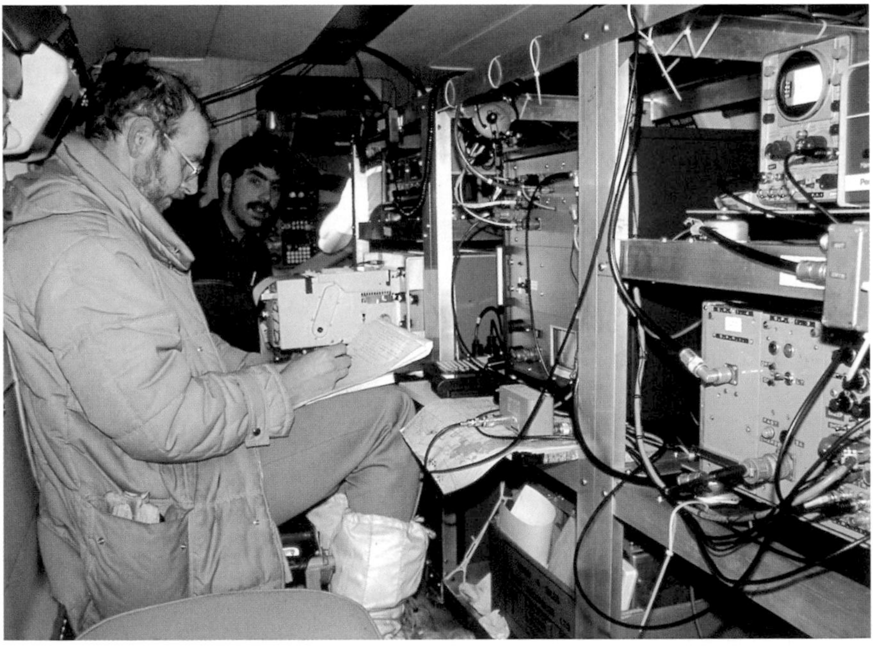

Plate 34 (fig. 16.4). SPRI RES equipment installed in BAS Twin Otter. Mike Gorman (left) and Paul Cooper (right) in the cramped operating area.

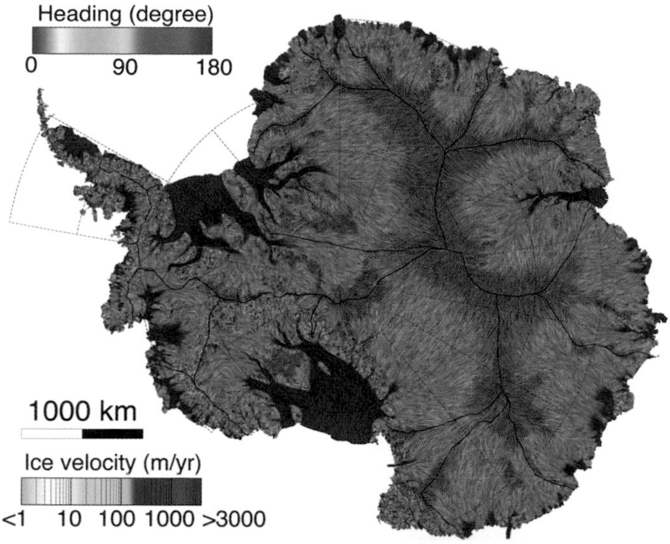

Plate 35 (fig. 17.2). Ice velocities and flow pattern over the Antarctic Ice Sheet using SAR interferometric phase data (interior) and speckle-tracking in fast-moving zones. The velocity is logarithmic. The principal ice divides, and hence major drainage basins are also designated. (From Mouginot et al. (2019); see footnote 360. (Courtesy the authors and American Geophysical Union).

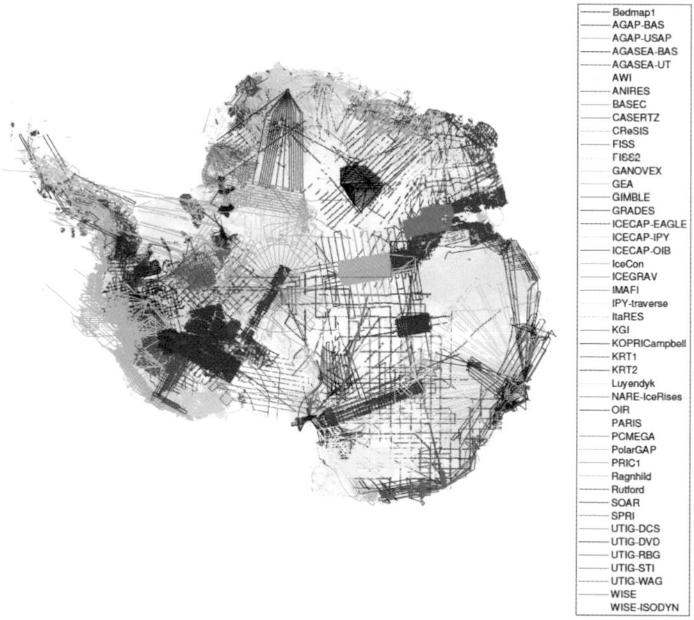

Plate 36 (fig. 17.3). Radio-echo sounding flight lines used in the Bedmap compilation to 2019. (courtesy BAS, UKRI).

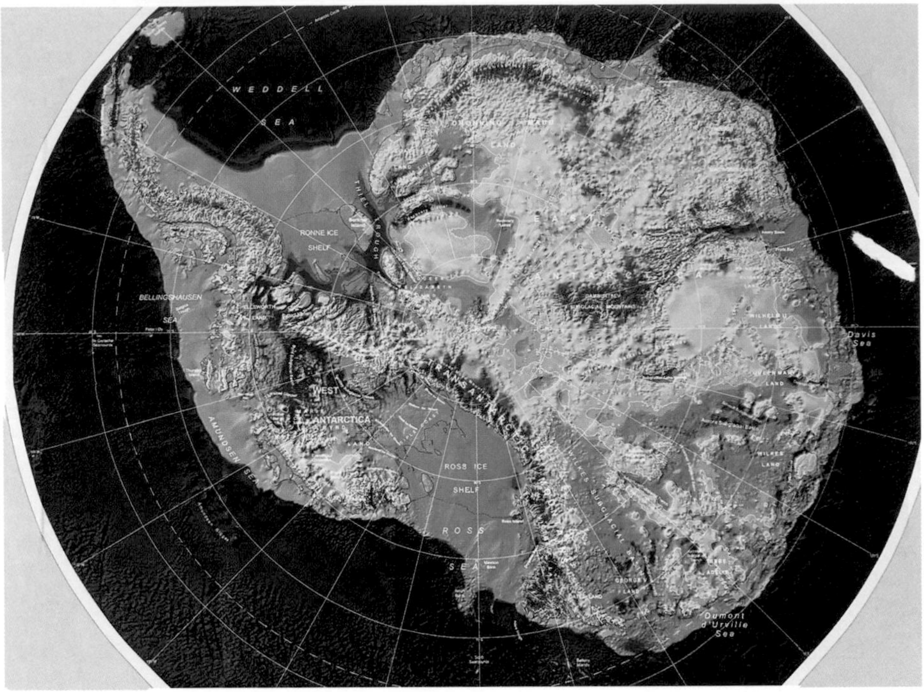

Plate 37 (fig. 17.4). Bedmap2 depicting subglacial relief of Antarctica. (Courtesy BAS, UKRI).

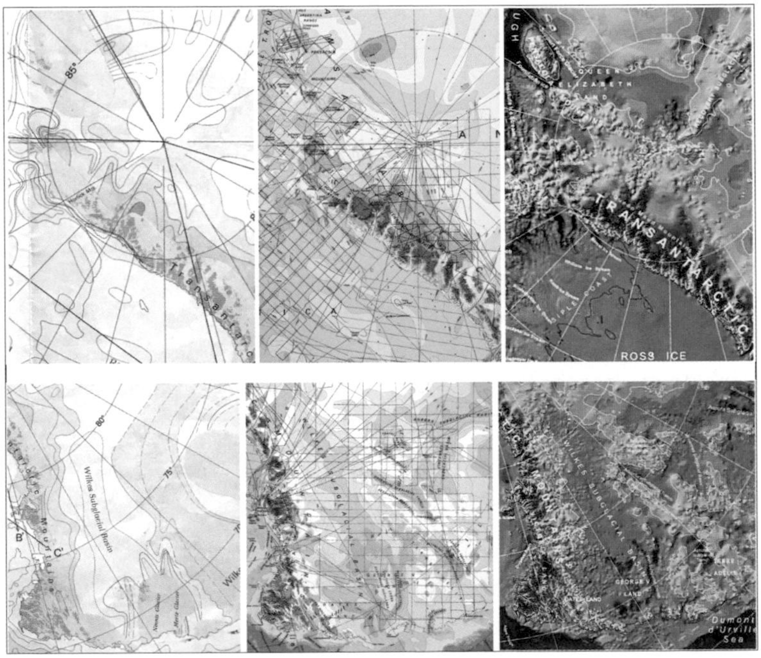

Plate 38 (fig. 17.5). Two comparisons of Antarctica bedrock relief. Left: (Bentley 1972); see footnote 368. Centre: SPRI Folio (1983). (Courtesy SPRI). Right: Bedmap2. (Courtesy BAS and Fretwell et al. (2013); see footnote 368).

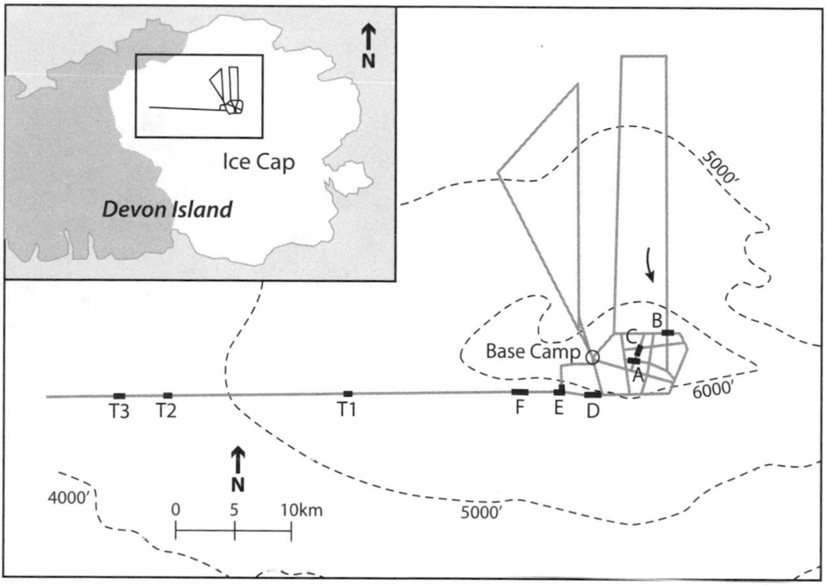

Figure 8.3. Location on Devon Island of the RES work in 1973 (see also Figure P.3). The contours are in feet (5000 ft = 1520 m). RES sounding lines are shown, and letters indicate the positions of various studies (referred to in the text). (Courtesy International Glaciological Society).

Chris Doake and Michael Gorman undertook a further project to measure ice flow rates based on a notion first proposed by John Nye.[179] This comprised conducting a detailed two-dimensional survey on the ice surface, of the fading pattern of echoes produced by roughness elements on the bed (Figure 8.4). This survey was to be repeated at a later date. Spatial correlation of the two fading patterns would give the displacement and thus the ice flow over the time interval. The survey was indeed repeated in the spring of 1975 along with the opportunity to perform some additional experiments.[180]

Robin's down-borehole experiments were aimed at making a precise determination of the velocity of electromagnetic waves within the ice cap by lowering an antenna down one of the existing boreholes (Figure 8.5). This specially constructed cylindrical aerial on a very long coaxial cable was lowered by winch. At the surface, radar pulses from a second antenna were

[179] Nye, J F; et al. (1972) Proposal for measuring the movement of a large ice sheet by observing radio-echoes, *Journal of Glaciology* 11 (63): 319–25.

[180] Doake, CSM; and Gorman, M R (1979) Performance of V.H.F. aerials close to a snow surface, *Journal of Glaciology* 22 (88): 551–53.

Figure 8.4. Doake (left) and Robin conducting RES ice movement experiments. The antenna was placed a few centimetres above the snow surface on fibreglass tracks in the direction of ice flow. It was then moved steadily to record the fading pattern. The traverses were repeated with the tracks shifted laterally by 20 cm to map an area of 8 m × 2 m. (Courtesy M R Gorman).

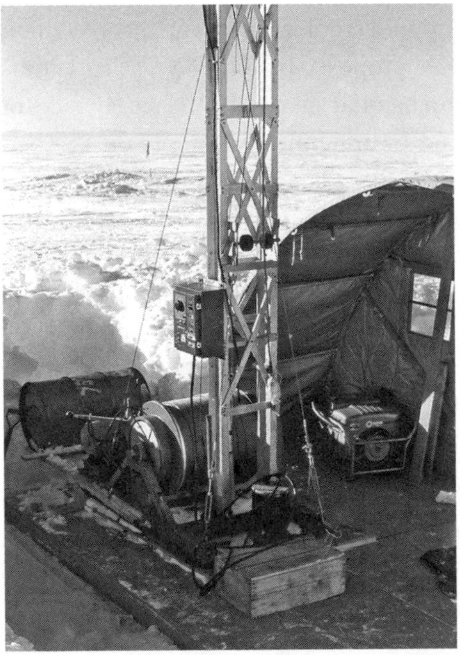

Figure 8.5. Winching mechanism located above the 300 m hole for the down-borehole radar experiments. (Courtesy M R Gorman).

directed obliquely downwards towards the borehole antenna using a phase-sensitive radar sounder operating at 440 MHz, newly constructed by Gorman. Very good results were obtained in defining the pattern in radio wave velocities with changing depth and density of the snow and ice.

Oswald noted an interesting encounter while on the traverse to the western edge of the ice cap:

> As we motored westwards, Alan Clayton was leading, with Oswald second, followed by the radar sled with Gorman across the wide-open snow surface. Alan . . . braked, and we all slowed to a standstill. He pointed at the surface. Polar bear tracks were embedded in the snow, well over a foot long, with claw marks in the forward edge, heading South. Amazing to see, but why would it have been there? 70km from the ice edge? A hundred metres further on, Alan braked again. I couldn't see why. Then I saw. Tracks of a lemming, parallel to the bear's. Hard to imagine that a ton of bear would have walked 50km up on to the ice in chase of a 20-gram lemming. . . . Maybe lemmings just taste really good?[181]

On completion of the field activities the SPRI team were picked up, in marginal weather conditions, by a Twin Otter that had landed some distance from the camp and taxied for an hour to collect them. Short on fuel, the ever-forgiving aircraft was tanked up with the camp's stock of heating and diesel oil and departed for Resolute by launching down the slope of the ice cap to escape the clutches of the increasingly wet and sticky snow—a dramatic end to a successful, highly concentrated Arctic campaign.

Almost 30 years later, in 2002, an SPRI team led by Julian Dowdeswell, with Gorman as the radar engineer, and in collaboration with Canadian colleagues were to fly a systematic 10 km grid of RES lines over the Devon Ice Cap. A total of 3,370 line-kilometres of data was acquired using a 100 MHz system constructed by Gorman and deployed from a Twin Otter aircraft. These data were used subsequently to calculate the volume of the ice cap and its rate of mass loss. RES revealed further surprises with the discovery of two hypersaline lakes beneath the ice cap, where basal temperatures are of the order of –10°C.[182]

[181] Oswald, GKA, personal communication, February 2021.

[182] Dowdeswell, J A; Benham, T J; Gorman, M R; Burgess, D; and Sharp, M J (2004) Form and flow of the Devon Island Ice Cap, Canadian Arctic, *Journal of Geophysical Research* 109, no. F02002, https://doi.org/10.1029/2003JF000095; Rutishauser, A; Blankenship, D D; Sharp, M; et al. (2018) Discovery of

8.4 The 'Keystone' of Gondwana

It was already well established in the early 1970s, through early tectonic plate reconstructions, that Antarctica had shared a common geological history with the other main components of the ancient super-continent of Gondwana—the southern element of Pangaea. Clustered around Antarctica, the land masses of Australia and New Zealand, South America, India and Ceylon, and Africa including Madagascar provided helpful analogues for the evolution of the southern continent (Figure 8.6[183]). It was considered that their known geology would assist in the interpretation of what lay beneath the ice as revealed by the topography and other geophysical measurements.

One aspect of my research was a comparison of the bed topography of East Antarctica with terrains in southern Africa. I was drawn particularly to the strong similarities between the Transantarctic Mountains, with their extension inland to the Wilkes Basin, and the landscapes and geological history of this part of Gondwana. Both regions, it was agreed, had experienced a long period of Late Palaeozoic and Mesozoic sedimentation forming the extensive sequences of the Karoo Supergroup in Africa and the Beacon Supergroup in Antarctica. These rock units were predominantly substantial thicknesses of near-shore, shallow marine and estuarine sediments which were uplifted and tilted without significant deformation. The gently dipping strata are clearly observed along the impressive escarpment or 'front' of the Transantarctic Mountains (the flank of the West Antarctic Rift) inland of and lying above the older 'crystalline' basement rocks[184] (Figures 8.7 and 8.8).[185]

The Beacon series was intruded by distinctive thick sills of dolerite during the Jurassic period at the commencement of the disassembling of Gond-

a hypersaline subglacial lake complex beneath Devon Ice Cap, Canadian Arctic. *Science Advances* 4 (4): eaar4353, https://doi.org/10.1126/sciadv.aar4353.

[183] Li, L; Lin, S; Xing, G; Jiang, Y; and He, J (2017) First direct evidence of Pan-African orogeny associated with Gondwana Assembly in the Cathaysia Block of southern China, www.nature.com, *Scientific Reports* 7:794, https://doi.org/10.1038/s41598-017-00950-x.

[184] Elliot, D H (2013) *The Geological and Tectonic Evolution of the Transantarctic Mountains: A Review, Geological Society, London, Special Publications* 381(1): 7–35, https://doi.org/10.1144/SP381.14.

[185] Craddock, CAM (ed.), 1969–70 *Geologic Maps of Antarctica*, American Geographical Society, Antarctic Map Folio Series, No.12; Elliott, D H (2013) (footnote 184); Miller, S C; et al. (2010) Cenozoic range-front faulting and development of the Transantarctic Mountains near Cape Surprise, Antarctica: Thermochronologic and geomorphologic constraints, *Tectonics* 29, no. 1, https://doi.org/10.1029/2009TC002457.

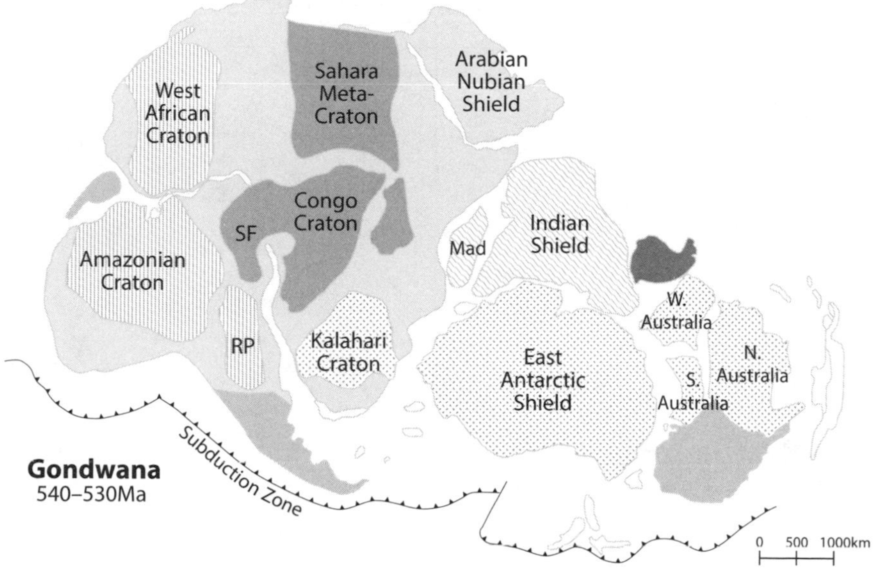

Figure 8.6. Typical reconstruction of Gondwana indicating the position of Antarctica and other ancient shield/craton regions. Mad = Madagascar; SF = San Francisco; RP = Rio de la Plata. (Adapted from Li et al., (2017); see footnote 183).

wana 177 to 183 Ma BP and, locally, possesses a capping of Cenozoic basalt lavas. In many places this combination gives rise to distinctive stepped morphology, and it was possible to identify virtually the same sequence of rocks in South Africa (Figure 8.9). Statistics of the terrain could be compared, as well as aspects of their gross morphology, with sub-ice areas of Antarctica using large-scale topographic maps and the current knowledge of geology from published works—and the similarities were considerable.

To advance this particular line of research I was awarded a fellowship offered within Cambridge University to travel to and study in South Africa between February and May of 1974.[186] This comprised visits to universities and their geology departments along with fieldwork to investigate particular formations to learn as much as possible about South African geology in the context of Gondwana.

The locations included the impressive Cape Ranges, Namaqualand, and the Karroo. Of particular note was the awe-inspiring escarpment of the Drakensberg Mountains with its close similarities to the east-facing

[186] Sir Henry Strakosh Fellowship.

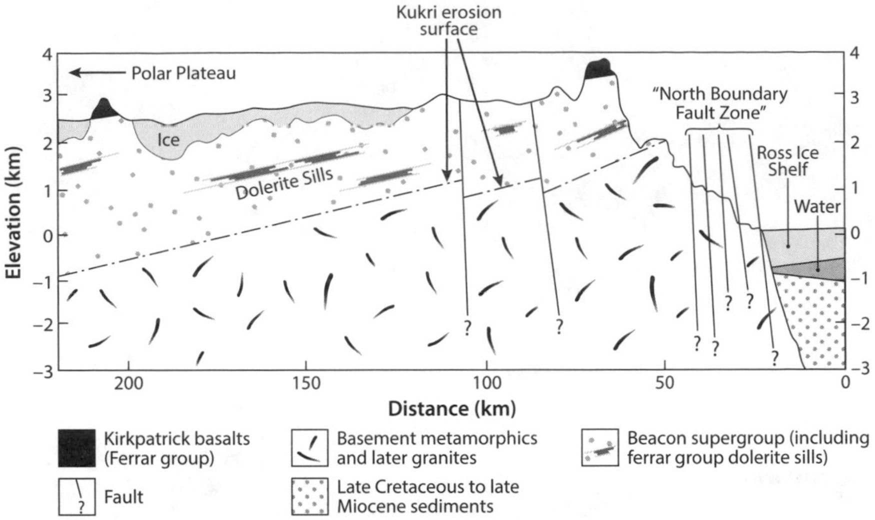

Figure 8.7. Representative geological cross section of the central Transantarctic Mountains from the Ross Sea/Ice Shelf in the north to the East Antarctic Ice Sheet (Polar Plateau). Inland of the ice shelf there is exposed basement of metamorphic and igneous rocks of Neoproterozoic to Lower Cambrian age, worn down to form the Kukri Erosion Surface and subsequently overlain by a Phanerozoic sequence of relatively undeformed sedimentary strata (Beacon Supergroup) intruded by thick sills of dolerite of Late Jurassic age and cappings of extrusive basalts (Ferrar Group). These latter mark commencement of the break-up of the ancient super-continent of Gondwana. The Ross Sea (part of the consequent West Antarctic Rift System) comprises a series of basins initiated in Cretaceous times and filled with sequences of later sediments (see also Figure 5.16). (Based on several sources; see footnote 185).

escarpments of the Transantarctic Mountains (Figures 8.8 and 8.9). At a gold mine at Evander on the High Veld I gained an impression of the geology in its third dimension. I returned to Cambridge with several portions of rock cores from Evander that had been drilled down through the Karoo sequences, including sections of the Dwyka Group glacial sediments of the Late Palaeozoic pan-Gondwana ice sheet.[187] The 'tillite' core has sat on all the desks I have ever occupied and remains facing me every day as I write this book.

[187] Visser, JNJ (1989) The Permo-Carboniferous Dwyka Formation of southern Africa: Deposition by a predominantly subpolar marine ice sheet, *Palaeogeography, Palaeoclimatology, Palaeoecology* 70 (4): 377–91. Visser, JNJ (1993) 'A Reconstruction of the Late Palaeozoic Ice Sheet on Southwestern Gondwana', in Findlay, R H; et al. (eds.) *Gondwana Eight: Assembly, Evolution and Dispersal*, Rotterdam: Balkema, 449–58.

Figure 8.8. Royal Society Range. Gently westerly (inland) dipping sequences of the Beacon Supergroup overlying crystalline basement rocks (foreground). Mount Lister is the highest peak (4025 m). Peak on left is Mount Rucker (3816 m) with Walcott Glacier lobe descending. An arm of Koettlitz Glacier is in the immediate foreground.

Figure 8.9. Sub-horizontal Karoo Supergroup sequence with its associated dolerite sills, at Mont aux Sources, Drakensberg Mountains, South Africa.

The sojourn in South Africa did much for my education on Gondwana geology and landscapes, and in later years, visits to Australia, India, and South America furthered an appreciation of this fascinating era in Earth history and the central role played by Antarctica. I was already seeing terrain in the Antarctic in a new light—structurally controlled tablelands and prominent mesas. There was much learning to absorb and to apply.[188]

8.5 A New Aircraft and New Instrumentation

The result of the various planning meetings had led the NSF to consider the development of a general remote-sensing aircraft for polar work that could, at an early stage, be constructed to take account of the sensor and

[188] It is worth recounting one memorable incident of this visit. In early May I was in Durban, where I met for the first time the well-known and somewhat irascible professor of geology Lester King. I had come across his research as an undergraduate when some of my lecturers had referred to him only as 'L C King'. I had misunderstood what they had *spoken* and thought this eminent geologist was a woman—'Elsie King'! It was only sometime later, when I obtained papers he had written, that I realised my gender mistake! It was, therefore, with enhanced interest that I had meetings with this, to me, legendary individual. But the story was to continue.

I had written to King well in advance announcing my visit and asked if he was willing to meet me and illuminate me on aspects of the geology and geomorphology of South Africa. Of all the senior individuals whom I contacted in this manner at various universities, King was the most welcoming and generous with his advance invitations and information—several full letters and a copy of his latest book. I rang him from the hotel in Durban, and he invited me and my geologist wife to his home for a 'working lunch'.

We arrived at the bungalow in the Durban suburbs to be met by a rather austere and formal individual. He sat us down in his lounge and more or less ordered his wife to prepare lunch: '[O]melettes will do, as we shall be engaged in a lot of discussion!' In due course the omelettes arrived, which we consumed during a long discourse by King on the geology, structures, and denudation history of the eastern flank of southern Africa. He was particularly keen for us to appreciate the cycles of uplift and down-cutting that had produced a distinctive stepped landscape, especially in Natal.

There was a great deal to take in, and, of course, with a copy of his book I was confident that we would be able to understand the complexities he was describing. After about two hours King finally drew his comments to a close. Then, to our great astonishment and consternation, he began a quiz—What is the age of the Intermission III planation surface? Which surface lies between Active Episodes D and E? It wasn't a quiz; it was an exam! By the skin of our teeth and with a bit of luck and pure guesswork we managed to respond with a reasonable amount of confidence and accuracy, but we could not wait to escape and feared he would go into another session after this revision. We made our excuses to leave but not before King had insisted that he take us into the field to show us some of the geomorphology of Natal—a two-day trip at least. Despite his clear erudition and research nothing could have been more uninviting. We demurred, indicating our itinerary could not accommodate such a generous offer. In the end, he was adamant that the next morning we accompany him to the top of the university tower, where he would point out as many features as could be distinguished from there. We acquiesced simply to get away. The next morning, I rang to say that unfortunately we had to change our schedule, and we would be unable to take up his kind invitation.

data collection requirements for a range of possible future science applications. There was a sense of déjà vu; such an airborne laboratory had been the vision of Bert Crary back in the early 1960s. The RES operations had partially fulfilled that aspiration, but now a fully fledged polar remote-sensing platform was being contemplated. There was little dissent from the view that this would have to be a new, ski-equipped LC-130 model to meet the vicissitudes of operations in the Antarctic.

The SPRI had lobbied to undertake RES flights from an aircraft with extended range. It was argued that little more of the 100 km grid planned to cover much of the ice sheet could be surveyed effectively by the 'F' model Hercules operating out of McMurdo Station without extensive and expensive refuelling facilities. The range superiority of the proposed 'R' model C-130 was very clear, and a striking map (Figure 8.10) showing the respective maximum ranges illustrated the argument.[189] Furthermore, we had reasoned it would be highly desirable to obtain other geophysical measurements in conjunction with the radio-echo sounding. Aside from the obvious scientific merit of providing for more powerful interpretations, it would reduce financial and logistic pressure to re-fly areas with different sensor systems. Initially, we maintained that the most useful geophysical measurement would be the geomagnetic field.

In late 1973, the SPRI was invited to contribute to a detailed discussion of a comprehensive data recording system for the new C-130. This was perceived to be central to any arrangement and mix of sensors and instruments that might be mounted on the aircraft and hence needed a comprehensive approach but also had to be practical from the point of view of those operations most likely to be pursued in future—such as RES. I was asked to represent the NSF at meetings in Hartford, Connecticut, in December 1973 with Kaman Aerospace, a designer and manufacturer of a range of military and commercial logging systems. My responsibility was to evaluate the proposed recording structure and its individual sensors, as well as those that would be most suitable for upcoming RES operations in Greenland and Antarctica.

Another person had been asked to join the liaison team representing a different set of interests, this time from NASA, whose priorities included validation of satellite remotely sensed data using aircraft. That person was

[189] Figure 3 in 'Radio-echo Sounding of Polar Ice Sheets: A Science Plan Compiled by the SPRI', November 1975, 34pp.

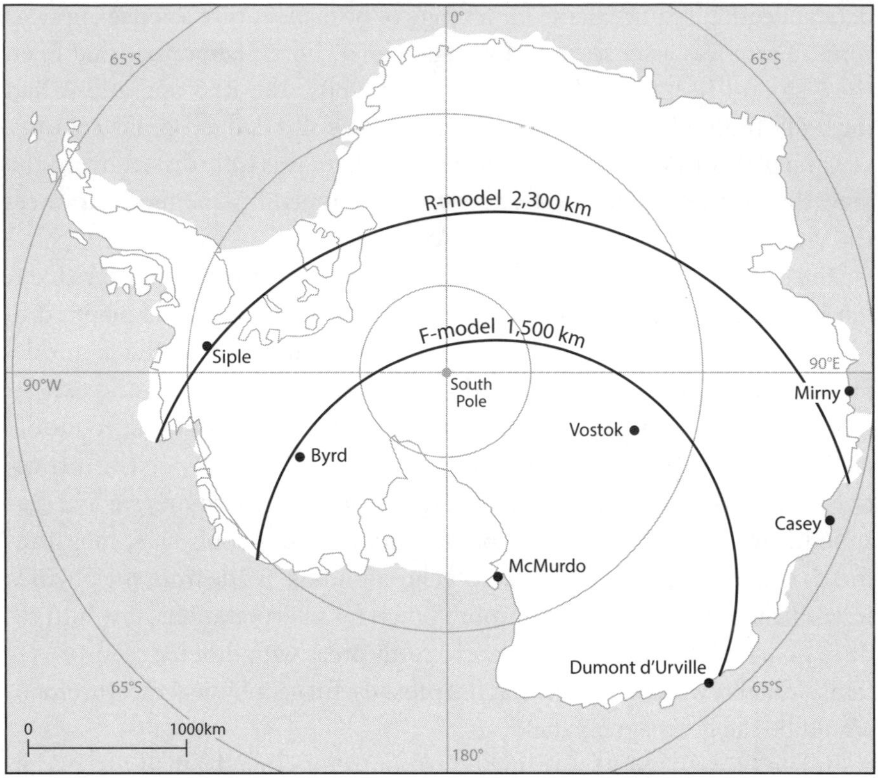

Figure 8.10. Maximum range of C-130 'R-' and 'F-' model aircraft for typical radio-echo sounding missions out of McMurdo Station.

Jay Zwally, a bright and eager physicist who was interested in ice and snow remote sensing. We struck up a friendship which has lasted throughout our respective careers. For three days Jay and I met with the senior Kaman engineers and system designers as we toyed with the specification, tolerances, and sensitivities of the many instruments that would comprise a comprehensive data recording package—pressure transducers, angle-of-attack indicators, groundspeed, track heading, pitch and roll angles, INS output, and so it went on.

One crucial feature was the recording medium. The 1971–72 RES season had used punched paper tape very successfully to record a sub-set of navigation data, and at the SPRI we were impressed by both its simplicity and reliability in Antarctic field operations. Kaman personnel were keen to have all the data recorded on magnetic tape. The dependability of this medium in the early 1970s was not high, and there were questions about data extraction, interference from other electrical systems, and cost, as the unit was to add an additional $50,000 to the bill!

In the end the data transfer rate was the critical factor. With the paper tape system the punch rate was 63 characters per second. There was a long discussion on the size and number of characters to be recorded. It was agreed that 100 characters were most probably needed, which meant that the original specification of one frame of data every second could not be met; sampling only every 3 seconds could ensure all the data were punched. There was debate, and several of the analogue inputs were sampled at 0.1-second intervals and averaged for 1 second and that was fine. The crucial INS output was to be read to 0.1′ of latitude, or approximately 185 m on the ground. Since the aircraft groundspeed was usually between 100 and 150 m per second, a faster cycling time would not be compatible with the positional resolution.

Zwally was very much in favour of the magnetic tape unit, especially for future programmes that might need 1 data point per second or finer. In the end, it was agreed to build in the magnetic tape recorder, despite the cost, and install a ruggedized high-speed paper tape punch as backup! Some years later these deliberations at Kaman formed the basis for the avionics component of the ARDS (Airborne Research Data System) facility for the new aircraft, which was overseen by the Applied Physics Laboratory (APL) at Johns Hopkins University in Baltimore.

Following the November 1973 meeting, Gudmandsen and his group at TUD were busy with the development of new radars and the aerials that would be attached to both existing and the new aircraft.[190] Liaison with these Danish colleagues was greatly assisted by arranging to meet periodically in either Cambridge or Lyngby/Copenhagen, enjoying each other's hospitality and sharing a common purpose. It was vital to settle on the design for the new C-130 as early as possible so that the hard fixing-points beneath the aircraft wings could be manufactured and installed during its construction. The schedule was tight if the new equipment was to meet the deadline for field operations in Greenland in northern summer of 1974 and thereafter be available for the SPRI Antarctic flights.

[190] Gudmandsen, P; Nilsson E; Pallisgaard M; Skou, N; and Sondergaard, F (1975) New equipment for radio-echo sounding, *Antarctic Journal of the United States* 10 (5): 234–36.

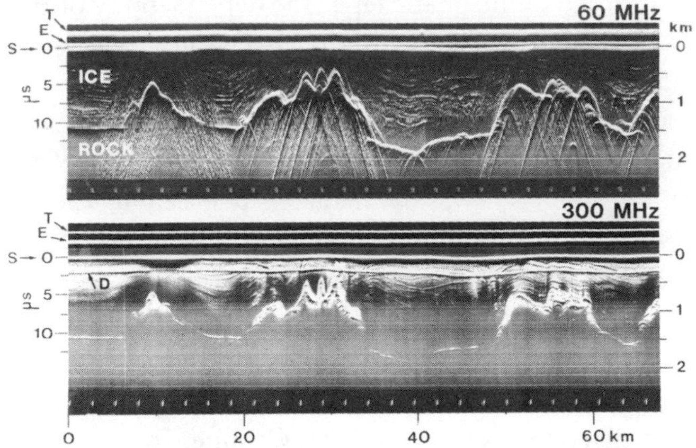

Figure 8.11. Simultaneous RES profiles inland of McMurdo Dry Valleys at 60 MHz (upper) and 300 MHz (lower). Left: echo delay time for one-way travel. Right: ice-depth scale. T = transmitter pulse; E = end of receiver suppression; S = ice surface return; D = 2nd multiple from ice surface.

TUD had worked through several concepts and the one that appeared to achieve the best results was a combination of two radars. There would be a radar operating at a centre frequency of 60 MHz with a peak power of 10 kW, a mean power of 250 W, and a good match in the bandwidth of 50 to 70 MHz. This would be the main system for sounding the deepest ice in the Antarctic and was already well proven. A second radar would be installed at a higher frequency, 300 MHz. The loss by absorption in the ice would be much greater, leading to an inability to penetrate deep ice; however, the narrower beam of the radar would give a much-improved definition of the bottom surface (Figures 8.11 and 8.12).

The antenna systems for the two radars would be mounted one beneath each of the main wings with the ability to operate in tandem. The 60 MHz configuration was to follow the general design used in the 1971–72 season but with several improvements. The fixture for the dipoles and the supporting struts were formed with a balsawood-fibreglass-epoxy technique commonly used in glider design and given a neutral aerofoil shape. This overcame the problems with drag and flexure experienced in 1971–72 (Figure 8.13). The 300 MHz system was more directional, with the antenna arrangement beneath the port wing comprising four 'backfire' elements producing a narrower beam 20° in the fore-and-aft direction (Figure 8.14). The resulting ensem-

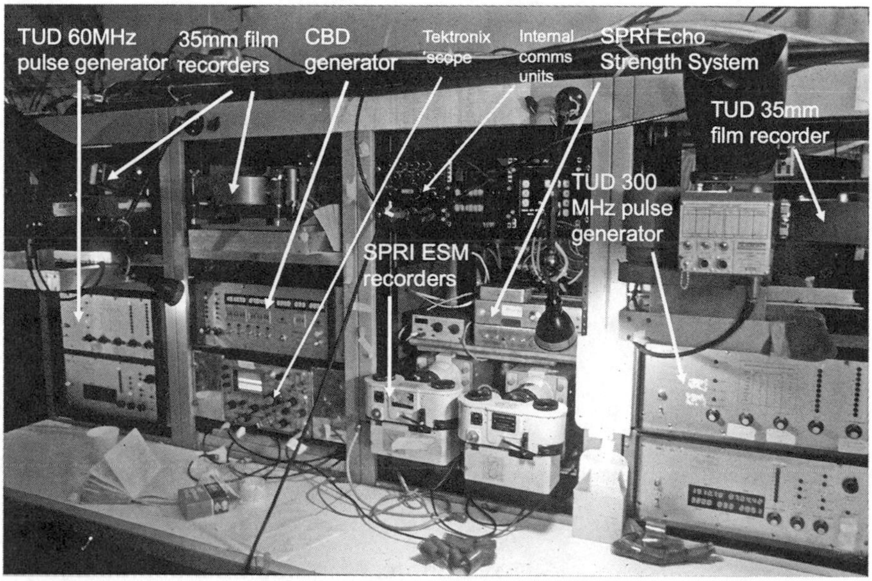

Figure 8.12. TUD RES systems with 60 MHz and 300 MHz radars, various photographic recording cassettes and cameras, and SPRI echo-strength device. All pallet-mounted inside C-130 cargo bay (1978–79 season).

ble provided a very powerful combined RES system for study of the ice and bed reflections.

At the SPRI data reduction initiatives were also under consideration. The increasing volume of RES film data and the new navigational information began to create handling problems. At the beginning of the RES programmes the film data were printed as long positive strips, some 15 cm in width, which could be scaled, and thicknesses were read off at a given sampling interval. This was satisfactory for the early ground and airborne experiments with small volumes of data. Following the first Antarctic season, in 1967, a digitising system was introduced in which the RES negative film was back-projected and enlarged onto a screen where an X,Y cursor would record the positions of the various features such as ice surface, bottom return, and internal layers. The data were then processed by computer to yield the ice thicknesses. The process was labour-intensive and required an 'intelligent' operator to undertake the digitisation and, especially, to interpret where echoes were weak, indistinct, or complex. All the research students took their turn to digitise, but later, funded by a NERC research grant, research

Figure 8.13. TUD 60 MHz aerial configuration in 1977–78 and 1978–79: array of four half-wave dipoles on C-130 #03. Magnetometer boom on the tail.

Figure 8.14. TUD 300 MHz antenna arrangement comprising four dipoles and parasitic elements in the 1974–75 season mounted on C-130 #320.

assistants were employed to support this task and allied aspects of data handling. The first to join the group at SPRI was Ann Fuzesy, followed by Joyce Whittington, who established the first RES database.

To improve both the speed and consistency of digitisation some form of automated film analysis was considered, and a local Cambridge computer-aided design company, Digico, designed and delivered a bespoke system. The equipment needed considerable nursing, and the task was probably too complex for the hardware and computing capabilities at the time. It was, overall, something of a disappointment. The luckless attempt to develop automated digitisation further is explored in section 10.6.

8.6 A Planning Dilemma

Despite the enthusiasm generated in respect of these various innovations and developments, the Lockheed-NSF project of building a new aircraft and its instrumentation would not be completed and the aircraft delivered for operation until later in 1975. There was, therefore, an emerging dilemma. It was important to maintain the momentum of the continental survey work following the success in 1971–72, and funds had been secured in Cambridge for another field operation as soon as possible, that is, 1974–75, but this could happen only if we accepted we would not have the new aircraft and its facilities. In addition, we had to consider our responsibilities to our research students who were expecting to work on both existing material and to share in the excitement and rigours of collecting new RES data in the field. Furthermore, other partners in the IAGP had new and pressing requirements for ice-thickness data.

The difficult decision was made to proceed with a further season in the older 'F' model C-130. It was argued it would still provide an opportunity to install the new TUD radar equipment and, in a timely manner, extend the grid and focus on areas that analyses to date had revealed as scientifically stimulating—the basal zone of the Ross Ice Shelf, the ice streams of Marie Byrd Land, sub-ice lakes, and the Wilkes Subglacial Basin.

Not all areas could be investigated by the small team, and there were excellent opportunities for several new projects, particularly for enquiring minds of doctoral students. The SPRI was successful in obtaining new studentships from the NERC. Chris Neal had joined the team as a PhD student in 1972 from Birmingham University, supervised by Gordon Robin, and was already at work on possible computer modelling of ice dynamics. Two more research students were added to the group for the 1974–75 season. Keith Rose came from the Department of Engineering at Cambridge to work with me on the ice streams of Marie Byrd Land, and Neil Hargreaves, also a Cambridge graduate, joined us to investigate electromagnetic effects in ice, particularly polarisation studies under Robin. I was especially pleased with Keith Rose's interest, as he and I had spent the summer of

1972 in East Greenland on the Roslin Glacier in the Stauning Alps. Fiercely outdoors-oriented, energetic, and with an engineer's practical bent, he was an ideal member of the team.

Preparations engaged the RES group throughout 1974. Michael Gorman had departed for another job in Cambridge—leaving a significant gap on the technical side—but, although unplanned, would be back in a couple of years' time. Soon we were joined by Dave Mackie, a lab technician who had been working in the Department of Zoology. With shoulder-length hair and a keen interest in popular music he added a new dimension to the 'academic' team but was no less professional or committed to the Antarctic programme. Gordon Robin, who had been unable to participate in 1971–72, was going to lead the next season. It would be his last before stepping back to let others take over.

8.7 Remote Sensing in Glaciology

In September 1974 the International Glaciological Society held a conference on 'Remote Sensing in Glaciology' in Cambridge. It provided an ideal opportunity for the SPRI RES group to present several aspects of its work and mingle with fellow scientists from around the world. Robin gave a wide-ranging review paper of RES discussing radio wave propagation, interpretation of returned signals, deconvolution/migration of profiles, the problems of sounding temperate glaciers, and the potential of RES to assist in determining ablation, accumulation, temperature, and ice movement.

Gordon Oswald reported on characterising the reflections from the ice base. He discussed further studies of the sub-ice lakes in East Antarctica and field results from Devon Island, Canada, concluding that beyond the defined boundaries of the lakes there were substantial areas of melting at the base of the ice sheet where the water was likely to be in thin layers. On Devon Island several traverses allowed Oswald to examine small-scale irregularities of the bed of the cold ice (at that time no indications of melting there), including the possible presence of morainal boulders. A difference in the calculated electrical permittivity along a traverse line across the western side of the ice cap was consistent with the geological boundary predicted to run beneath the ice cap. Oswald completed his PhD thesis in 1975 and moved on to a position at an electronics company in the Cambridge area. In later years he returned to some of these RES interests with studies of data gathered in Greenland and Antarctica.

Robin presented a further paper from Devon Island fieldwork, determining the velocity of radio waves in ice. This parameter, crucial to converting travel time to ice depth, is subject to the effects of ice density, temperature, and crystal orientation. Almost all previous studies of velocity had been measured under laboratory conditions; there were virtually no comprehensive field measurements. Robin's borehole experiments yielded a valuable log of the changes in propagation velocity in steps down the hole. Below 50 m the results showed a high level of reliability and consistency and gave a value of 167.7 ± 0.3 m μs^{-1} at −20°C.

I gave a paper on comparing the ice-thickness measurements by radio-echo sounding with those from seismic shooting and gravity observations. RES was coming of age, and for it to be seen as the most rapid and continuous method of ice-depth measurement, it had to have an accuracy equal to if not better than existing acoustic techniques. From the grid of flights conducted in 1971–72 with the improved navigational systems, the RES data were shown to be internally consistent to better than 1%. Sixty comparisons were made of thicknesses measured on eight earlier oversnow traverses. Fewer than one-third of the seismic depths fell within a 3% limit suggested earlier by Robin, and 21 comparisons, obtained during three traverses, fell outside a 10% difference that allowed for disparities caused by extreme errors in position. It was argued that because of the high reliability of RES measurements and the confidence in interpretation given by the continuity of the reflections shown on the records they should be used as a standard reference for all other geophysical measurements of polar ice sheets.

8.8 Preparations for the New Season

There was considerable discussion within the SPRI regarding the principal research aims and priorities for the upcoming season. These had to meet a variety of objectives—the ongoing research interests of the staff, potential projects for new research students, the requests from colleagues in other countries, the requirements of the now very active IAGP, and, finally, the overall intellectual drive to gain as wide a coverage as possible of the unknown regions of Antarctica.[191]

[191] Bentley, C R; et al. (1972) (footnote 141).

8.8.1 Dome C and East Antarctica

The IAGP was now identified as the primary coordinating mechanism for the activities of several national programmes. The 1971–72 season had produced reconnaissance mapping and other insights into the glaciology of a large tract of East Antarctica, which had been a valuable input. Now there was a need to fill in gaps, particularly around the coastal margin of Wilkes Land and Terre Adélie, and to conduct more detailed surveys in specific locations where new campaigns were to be undertaken. Coordination was discussed at the IAGP meetings attended by Robin. Of high importance were the developing plans for deep ice core drilling at Dome C. Cores recovered in northwest Greenland at Camp Century and in the centre of West Antarctica at Byrd Station in the 1960s had pioneered the techniques of drilling and provided tantalising glimpses of the climate records that could be retrieved. Both these locations for drilling, however, had been determined by the presence of a well-established logistics facility rather than on scientific criteria. This selection process was to change. Led by Claude Lorius, scientists at the Laboratoire de Glaciologie in Grenoble were considering carefully where to locate a deep hole in East Antarctica (Figure 8.15). In preparation they had undertaken surface sampling and shallow drilling on traverses inland of the French research station, Dumont D'Urville.[192]

From these forays Lorius' team had established a definitive correlation between deuterium isotopic values and average temperatures, which could then be used to interpret the isotope signatures in deep ice cores in terms of changing climate conditions in the past. Lorius had plans for a variety of other measurements on ice cores, including determining the composition of past atmospheres by recovering air trapped in tiny bubbles within the ice. The trace chemistry of the ice would also shed light on atmospheric processes and interactions with the nearby Southern Ocean. All this potential research would feed the growing and exciting study of climate change which was seen to be of increasing concern.[193]

Selecting an appropriate site for the deep drilling was critical and had to be based on a clear scientific rationale. The SPRI RES maps had shown the presence of a well-defined but subdued surface dome halfway between the

[192] Jouzel, J; Lorius, C; and Raynaud, D (2013) *The White Planet* (English translation), Princeton, NJ: Princeton University Press, 306pp.

[193] Johnsen, S J; et al. (1972) Oxygen isotope profiles through the Antarctic and Greenland ice sheets, *Nature* 235:429–34.

Figure 8.15. Claude Lorius, circa 1996.

main ice sheet crest in East Antarctica (Dome A) and the coast—this was Dome C, referred to earlier. Domes are ideal for recovering ice cores and interpreting the results. The snow falling on the surface travels downwards in a more or less vertical direction, thus retaining its geographically specific signature throughout and avoiding the complications of lateral ice flow. Whilst it is possible for the surface configuration to change over long time periods and during expansion and contraction of the ice sheet during ice ages, it was considered that Dome C would be a sufficiently stable location to warrant investigation. The RES showed the ice depth was more than 3000 m, varying by a few hundred metres over distances of 50 km or more. Modelling of the age of the ice suggested it could be more than half a million years old at its base, which would provide an outstanding record of climate.

The French planned a drilling campaign but first needed further data for Dome C—surface chemistry, estimates of the annual snow accumulation, and detailed ice thicknesses. It was agreed that the SPRI would fly a more tightly spaced grid around the projected drill site to gather the necessary depth and bedrock topography information, as well as conduct a sounding flight along the surface traverse route the French teams were taking from Dumont D'Urville to Dome C, more than 1000 km. This would assist them in their modelling of the ice flow regime. Charles Bentley from the University of Wisconsin had also expressed an interest in this area for various experiments and was keen to have a finer grid of soundings. These would be used to refine and later assist in the interpretation of the ice core results and provide another US scientific contribution to IAGP.

Significant gaps in the grid over East Antarctica were in the coastal zones between 135°E and 155°E. Accordingly, lines were planned to fill in these areas, albeit at extreme range from McMurdo. Other tracks were planned around the Soviet station at Vostok where a deep coring programme was already underway, and from there to the South Pole along or parallel to the 90°E meridian. One flight was to be devoted to the special study of sub-ice lakes.

8.8.2 Marie Byrd Land

The West Antarctic Ice Sheet was emerging as the focus for future research into its long-term stability. In the late 1970s John Mercer at the Institute of Polar Studies at The Ohio State University had postulated that under a warmer climate the great ice shelves in the Ross and Weddell Seas could break up, resulting in the rapid collapse of the adjacent ice sheet in West Antarctica, which is grounded below sea level and is therefore potentially unstable. He calculated the loss of the grounded ice could raise world sea levels by as much as 5 m.[194] His hypothesis was based on an examination of the higher sea levels during the last inter-glacial about 120 to 125 thousand years ago, for which the only explanation seemed to be loss of West Antarctic ice. Mercer's ideas generated considerable interest and concern that contemporary greenhouse warming could induce a similar reaction. Nevertheless, it would take another decade for the reality of global climate warming to become mainstream.

Mercer was a periodic visitor to the SPRI, and this lanky, slightly hesitant but energetically minded researcher would engage us with his latest ideas. He readily shared his passion for scientific exploration in remote places, especially in the Andes of southern South America, and Antarctica.

One of the related notions at the time was that the recently identified ice streams along the Siple Coast would be the principal conduits for the increased ice discharge from the central region of West Antarctica should the ice sheet begin to collapse. Other ideas being promulgated suggested the ice sheet could undergo periodic 'surging' in which it would advance at a rate an order of magnitude greater than its normal mode.[195] Together these concepts demanded further investigation for which only rudimentary gla-

[194] Mercer, J H (1978) West Antarctic Ice Sheet and CO_2 greenhouse effect: A threat of disaster, *Nature* 271:321–25.

[195] Wilson, A T (1978) 'Past Surges in the West Antarctic Ice Sheet and Climatological Significance', in van Zinderen Bakker, E M (ed.) *Antarctic Glacial History and World Palaeoenvironments*, Rotterdam: A A Balkema, 33–39.

ciological information existed, particularly better descriptions of the present-day configuration of the ice sheet surface, the sub-ice bedrock topography, as well as definition of the conditions at the base of the five known ice streams. Terry Hughes at the University of Maine, a large, gregarious, and energetic freethinker, was especially active in promoting a programme of research of the whole ice sheet which he called WISP—the West Antarctic Ice Sheet Project.[196] This would be tailored to coordinate with the already developing programme of science on the Ross Ice Shelf (see section 8.8.3).

At the SPRI a campaign was planned to provide systematic RES data for the area extending from the grounding line of the Ross Ice Shelf inland across the ice streams to the ice divide, the region separating ice flow into the Ross Sea drainage basin from that flowing east into the Ronne and Filchner Ice Shelves of the Weddell Sea. In practice this comprised a 50 km–spaced grid from the Transantarctic Mountains to the Rockefeller Plateau, and between the Ross Ice Shelf and Byrd Station. The region with its exciting scientific prospects was to be the focus of Keith Rose's research.

8.8.3 Ross Ice Shelf

The Ross Ice Shelf had already received considerable attention from numbers of RES flights radiating out from McMurdo Sound during the 1969–70 season and an additional flight of parallel lines in its southeast corner conducted in 1971–72. Robin was working on the data from these missions, investigating the patterns of ice thickness and the flow regime. Because the behaviour of ice shelves is also associated with the stability of the West Antarctic Ice Sheet, the US NSF was sponsoring research in the form of the Ross Ice Shelf Project (RISP).[197] One of its primary aims was to drill through the ice shelf to gain access to the water column, the sea floor, and its sediments below, as well as to conduct a variety of physical and chemical measurements down the hole. These it was postulated would likely reveal past changes in thickness, extent, and dynamics of the ice shelf such as grounding and floatation cycles, as well as afford knowledge of oceanic currents and circulation beneath the ice shelf (important for understanding melting or freezing at this interface). In preparation and to understand

[196] Hughes, T J (1973) Is the West Antarctic Ice Sheet disintegrating? *Journal of Geophysical Research* 78:7884–910.

[197] Zumberge, J H (1971) (footnote 118).

the broader aspects of the ice shelf, a programme of surface-based glaciological and geophysical investigations was instigated in the 1973–74 season. This latter sub-project, the Ross Ice Shelf Geophysical and Glaciological Survey (RIGGS), led by Charles Bentley at the University of Wisconsin, undertook measurements of ice thickness and water column depth using seismic and gravity techniques at regular grid points spread out over the ice shelf.[198]

It was obvious that the SPRI plans for RES in 1974–75 could provide additional information to RISP and RIGGS, but clearly it was inappropriate to duplicate costly ground-based research; there needed to be an 'added-value' component. For any floating ice shelf away from the drag effects of neighbouring land, the principal process that determines its form and behaviour is the flow under its own weight. Critical parameters in determining flow are the surface mass balance (typically net accumulation) and the melting or accretion of ice at the base, where the bottom of the ice shelf is in contact with the sea. Of these the trickier and more problematical is the basal melting or freezing, and in the mid-1970s little information existed to provide even an order of magnitude estimate of this quantity, never mind its sign!

Chris Neal was convinced that the problem could be studied using the changing nature of the radar reflections from the ice/water interface. If there was melting, then clear ice would be in contact with the ocean and would produce a very strong reflection owing to the significant dielectric contrast between ice and seawater. If, however, there was a layer of accreted ice at the base, this would most likely contain brine, and the electrical properties would be different and less contrasting.

Gordon Oswald had shown it was feasible to determine some statistics that could help interpretation from the shape of the returned pulses. He had used A-scope images every metre along a track on the surface of the ice cap on Devon Island to derive such values.[199] Neal worked out that for the average conditions on the Ross Ice Shelf he would need samples at least every 3 m along a very much longer line, which he calculated as 1 million frames per flight! A fresh look and some innovation were needed.

[198] Rutford, R H (1974) The Ross Ice Shelf Project 1973–74, *Antarctic Journal of the United States* 9 (4): 157.

[199] Gordon Oswald has commented (personal communication, January 2021) that recording of the complex ('I/Q') received waveforms would have been better still. At the time that was beyond the state of the art.

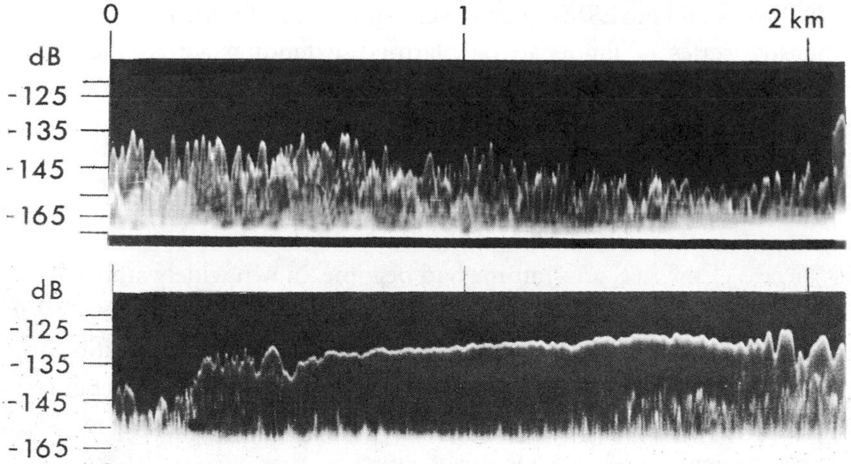

Figure 8.16. Echo-strength measurements (ESM) exhibiting the power reflection returns from two different sub-ice surfaces. Top: rough ice/rock interface with high-frequency variations in the returned power. Bottom: subglacial lake showing high and persistent returned echo strengths.

Neal came up with a clever solution, which was to record the peak power of the returned radar signal on continuous film, in a manner similar to recording the ice-thickness data. The system was termed ESM (echo strength measurement), and it proved highly successful.[200] The procedure required that the much stronger return from the ice sheet or ice shelf surface be blanked out to prevent it from obscuring the weaker return from the base. To accomplish this the bottom reflection (or any other return of interest) was 'gated' (Figure 8.16). The ESM recorder was duly constructed by Neal with the assistance of David Meldrum. In the first season the 'gate' had to be manually positioned to track the fluctuating range of the returned pulse and was somewhat labour-intensive. Nevertheless, the system worked extremely well and with the generally slowly changing thicknesses exhibited by the ice shelf the task was not excessively arduous. The technique could also be used over the ice sheet to examine the reflection properties of prominent layers or the ice/bedrock interface, although the much rougher topography placed exhausting demands on the ESM operator.

[200] Neal, C S (1976) Radio-echo power profiling, *Journal of Glaciology* 17 (77): 527–30.

The theory and the ESM concept were sufficiently compelling for a comprehensive series of flights to be planned, extending across the whole breadth of the Ross Ice Shelf, 50 km apart and parallel to the ice front to record the radio signal returned from the ice/water interface.

8.8.4 Other Projects

The success of the RES programme had become known widely amongst the glaciological and geophysical Antarctic fraternity, and requests for SPRI to obtain data for other investigators began to increase. The Antarctic community is generous in its willingness to assist other scientists, particularly as the costs of operations are so high; combining forces wherever possible has always been considered positively.

Ian Whillans, a bubbly, enthusiastic and gifted glaciologist at The Ohio State University, was working on determining the long-term changes to the West Antarctic Ice Sheet. He had established an array of surface markers up the flowline from Byrd Station for about 170 km and had obtained data on surface accumulation and movement. Whillans was keen to acquire ice thicknesses along his survey lines and any information from internal reflections (layers). These he argued, if they were truly representing time-coincident horizons formerly at the ice sheet surface (i.e. isochrones), could be compared with time horizons calculated from glaciological theory, using the surface data alone. The results might hint at the stability or otherwise of the ice sheet. It was agreed that we would do our best to meet his request, although the flying along his narrowly spaced lines would be very 'tight' and demanding.

The revelation of substantial mountainous ranges buried beneath the thick ice of East Antarctica was raising intriguing questions regarding their role in the formation and early development of the ice sheet. A wide range of direct and indirect evidence for the initiation of the ice sheet in Antarctica—including marine glacial sediments, global sea-level changes, and palaeontological investigations—suggested strongly that the growth of a full-bodied ice sheet would have commenced in the Late Cenozoic, perhaps some 30–40 Ma BP.[201] RES studies of the inland flank of the Transantarctic Moun-

[201] Drewry, D J (1975) Initiation and growth of the East Antarctic Ice Sheet, *Journal of the Geological Society of London* 131 (3): 255–73.

Figure 8.17. Ferrar Glacier Valley, southern Victoria Land, looking west (the glacier is 6 km wide). Taylor Glacier is in the upper part of the adjacent valley (right).

tains had mapped inland-trending glacial valleys that predated the present ice sheet, and similar details were emerging of the topography of the interior Gamburtsev Mountains. These were now strong contenders as highland centres for the initiation and expansion of the ice sheet. Such models also implied the Transantarctic Mountains were probably 1500–2000 m lower in elevation at the commencement of glaciation. I would return to this last-mentioned theme with a colleague from the Department of Earth Sciences in Cambridge, Alan Smith, some 10 years later.

Linked to these considerations was the emerging detail of the ice surface configuration in East Antarctica. This enabled the drawing of ice flowlines with some confidence. The ice emanating from the Dome C region and flowing east to the Transantarctic Mountains was observed to diverge inland of the McMurdo Dry Valleys, being directed towards the major outlet glaciers in the north (Mawson and David Glaciers) and in the south (Skelton and Mulock Glaciers). This pattern could have implications for the interpretation of the glacial history of much of East Antarctica, which was being worked out in meticulous detail in the valleys by George Denton and his colleagues from the University of Maine. It was considered that certain moraines, glacial drift, and erosional features reflected changes in the size and extent of ice sheet. A detailed RES survey was planned extending inland from the western edge of the mountains at the 'back' of these valleys (Figure 8.17). I had a hunch that such a survey might throw up some unexpected features and trigger further research. In the event, I was not to be disappointed, and the project led to a series of studies that resulted in some occasionally heated discussions with Denton.

9

1974–75

Radio-Echo Sounding Comes of Age

The team for the next—the fourth—season was taking shape. Despite a gap of more than two years, activity was building once more. Robin would be in charge overall; Meldrum would head the SPRI technical side with Mackie in support. I was to be responsible for the flight planning and in-flight glaciological observations, shared with Robin. Chris Neal and Neil Hargreaves would assist in various ways as their experience and competence developed. Because we would be using new TUD 60 MHz and 300 MHz equipment with the SPRI MkIV in reserve, the Danes would send two of their best electronics staff to assist with the programme—Nils Skou and Mogens Pallisgaard—working directly with Meldrum. They were already familiar faces through the excellent modus operandi established with Gudmandsen and his team in Lyngby. The 60 MHz radar had previously been tested very successfully in Greenland during the northern summer of 1974.

Owing to the new configuration of the RES system and modified aerials, installation and airborne trials were undertaken just prior to deployment to Antarctica.[202] Meldrum, Neal, Rose, and Mackie flew to America for these operations in October and joined Skou and Pallisgaard at the US Naval Air Development Center at Warminster, Pennsylvania. Two weeks spent there involved the fitting-out and flight testing. A highly effective relationship developed between the SPRI/TUD personnel and the aircrew, fostered by the commander of the aircraft, Lt Cmdr Arthur 'Art' Herr Jr. He purchased yellow crew-caps emblazoned with the aircraft bureau number,

[202] Gudmandsen, P (1975) (footnote 190).

320, that, although seemingly rather trite, had a powerful effect in bringing the #320 team together. Chris Neal observed a DYMO strip on the flight deck that read 'All the good guys wear yellow caps'.[203]

In late September 1974, a plethora of glaciological events were hosted by the SPRI capitalising on the presence of several key players attending the International Glaciological Society Symposium on Remote Sensing in Glaciology. This enabled the SCAR Working Group on Glaciology, as well as the Council of the IAGP, to hold discussions. It was a busy time. The SPRI maps of East Antarctica had been printed and were made available at the meeting. The plans elaborated in the previous chapter for RES in 1974–75 were set out in general terms, but a separate session was organised to specify particular third-party requests for sounding coverage. That meeting included Robin and Drewry (from SPRI), Preben Gudmandsen (TUD), John Clough from Madison, Ian Whillans from The Ohio State, Terry Hughes from Maine, and Bob Thomas from the University of Nebraska. In total the NSF had allocated 350 hours for the SPRI project, which was a generous slice of its air capability. But little did we know at the time we would never again be the recipients of such generosity. The various requests and our own plans were plotted on air navigation charts in draft form so that we could move quickly once we commenced the RES season. They were also necessary for preliminary discussions with Art Herr and the aircrews that were to fly the lines.

9.1 Initial Deployment

It was planned that I would route to New Zealand via Sydney using commercial airlines. I was to liaise with the NSF and Navy officials in Christchurch as the 'advance party' to ensure we lost no time in preparing for departure to Antarctica. But all those plans fell apart. I arrived on 4 November and had to wait two weeks to be joined by the majority of SPRI colleagues, the trusty Hercules aircraft, #320, and its crew. I was able, however, to spend more time with the New Zealand Antarctic Division and with scientists at Canterbury University involved in Antarctic work, as well as to finish a paper on the production of the large-scale map of Antarctica. Margaret Lanyon, assistant to the NSF representatives in the USARP office, was holding the fort on a temporary basis—a good move capitalising on her charm and

[203] Neal, C S, audio interview SPRI RS2020–8.

efficiency. The office itself had been moved and was now housed in the US Navy Operations building. Lanyon proved to be very helpful in providing situation updates.

Art Herr flew #320 down from Warminster, along with Meldrum, Neal, Rose, and Mackie, calling first at Memphis, Tennessee, and thence to Point Magu, close to Los Angeles, the home of the VXE-6 Squadron. The SPRI team was thereafter scheduled to deploy across the Pacific to New Zealand. It should be recalled that #320 was one of the oldest C-130 aircraft and had a reduced fuel payload and hence range. Owing to strong headwinds, two attempts to fly to the first refuelling stop in Hawaii were aborted. On the third take-off from California, they made it to the most easterly air facility (General Lyman Field) at Hilo on the Big Island with 'dry tanks'. The aircraft refuelled and flew on to the Naval Air Station Barbers Point, on Oahu, where there was a two-day stopover before continuing via Canton Island (another US air base in the Phoenix Islands of Kiribati),[204] and Nandi in Fiji, not arriving in Christchurch, New Zealand, until 21 November at the unsocial local time of 4.00 a.m.! 'Flight-following' had failed to inform Harewood Airport of the arrival of #320, and so it was not until 6.00 a.m., after New Zealand Health, Customs and Agriculture had cleared the plane, that the crew and SPRI and TUD contingent were able, wearily, to leave the plane. It had been some journey! Now that the team was together, there ensued a busy period involving aircraft checks, preparing equipment, drawing cold-weather clothing, and briefing the flight crew (Figure 9.1).

[204] Chris Neal (footnote 203) records that as part of the briefing for travel in the C-130 the SPRI team had been talked through a safety drill in the case of possible emergency ditching in the sea. Given the several attempts to fly to Hawaii with barely sufficient fuel, the team realized this was not an entirely theoretical situation.

According to Neal, the drill instructions for ditching were that the aircraft loadmaster would order everyone to don lifejackets and strap into the webbing seats; the escape hatches fore and aft were pointed out. The aircraft would descend toward the sea surface. At 100 feet above the water the plane would level out, and there would be six blasts on the internal alarm system—the signal that the aircraft was about to hit the sea.

After the stopover at Barbers Point the aircraft left Oahu on the next leg of the journey. Neal recalls that after some time the aircraft started to descend towards Canton Island. As the plane descended further suddenly and to much consternation there were six blasts on the alarm system! Neal looked out of one of the aircraft porthole windows and there was nothing but sea—so a ditch was imminent. And then a strip of land appeared below, and the plane landed safely on the small island! Apparently, there had been a fire warning alarm, and the pilot had decided it was as good a time as any to undertake a ditching drill without telling the crew or SPRI team! Neal notes that it added a certain 'frisson' to the journey.

Figure 9.1. Trying out the cold-weather clothing issue in the USARP store at Christchurch airport.

The increasing incorporation of the SPRI team into the aircrew of #320 was put to the test on 23 November. Art Herr considered that any method of reducing drag on the aircraft would assist in improving its range and so he requested that the SPRI team and the aircrew muster at 2.00 p.m. at the hanger to wash down the airplane. This involved stripping to the waist with a pair of shorts and, with long-handled brushes, scrubbing off all the grease, oil, and dirt from the fuselage using a powerful solvent and high-pressure water hoses. It was a filthy, messy, and odorous job that also entailed a few high jinks with the hoses, all of which enabled us to bond better with the crew. As these various tasks progressed satisfactorily, it was possible to relax a little, and a party was held for all the personnel, aircrew, and NSF staff involved in the SPRI programme before #320 departed for McMurdo on 27 November. Even this departure had been delayed by two days owing to strong and persistent headwinds.

On arrival at McMurdo, the team were housed in a medium-size Jamesway hut (see Figure 7.2). It did not possess quite the same status as the infamous 'Vermin Villa' encountered in previous seasons but was certainly not luxurious. Chris Neal commented that the place was dark and dingy, sleeping arrangements were on bunk beds, and in the middle of the building there was an oil-burning stove that gave off fumes.[205] The upside was that we had the whole building for the SPRI-TUD group of nine, plus it of-

[205] Neal, C S (footnote 203).

fered a 'lounge area' and small bar (but without contents). After a couple of weeks, as space became available in more congenial accommodation, the team moved out with few regrets.

The immediate impression of our NSF colleagues, based in the 'Chalet' (the NSF HQ), was not hugely positive. They appeared to have little interest in our programme and seemed focused on viewing movies and consuming large quantities of food and drink. In contrast, the Navy and VXE-6 personnel proved exceptionally helpful. This lack of engagement by NSF staff was compounded by the visit of several 'DVs' (distinguished visitors) in mid-December who had been flown from Christchurch to witness the US operation. McMurdo was 'in chaos' for a week during the visit, as the DVs' schedule took precedence. They received briefings on the research programmes being conducted 'on the ice'. Of course, many scientists were at remote stations or in the deep field. Our SPRI group was one of the few operating out of McMurdo, so we were called upon to give presentations and entertain the DVs. This was somewhat incongruous, as the visiting group were mainly high-ranking officials and politicians on a trip to see how their country's dollars were being spent, and here were Brits and Danes enjoying American munificence and briefing them on their science expenditure. The visit succeeded in demonstrating the hugely important dimension of international cooperation that is a hallmark of Antarctic science. We made certain we gave shedloads of credit to our US colleagues and the NSF.

We were extremely lucky with the selection of the aircraft commander and aircrew assigned by the Navy to our project. Lt Cmdr Art Herr Jr, in overall charge, was a highly professional individual with considerable flying experience in the Antarctic. He commanded respect from his crew as well as the SPRI team, and his interest in the science was very encouraging (Figure 9.2). He quickly embraced the RES work and viewed it as one of his most exciting and challenging flying operations. He badgered the VXE-6 operational staff on our behalf for repairs, fuel deployments, and equipment. In this way, he ensured that the Navy gave sufficient priority, time, and general support to the programme. The extensive and valuable sounding work we achieved during this season was in large measure due to his enthusiastic drive.

The established scheme for airborne activity was to divide into two teams to enable the flying to operate as continuously as possible, with one crew taking over as the other completed its mission. Art Herr led one section of the aircrew; his colleague Lt Cmdr Robert 'Bob' Nedry led the other. More

Figure 9.2. Left to right: Lt Cmdr Arthur 'Art' Herr Jr with Mogens Pallisgaard (TUD) and David Meldrum (SPRI) at McMurdo.

puckish and somewhat younger, Nedry nevertheless also embodied the focused approach that was needed for a successful season. Between the two crews we could fly back-to-back missions whenever the weather permitted and the aircraft and RES equipment were functioning satisfactorily.

We had learned that it was preferable to have close and regular liaison with the US Navy staff and the VXE-6 Squadron, as well as to have rapid access to the meteorological forecaster. In previous years we had established our activities in the Earth Science Lab at McMurdo (see Figure 7.2). This time, through Herr's intervention, we had the ability to occupy a suite of offices for general planning and 'laboratory'-type work in the dedicated Navy HQ building, which was located closer to the centre of McMurdo (Figure 9.3). This was to prove of great benefit and convenience, as well as helped establish good rapport with the wider Navy squadron personnel under the command of Capt F C Holt.

9.2 Operations Commence

An important restriction on the flying programme was the weight of the aircraft, an aspect already encountered in the positioning flight across the Pacific. Owing to its age and in consideration of safety factors, #320's all-up

Figure 9.3. Building No.165, the US Navy HQ in McMurdo. (Courtesy NSF).

weight was reduced from previous seasons by 4,500 kg. This directly affected the quantity of fuel taken onboard and resulted in flights to and from the principal sounding area needing to be made at high altitude, where the fuel burn rate was lowest. The first test flight took place on 28/29 November, followed by a flight into East Antarctica to provide a full shake-down of the equipment. No significant problems were encountered, and the season was primed.

The flight operations under Art Herr worked extremely well. Should the aircraft prove reliable, two aircrew teams would fly in rotation. This placed significant demands at the end of each flight to ensure notification of any faults with the RES equipment (sometimes in advance by radio), reloading of the cassettes for the film cameras, and other housekeeping chores. Quite often, minor faults would require someone—mostly in rotation as well—to work back at the laboratory to fix the problems. Nevertheless, throughout the season it was rare for an RES equipment fault to abort a flight; for the SPRI team this was a matter of principle! A cancelled mission was typically the result of an aircraft malfunction or bad weather.

Early in the season it became apparent that the two aircrew teams saw themselves in competition to log the most hours of flying. Given that we had previous experience of how a season could come to a rapid and unexpected termination, it was vital to undertake as many missions as early as possible, so this competitive spirit worked to our advantage. A 'how goes it' chart was devised that plotted the cumulative hours for each of the teams. This was produced in large format and displayed prominently in the Navy building. We kept it up-to-date, and it proved a wonderful spur to the flyers who would consult it to see whether they or the rival team had achieved

Figure 9.4. Meldrum (left) and the author discuss flight routings in VXE-6 Squadron offices where SPRI had its operations. The map shows a compendium of flight lines from which, according to range, refuelling, and weather considerations, several could be selected for a mission. (Courtesy US Navy, US Antarctic Program).

more time in the air. We also kept a large map on display on which were reproduced all the planned flight lines for the season (Figure 9.4) This, too, we marked up regularly, showing the lines we had accomplished in a prominent colour.

Some of the first flights were out to the coastal fringes of Wilkes Land to fill in gaps in the East Antarctic grid. Several missions took us to the French station, Dumont D'Urville, which we overflew at low level, and to the impressive ice stream outlets of the Astrolabe, Mertz and Ninnis Glaciers that fed staggering quantities of icebergs into the Southern Ocean. In recent years these outlet glaciers have become important features for study, as they link the ocean to the ice sheet and are conduits through which climate change responses are most likely to be transmitted.[206] The number of flights and

[206] The Mertz Glacier flowline was modelled by Len Watts, an SPRI graduate student ('Finite Element Simulations of Ice Mass Flow', PhD thesis, University of Cambridge, 1988, 249pp). His results predicted that in the transition zone between the ice sheet and ice tongue, as well as in upstream basins, the bed was at the pressure melting point. Watts ran the model forward for 30 years with a CO_2 warming scenario (temperature increase of 6.5°C, surface accumulation increase of 30%). For the lower part of the ice stream (100 km from the floating ice front and some 500 m in thickness) the basal

hours stacked up steadily—70 hours on 8 December; 108 hours on 15 December; and 140 hours on 19 December, by which time we estimated we were about a third of the way through the programme. We were developing a rhythm as long as the airplane continued to perform without any major faults; it was hectic flying all hours, day and 'night'.

Although operations were proceeding smoothly, there were moments of high drama. Chris Neal recollected a tricky incident. During briefings on operational procedures, the SPRI team were told the two worst situations on an aircraft were, first, a fire onboard, and second, personnel disobeying orders, and this applied to the injunction to remain strapped into your seat during what were often quite bumpy landings. During the season and whilst on a long mission, #320 was flying into Byrd Station. In the descent Neal noticed that smoke was pouring out of one of the RES oscilloscopes! Neal remarked that he was in a dilemma—which of his instructions took precedence—stay strapped in or deal with the fire! As the aircraft was on its final approach he unstrapped, jumped up, and shouted, ''scope on fire!' In the ensuing melee the loadmaster came across and extinguished the fire. All was satisfactory. Neal had made the right decision, but the offending oscilloscope was removed from the rack and for the rest of the return flight to McMurdo was strapped down on the back ramp of the aircraft like a truculent animal lest it reignite.

On 19 December the RES operation was moved to the McMurdo Skiway, which reduced the take-off weight further, but the skiway's rougher surface caused shocks to the antenna supports, so much so that on one flight struts broke loose, and the sortie had to be aborted. These shocks produced other effects, as well. On one later flight at high altitude that required pressurisation the cabin pressure dropped steadily. The loadmaster believed it was caused by a crack in the fuselage. He lit up a cigarette and proceed to walk through the aircraft hold, bending down frequently to blow smoke around the sides of the plane seeking the location of the crack, which would suck out the smoke. And he found the spot! It was on the rear part of the aircraft belly. The plane commander radioed back to Williams field for the 'tinsmith' (in less prosaic terminology this translated, and inflated, to a structural engineer) to meet the aircraft on landing and fix the problem.

---

temperatures rose by 5°C, the area of bottom melting expanded, and the melting rate doubled; ice velocities increased by 10 m a^{-1}. These changes would result in some thinning and advance of the ice sheet, with a consequent increase in calving of ice at the ice front. Further inland, beneath the much thicker ice sheet, sliding was predicted to be minimal.

When #320 was parked after the mission, a tall, gangly man in grubby overalls walked out to the plane—the tinsmith. His repair consisted of running several long strips of duct tape along the base of the fuselage where the crack was surmised to be located. After that none of us worried too much about 'minor' structural defects. It was rather a coincidence that our aircraft developed some mechanical faults around 25 December. And we were not the only C-130 with problems. The oldest of the airplanes, another 'F' model, #319, had been grounded for so long at Williams Field air facility it was regularly called 'Building 319'.

9.3 Siple Coast—Domes and Ice Streams

A rectangular grid was being flown across western Marie Byrd Land to better define the ice flow into the Ross Ice Shelf. Normally, missions over the ice sheet were somewhat routine in as much there are rarely significant surface features to observe. The southern ends of the flight lines in this region were more interesting as they approached the frontal ranges of the Transantarctic Mountains. However, as the flight lines crossed the ice inland of the Ross Ice Shelf, zones of heavy crevassing were spotted, caused by the entry of fast flowing ice into the more sluggish ice shelf. In discussion with the Navy pilots we discovered they knew these features well and used them as 'guides' to navigation when en route to Byrd Station. They called these areas by colourful names that mostly related to their shapes as seen from the air—'boomerang', 'steer's head', 'the skua', and 'broken arrow'.

At the regional scale these features resolved themselves into long linear bands, which we quickly interpreted as the margins of ice streams (see Figures 9.5 and 10.2). As deduced from the earlier flight in 1969–70 (section 5.7), the ice from the high inland region was moving down toward the ice shelf and self-assembling into a series of fast-flowing streams embedded within the ice sheet. These would typically be several hundred kilometres long and up to 60 km wide. The difference in velocity between the fast flow of the streams (several hundreds of metres per year) and the slowly moving inter-stream zones (typically only a few tens of metres per year), caused considerable shearing at the margins, and the ice being torn apart created crevasses and rifts. These fragmented regions stood out clearly as we flew across them and on the edge of the ice shelf formed zones of chaotic broken ice (see Figure 5.19).

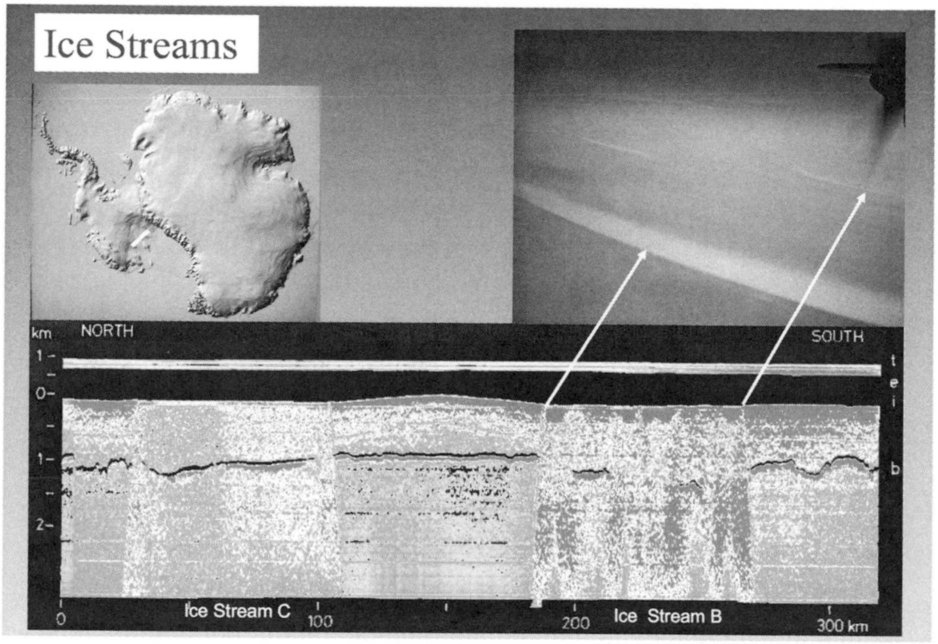

Figure 9.5. Ice streams in Marie Byrd Land (Siple Coast). Bottom: RES record, in false colour (located on the map). Margins of B and C delimited by the deep clutter echoes from crevassed shear bands depicted in the photograph of Ice Stream B. The ice stream flow is left to right.

The heavy crevassing scattered the radar signal, producing a region of intense clutter echoes on the RES records. These would be both strong and persistent, often obscuring the return from the bottom of the ice. The onset of the clutter could be very marked, and it was possible to use this characteristic to define quite precisely and map the margins of the ice streams. We deduced that the fast flow and the low gradient of the ice streams was the result of basal melting and water saturation lowering the friction at the bed (also termed the basal shear stress). The zones in between, we discovered, were elongated domes or ridges, where the ice was probably frozen to the bed and the ice flowing only slowly down the slopes with the regional gradient. A pattern began to emerge of five principal ice streams characterising the region. The RES in this area, its mapping, and interpretation formed a core part of the studies by Keith Rose, and some of his conclusions are explored in section 10.1.

9.4 'Lake' Vostok

Two or three flights were planned in East Antarctica to the remote central dome, the highest part of the continental ice sheet. These would augment the forays made in 1969–70 and 1971–72, although it was accepted that they were logistically at maximum range for our C-130 out of McMurdo, even with refuelling at South Pole Station. It was already known that the subglacial Gamburtsev mountain range in this region was impressive and enigmatic—there are few intra-craton mountain ranges around the world, so its further delineation would be important.[207] The flights tended to route over the Soviet Vostok Station, as this provided an excellent geographical control point in this far-flung region. It was on one of these flights that as soon as the aircraft descended from altitude to the normal sounding elevation, a very strong return from the base of the ice sheet was picked up despite the depth being nearly 4000 m, and it continued for well over 100 km. From Oswald's work the presence of sub-ice 'lakes' had been identified, but these tended to have lateral dimensions approximating the ice depth of a few kilometres. What we were seeing here seemed extraordinary—an order of magnitude larger in its size. Robin's excitement is seen in his entry in the glaciologists' logbook (Figure 9.6). Subsequent flights into this region confirmed the considerable extent of basal water at this location (Figure 9.7).

[207] Extensive geophysical soundings (RES, magnetic, and gravity observations, both ground- and airborne-based) undertaken during the AGAP campaign have enabled a better understanding of the presence and origin of this extensive mountain area (see footnote 135). The Gamburtsev chain is similar in areal extent to the European Alps and displays 'alpine' topography with high peaks (up to 3000 m asl) and deep valleys (this type of terrain was identified from the earliest SPRI RES records (see footnote 89)). Most recent tectonic hypotheses consider that the proto-Gamburtsev Mountains resulted from mountain building in the Late Proterozoic (some 1 Ga BP) during the coalescence of the ancient super-continent of Rodina. The continental collisions would have created an elevated mountain chain with a deep crustal 'root'. Only this root would have been preserved following a long period of denudation. It is proposed that further crustal shortening took place as the super-continents of Gondwana and Pangaea assembled in Late Proterozoic and Palaeozoic times (Veevers, J J; et al (2008) Provenance of the Gamburtsev Mountains from U–Pb and Hf analysis of detrital zircons in Cretaceous to Quaternary sediments in Prydz Bay and beneath the Amery Ice Shelf, *Sedimentary Geology* 211:12–32). Subsequently, the break-up of Gondwana in the Mesozoic (200-100 Ma BP) remobilised an old rift structure and raised the ancient Gamburtsev lithospheric root. Extensive rift structures remain around the Gamburtsev Mountains, the principal ones being the Lambert Rift and the rifts between the Gamburtsevs and the Vostok Sub-glacial Highlands (VSH); the latter rift contains Lake Sovetskaya and another large lake, 'L90E'. Lake Vostok sits within a further rift on the eastern flank of the VSH. The elevated mountain root of the Gamburtsev Mountains was subject to a fresh cycle of sub-aerial fluvial and later glacial erosion that produced its present but hidden landscape. Only deep drilling into bedrock will assist in advancing our geological knowledge of this intriguing region (Talalay, P; et al. (2018) Drilling project at Gamburtsev Subglacial Mountains, East Antarctica: Recent progress and plans for the future. *Geological Society, London, Special Publications* 461 (1): 145, http://dx.doi.org/10.1144/SP461.9).

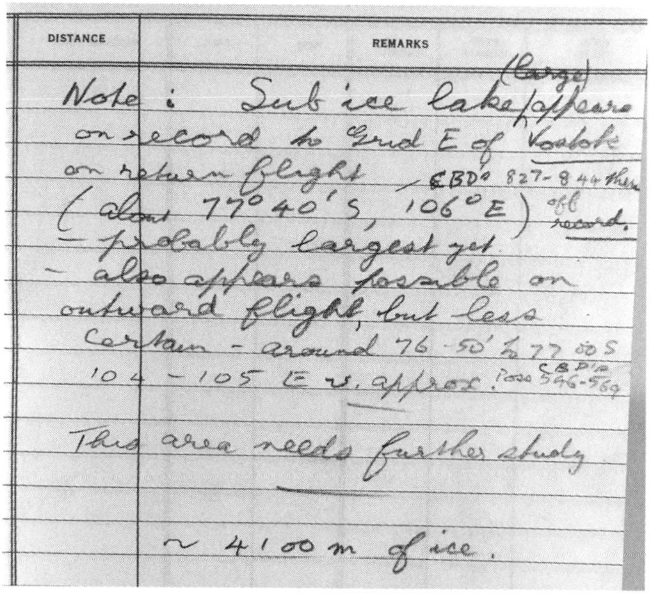

DISTANCE | REMARKS

Note: Sub ice lake (large) appears on record to Grid E of Vostok on return flight, CBD° 827–844 then off record (about 77°40'S, 106°E) – probably largest yet.
– also appears possible on outward flight, but less certain – around 76–50' to 77 00S 104–105 E v. approx. Posn CBD/a 546–589

This area needs further study

~ 4100 m of ice.

Figure 9.6. Robin's logbook entry on 23 December 1974 indicating his clear recognition of the emerging scale of what was to become Lake Vostok.

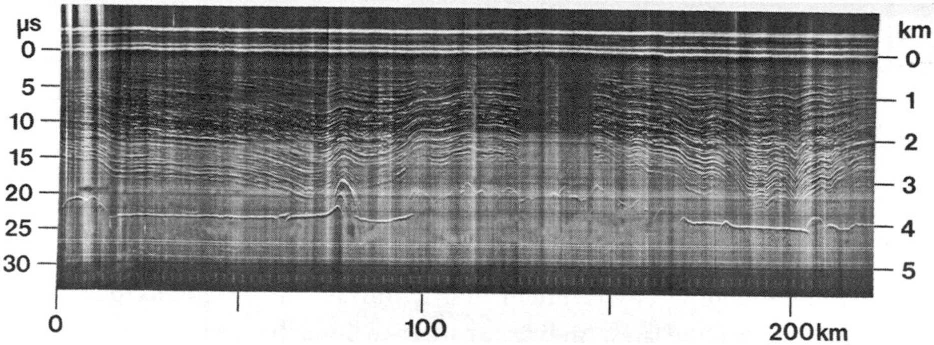

Figure 9.7. RES profile showing two crossings of Lake Vostok (shown in Figure 9.8 at track X-Y).

These occurrences were compiled into a sketch map showing how the water reflections joined to make up a much larger water body or lake close to Vostok Station (Figure 9.8).[208] Once again the RES results had astounded us with a new and challenging discovery.

[208] Robin, G de Q; Drewry, D J; and Meldrum, D T (1977) International studies of ice sheet and bedrock, *Philosophical Transactions of the Royal Society of London B* 279:185–96.

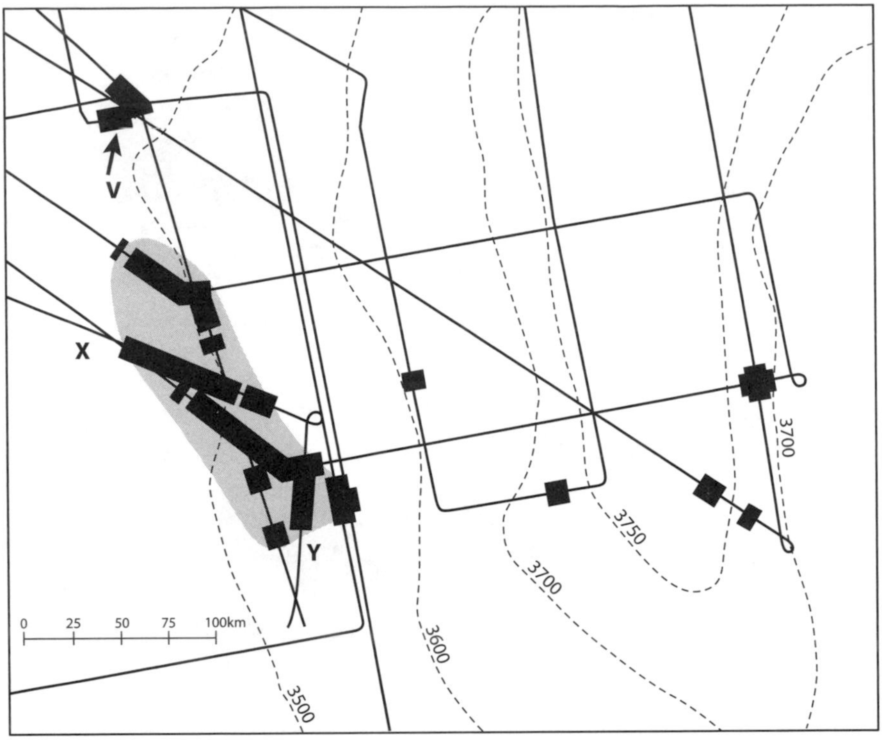

Figure 9.8. RES flights in vicinity of Ridge B and Vostok Station (at V), from the 1971–72 and 1974–75 seasons. Ice sheet surface contours are in metres above sea level. Black blocks denote areas where sub-ice water was detected. Shaded area was the sketch of the Vostok 'Lake'. The 'lake' was not extended to beneath the station itself; this was accomplished by subsequent geophysical fieldwork. (Adapted from Robin et al. (1977); see footnote 208).

The often-febrile environment at McMurdo with scientists passing through enabled the latest findings or ideas to be exchanged. Igor Zotikov, from the Institute of Geography at the Academy of Sciences of the USSR in Moscow (I recall him as a short, stocky, convivial and highly competent glaciologist-cum-engineer), was working with the Ross Ice Shelf Project on drilling through the shelf to the seawater beneath. Some of us had visited his drill camp on the ice shelf earlier in the season and were impressed by this venture. He was in McMurdo for a short break, and Robin engaged him in conversation about our discovery of a large lake near Vostok. It should be recalled from section 7.8 that Zotikov's earlier work had been on the thermal regime of the Antarctic Ice Sheet, and his temperature models had suggested that in places the base of the ice sheet could be at the pressure melting point, so this news was of considerable interest to him. In discussion Zotikov recalled a report by a senior aircraft navigator with the Soviet Antarctic Expedition of 1959 (R V Robinson) who described a shallow but extensive *surface* depression in the Vostok region. In his short paper he wrote: 'Natural landmarks in the interior of the continent include, in addition to individual mountains and mountain ridges, oval depressions with gentle "shores" which are visible from an airplane over the plateau. The depth of these depressions usually does not exceed 20–30 m and their length 10–12 km. These unusual depressions are sometimes called "lakes" by pilots.'[209] Our SPRI flights confirmed this 'shoreline' feature.

We calculated that the 'true' lake *beneath* the ice would be about 180 km long and with a typical width of 45 km, about two-thirds the size of Lake Ontario in North America and two and a half times the area of Lake Geneva in Europe.[210] The ice thickness was 3.95 km, with the water at a relatively uniform level of 500 m below sea level—clearly occupying a basin or depression in the underlying rock surface. We argued that over such a large area of water the friction at the base of the ice would be reduced to zero, and the ice above would most likely be in hydrostatic equilibrium, (i.e. 'floating'), giving rise to a relatively flat upper surface, as had been observed by the Soviet aviators. Examination of satellite imagery that was available a few years later confirmed the smooth surface and outline of the lake below (Figure 9.9).

It was fascinating to reflect on these findings. Just as the smaller subglacial lakes uncovered by the RES missions in 1971–72 had spurred all types of scientific investigation, the existence of a massive lake beneath Vostok spawned a whole industry of scientific endeavour—chemists, biologists, geologists, geophysicists, glaciologists, of course, and many others were intrigued by the questions its presence raised in the late 1970s. Why was it there? How deep was the water? What lay below—sediments or rock or both? How old was the water? How old was the lake? What was the composition of the water, and what might it indicate? Was it fresh or saline? Was there life in the water? The questions were endless, triggering a rash of further scientific exploration of the lake.

[209] Robinson, R V (1960) From the experience of visual orientation in Antarctica (English translation), *Information Bulletin of the Soviet Antarctic Expedition* 18:8–29.

[210] The dimensions from the most recent surveys indicate the size as 250 km (160 mi) by 50 km (30 mi) at its widest point. It possesses an area of 12,500 km^2 (4830 sq mi). The average depth of water measured by seismic sounding is 432 m (1417 ft). The volume is estimated as 5,400 km^3 (1300 cu mi).

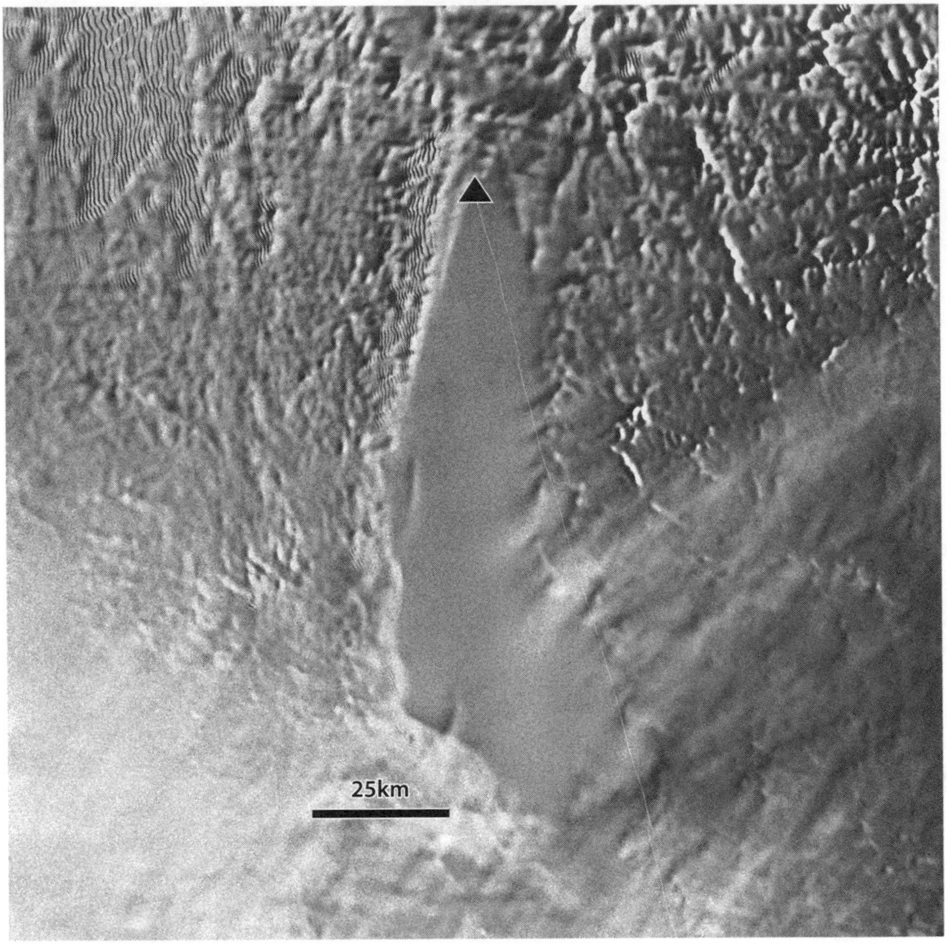

Figure 9.9. Satellite image of the ice sheet surface above Vostok Lake showing the location of the Soviet Station (triangle). The thin white line from Vostok Station across the right-hand-side margin of the lake is the tractor resupply trail to the coast, at Mirnyy (see also Figure 5.26). (From NASA MODIS instrument, courtesy NASA NSIDC DAAC).

The Soviets had for some years been drilling cores at Vostok for climate studies. Their intention was to retrieve samples from the base of the ice at about 3750 m, but it became very clear that the station was situated over one end of the lake, and boring to the bottom would involve the drill entering the lake and consequently contaminating it.[211]

[211] Drilling often involves the use of various fluids to fill the hole. Because ice is a 'plastic' material, a hole at depth will commence closing owing to the ice overburden pressure. Filling the hole with a

An international group was formed by SCAR to discuss these issues at a first meeting in Cambridge in 1994, attended by Robin and Martin Siegert, an SPRI research student and postdoc, who was later to take a leading role in studying these lakes. It was agreed that the drilling should be halted some 10 m above the lake and should not proceed deeper until the environmental issues posed by the operation had been resolved satisfactorily and a drilling protocol agreed. After a hiatus of more than 30 years the Soviets did recommence drilling in 2012 and 2015 to a depth of 3768 m. Unfortunately, the retrieved basal ice core was possibly contaminated, so that interpretations of the organic contents and other chemical signatures have been suspect.

A dominant player in the exploration of Lake Vostok was the Soviet scientist Andrei Kapitza (see section 2.4). Kapitza was a larger-than-life character, large in size and large in ego. He had been born in Cambridge in 1931, the son of the Soviet academician Piotr Kapitza, who worked with Ernest Rutherford at the Cavendish Laboratory and shared the Nobel Prize in Physics (1978) for his low-temperature research. His father's achievement was a success that the younger Kapitza constantly attempted to emulate. It cast a shadow on his own career from which he vigorously attempted to escape and, combined with the difficulties of pursuing a scientific profession in the Soviet Union, was perhaps why he took any opportunity for self-promotion.

To be fair, Kapitza had served with the Soviet Antarctic Expeditions in the IGY between 1955 and 1964. On four expeditions he conducted seismic sounding on the interior plateau in and around Vostok Station and out to the Pole of Relative Inaccessibility. At the time these were tough and demanding journeys given the constraints of high altitude, severe temperatures, and rudimentary equipment. Kapitza's reconnaissance sounding work was certainly valuable in giving early indications of the ice thickness in these central regions and the presence of the Gamburtsev Mountains. We have also seen that there were significant differences between some of Kapitza's depth determinations and the SPRI radar records.

Kapitza made seismic reflection measurements at Vostok but did not interpret, at the time, any of the returns on his seismograms as indicating the presence of water. Indeed, he made claims regarding the presence of

---

liquid of the same density as the ice (0.92 kg m^{-3}) will prevent this. In the early days the liquid was a combination of diesel oil and alcohol—containing many contaminants that would be unacceptable should they enter a sub-ice water body.

sedimentary rocks beneath the ice below Vostok. This was of particular interest to me, and I wrote to him seeking information about his seismic work. Only a refraction profile yielding P-wave velocities would provide convincing data upon which to make such a supposition regarding the subglacial materials. Kapitza replied that he had used the method of 'reflected' waves and had no data on P-wave velocities.[212] No water had been mentioned, and his inference of sedimentary strata had little basis.

Only later, at the 1994 meeting, was it agreed that he and Robin, with his earlier experience of seismic work in Antarctica, should discuss his seismograms to ascertain whether certain reflections were considered to have come from the lake floor. A joint paper was published by the group but only after considerable disagreements had been resolved over who had or had not 'discovered' Lake Vostok.[213] This intriguing debate is admirably described by Martin Siegert.[214] Despite the clear case that its discovery remained equivocal,

[212] I wrote to Kapitza asking: 'In your paper (*Info. Bull. Soviet Antarctic Expeds.* No. 19 1960, 10–14) you indicate that seismic reflections from the deepest geological layers indicated the presence of sedimentary rocks beneath Vostok Station. Would it be possible for you to tell me whether this result came from a seismic *refraction* profile shot at Vostok and if so what compressional ("P") wave velocities were encountered in the sub-glacial rocks? If it was not a refraction profile, what method was used to obtain this result?' (Letter, 27 November 1975). Kapitza's reply was very brief (seven lines) that included the following: 'As regards your question on the seismic measurements made at Vostok Station, I want to inform you that the method of reflected waves was used and we have no data on "P" wave velocities.' (undated letter but probably December 1975 or early January 1976, as I replied to thank him on 20 January 1976).

[213] Kapitza, A; Ridley, J K; Robin, G de Q; Siegert, M J; and Zotikov, I (1996) Large deep freshwater lake beneath the ice of central East Antarctica. *Nature* 381:684–86.

[214] 'The actual writing of the paper by Kapitza et al. (1996) (footnote 213) was a challenging process, owing mainly to the different views between Kapitza, Robin, and Zotikov on who was the first to discover subglacial lakes. Robin argued that his team were the first to publish evidence of subglacial lakes, which was undeniable. However, Kapitza claimed that his 1961 PhD thesis included the first mention of subsurface water (referred to as "meltwater lenses") in Antarctica. Zotikov disagreed with Kapitza, however, because his analysis lacked appreciation of ice-sheet thermal dynamics [Zotikov (1963) (footnote 164)] and was, consequently, unsubstantiated. Kapitza had obtained the seismic data, however, and within those data was the evidence of basal water. Unfortunately, Kapitza had not spotted the lake-floor reflections . . . and so never published. . . . He often referred to his failure to tie his ideas on basal water with the seismic data as a scientific regret. At one point in the writing process, this disagreement resulted in the abandonment of the paper until a precise wording of Kapitza's early ideas was eventually accepted by all [footnote 213], Siegert, M J (2018) 'A 60-Year International History of Antarctic Subglacial Lake Exploration', in Siegert, M J; Jamieson, SSR; and White, D A (eds.) (2018) *Exploration of Subsurface Antarctica: Uncovering Past Changes and Modern Processes. Geological Society, London Special Publications* 461:7–21.

Kapitza always claimed thereafter that he had discovered the lake. A later, more balanced, appraisal by Russian colleagues was given by Kotlyakov and Krenev in 2016.[215]

9.5 Inland of Dry Valleys

As recounted in chapter 8, interpretation of the landforms revealed by the RES surveys had stimulated my interest in the glacial history of Antarctica. The fluctuations and interplay of the East Antarctic Ice Sheet, of the Ross Ice Shelf and so-called alpine glaciers over the last several million years were admirably recorded in and around the Dry Valleys of southern Victoria Land close to McMurdo Station. The sequence of events had been developed in great detail by George Denton, whose contributions have been of considerable significance and widely recognised. The plan was to run closely spaced radio-echo sounding flight lines on the inland flank of the Transantarctic Mountains west of the Dry Valleys. Parker Calkin, professor at the Department of Geological Sciences, State University of New York at Buffalo, had spent a year's sabbatical at the SPRI working on earlier RES data in and around these ice-free valleys and had already made some observations from the 1969–70 season, but the dataset was not very comprehensive. Flight No. 142 would survey this area at low level commencing at the northern end of Skelton Névé and extending north to the head of Mawson Glacier. The flight comprised 10 sub-parallel lines approximately 10–15 km apart and a final cross-tie line that descended down Wright Valley. Figure 9.10 shows the lines and the surface elevations derived from the soundings revealing the presence of a small dome.

The details hinted at some complications to the simple coupling of moraines and erosive features in Taylor Valley to variations in the thickness of the main ice sheet. The 'Taylor Dome' was likely to have a blocking effect except for the very largest fluctuations. This survey set in train a small but rather interesting programme of investigation that unfolded in the margins of later seasons and is explored in chapters 12 and 13.

[215] Kotlyakov, V M; and Krenev, V A (2016) Who discovered the Lake Vostok? (in Russian) *Ice and Snow* 56 (3): 427–32.

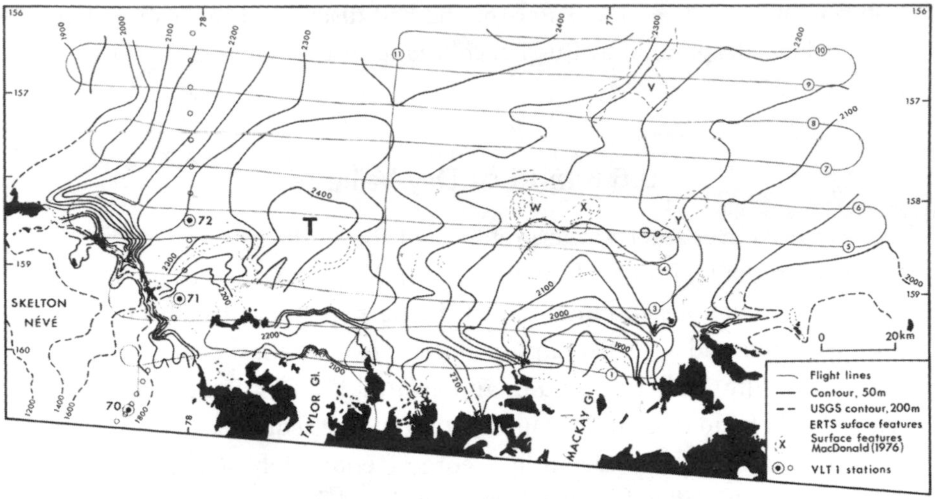

Figure 9.10. Ice sheet surface contours inland of McMurdo Dry Valleys. Flight lines are shown. Exposed rock is in black. T marks the ice dome at 2400 m inland of Taylor Glacier.

9.6 Disaster at Dome C!

By the beginning of January 1975, the team had flown 225 hours. The flights over Dome C had been completed, as had all lines in Terre Adélie. There was one more line to undertake on the East Antarctic plateau towards Dome B, as well as two 'research flights'. Three missions had still to be flown in the vicinity of the Transantarctic Mountains, and Chris Neal required three more sorties over the Ross Ice Shelf. However, the main concern was Marie Byrd Land, where the survey programme was badly behind schedule and required at least six flights with refuelling at Byrd Station. The pace did pick up, and by 9 January another 65 hours had been added; we were very pleased with the steady progress.

The flights to Dome C were particularly important, as they were to provide the site survey data on ice thicknesses and bedrock topography for the deep core drilling by the French under Claude Lorius. Flights were completed in this area with a minimum grid spacing of 50 km. We flew at low level over the virtually level region and were able easily to spot the camp that had been established for the initial surface reconnaissance. At an elevation of over 3200 m it would prove a challenging location for the various drilling parties in later years and the erection in 2005 of a new research station—'Concordia'—operated by French and Italian scientists.

Figure 9.11. JATO take-off from the Shackleton Glacier, central Transantarctic Mountains. (Courtesy Charles Kaminski and NSF, USAP Photo Library).

On Wednesday, 15 January, we were alarmed to hear that there had been an aircraft crash at Dome C. The oldest of the Hercules, #319, had been sent out to pick up Lorius and his French party, who had been undertaking the glaciological programme. Owing to the height of the camp and the rough surface, the pilot had ordered a jet-assisted take-off (JATO). A C-130 had already crashed in Antarctica a few years earlier using JATO and, similarly, picking up a French surface party, on that occasion closer to the coast (chapter 7). The use of JATO is a dangerous as well as spectacular procedure which involves attaching several JATO 'bottles' to brackets on the fuselage to the rear of the aft paratroop deployment doors. Typically, three or four such bottles are locked into place on either side of the aircraft. The airplane commander commences a full speed take-off over the ice surface. When he judges the speed is approaching rotation, he fires the JATO, which provides a significant additional thrust to the airplane, launching it into the air (Figure 9.11).

Participating in a JATO take-off is both an exhilarating and alarming experience in which a number of the SPRI team shared. Because the ice surface is frequently rough with sastrugi, the airplane will shake, bounce, and lurch alarmingly as it gathers speed. When the JATO bottles are fired, the acceleration is akin to being punched heavily in the back! Then the plane is airborne, the bouncing ceases suddenly, and the noise dies away as the spent bottles are jettisoned. There is relief all around.

Figure 9.12. Crashed C-130 #319 at Dome C. Much of the starboard wing and engine has been lost. (Courtesy US Navy, US Antarctic Program).

In the crash at Dome C a JATO bottle broke loose and, with no guidance system of its own, shot forward, passing through part of the inboard starboard engine. Fortunately, the aircraft had not gained much altitude, and the pilot was able to land immediately. Fire then spread to the outboard engine, and the wing burnt completely and fell off (Figure 9.12).

By a miracle, no one was injured thanks to the professional and skillful actions of the pilot. Lorius and the other scientific personnel and aircrew returned to the camp. Mac Center (air traffic control at McMurdo) immediately diverted others of its C-130 aircraft that were already in the air to Dome C. Planes #129 and #131, both new 'R' models, arrived at the scene of the crash a few hours later. Personnel were transferred to #129, which took off without JATO, fearing a repeat problem should there be further faults with the JATO bottles. During the long high-power take-off over the rough surface, the plane's nose ski dug into soft snow on the crest of sastrugi, just before rotation. As a result, the nose wheel/ski assembly buckled and tore apart, and the aircraft was brought to a rapid halt.

The situation at Dome C was turning into a disaster! Luckily, once more, no one was injured. Lorius and all the #319 crew had now experienced two crashes in less than 4 hours and were clearly shocked. They were transferred to #131, which had been rerouted from the South Pole Station and already had a number of personnel onboard and a full fuel tank in the fuselage. With

43 passengers and a large quantity of equipment, the take-off was going to be difficult, but using JATO, they got airborne and arrived safely back in McMurdo.

Lorius vividly describes this last exit from Dome C in his book *Glaces de l'Antarctique*: 'We all moved back to the back ramp/rear cargo door to provide maximum lift for the front skis. Full throttle . . . one minute, maybe two that seem like forever, firing rockets and the plane rushes up into the air. A general "Hurray" was followed by "F . . . you Dome Charlie". We were on the verge of disaster—but at 4.30 a.m. we landed in the mist at McMurdo.'[216]

The situation was desperate. With two wrecks now lying 1000 km (600 mi) from base, the US air support in Antarctica had been crippled, cut from five to three airplanes.[217] For two days all operations ceased whilst the Navy, VXE-6, and NSF convened meetings to discuss the crisis and liaise with Washington and other centres. I was able to talk with my friend Claude Lorius a few days later, and he was still very shaken by the experience but so glad that nobody had been seriously hurt.

9.7 The Season Concludes

We were within an ace of completing our full programme and needed just a further handful of flights. Our dilemma was that the NSF required all the remaining aircraft to evacuate personnel from McMurdo, other bases, and field stations. What hope was there for us to complete our planned operations? Perhaps it was the influence of Art Herr; maybe the excellent rapport we had achieved with the VXE-6 and Navy commanders; possibly the fact that we needed only two more flights, which would not be a big demand, and #320 was unable to undertake alternative missions without the time-consuming task of removing all the RES equipment and the antennas that led to the agreement for us to undertake these last sorties. We were extraordinarily grateful given the difficult circumstances.

The problems caused by the crashes had further impacts, namely, on the return of both cargo and personnel; here were echoes of 1971–72. RES

[216] Lorius, C (1991) *Glaces de L'Antarctique, une mémoire, des passions*, Paris: Éditions Odil Jacob, 301pp.

[217] Two Hercules damaged (1975) *Antarctic Journal of the United States* 10 (2): 61–62.

Figure 9.13. RES team 1974–75. Back row: David Mackie, (Admiral Richard E Byrd), David Meldrum; Middle: Neil Hargreaves, Mogens Pallisgaard, David Drewry, Bob Nedry (pilot); Bottom: Chris Neal, Art Herr (plane commander), Gordon Robin, Keith Rose.

equipment was to be shipped back in containers on the Navy supply ship USNS *Private John R. Towle*, so following completion of our flights we spent three days of hectic packing of two 8-ft containers. The SPRI crew flew back, thankfully, to Christchurch on 6 February; it was a long flight, just over 9 hours due to very strong (100 kts) headwinds.

Despite a tumultuous end to the campaign, we had accomplished a great deal. It had been a long and intense season, but the whole SPRI team, including our TUD colleagues, had worked with great diligence and persistence (Figure 9.13). The first use of TUD radars in Antarctica had demonstrated their effectiveness. The grids and lines over Marie Byrd Land and the Ross Ice Shelf formed a dense mesh, and in East Antarctica the additional flights filled out the pattern we had commenced in

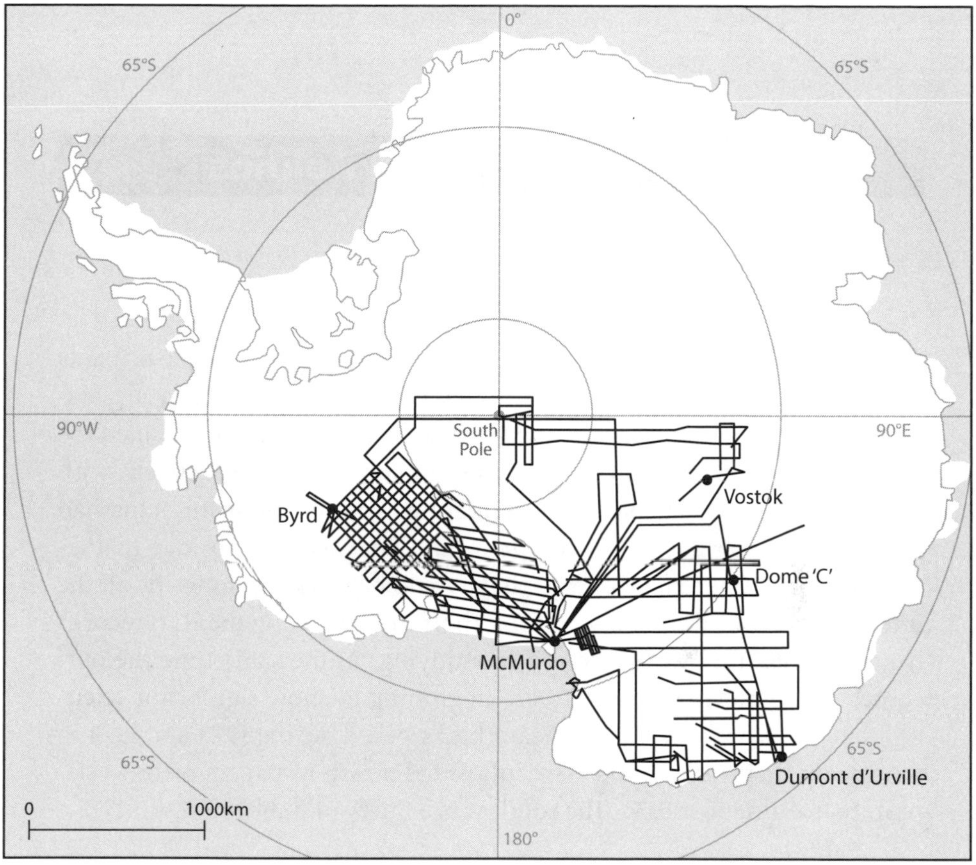

Figure 9.14. RES flight lines conducted in the 1974–75 season.

1971–72. We had flown Dome C and lines around South Pole and Vostok, as well as discovered its immense subglacial lake, and had undertaken other 'projects'. Fifty flights had been completed with 320 hours of flying (Figure 9.14). It was a colossal effort; we would not see the like of this season again.

10

Data, Research, and Politics

A mammoth undertaking of data reduction confronted the field participants on their return to Cambridge in February 1975, the downside of a successful and lengthy field season. Ian Holyer and Lina Talbot had now joined the group, taking over from Joyce Whittington and tasked, respectively, with enhancing the RES database and performing routine digitising. Lina had worked for a pharmaceutical company, and Ian was a Cambridge mathematician from Churchill College. They would be working alongside all the other members of the team who took their turns tackling the RES records from the areas they were involved in studying. At the same time the outcomes of the previous seasons were beginning to show significant scientific returns. Several of the RES team had worked during 1973 and 1974 in writing up their results, and they submitted a rash of papers prior to departure in November 1974. The result was a flurry of publications in 1975.

10.1 Ice Streams of Marie Byrd Land

Keith Rose was working on the impressive network of lines flown in the 1974–75 season; the date from their reduction enabled the first thorough delineation of ice streams and grounded areas, and assisted the investigation of the form and flow of the ice sheet in this region.[218]

Rose prepared a series of studies with accompanying maps showing in considerable detail the pattern of ice streams and intervening domes and ridges from the edge of the Transantarctic Mountains to the coastal Rockefeller Mountains (Figure 10.1). The bedrock revealed, perhaps not unsurprisingly, that the ice streams are mostly located above channels in the bedrock. He made a number of glaciological calculations—basal shear stresses ('friction' at the bed), temperatures through the ice, and balance velocities.

[218] Rose, K E (1979) Characteristics of ice flow in Marie Byrd Land, Antarctica, *Journal of Glaciology* 24 (90): 63–75.

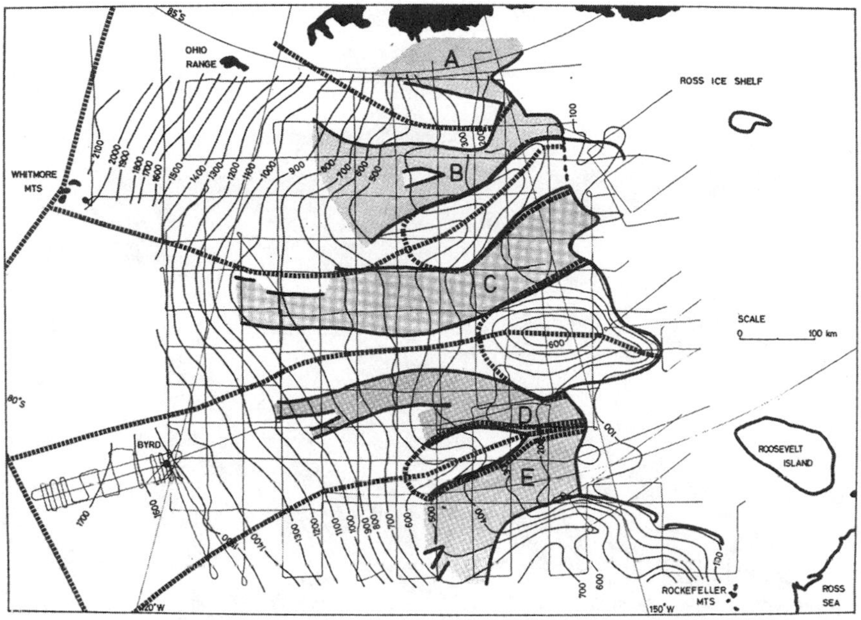

Figure 10.1. Ice streams (shaded, A–E) and intervening ice rises and ridges in Marie Byrd Land. The heavy dashed lines delimit the inferred catchment of the ice streams. 1974–75 flight lines are shown. (Lines inland of Byrd Station were flown in support of Ian Whillan's surface strain network.) (After Rose (1979); see footnote 218). (Courtesy International Glaciological Society).

This last quantity is the average speed of the ice across any section of an ice stream (or glacier) that is necessary to discharge all the snow accumulating in the drainage basin higher up, assuming that the whole system remains in a steady state. Rose observed that the basal shear stresses decreased to very low values as the ice streams approached the grounding line, where they come afloat and form part of the ice shelf, suggesting the presence of a zone of water. The temperature calculations implied a pattern of melting at their base, while the intervening domes and ridges appeared to be frozen to the substratum.

Ice Stream C seemed anomalous. Rose commented that although the typical clutter echoes associated with ice streams (Figures 9.5 and 10.2) extended well inland, the surface of this ice stream was featureless. He noted that very low ice velocities measured on the Ross Ice Shelf beyond the grounding line of Ice Stream C (by the RISP) suggested it was stagnant. His calculated balance velocities were, in contrast, of the order of 200 m a^{-1} (metres per annum). To reconcile the various features from the records and the adjacent

Figure 10.2. Shear margin of Ice Steam B. (Courtesy US Navy, US Antarctic Program).

measurements, Rose suggested that Ice Stream C is presently in the quiescent phase of a cycle of surging. More recent studies have shown that C is stagnant and that Ice Streams A and B are also on the cusp of stalling, with significant areas of the bed frozen.[219]

10.2 The Ross Ice Shelf Revisited

Gordon Robin had assembled the ice-thickness data for the Ross Ice Shelf from the many flight lines radiating out of McMurdo in the previous seasons (some 35,000 line-kilometres) and his article in *Nature* came out with

[219] Joughin, I; Tulaczyk, S; and Macayeal, D R (2004) Melting and freezing beneath the Ross ice streams, Antarctica, *Journal of Glaciology* 50 (168): 96–108.

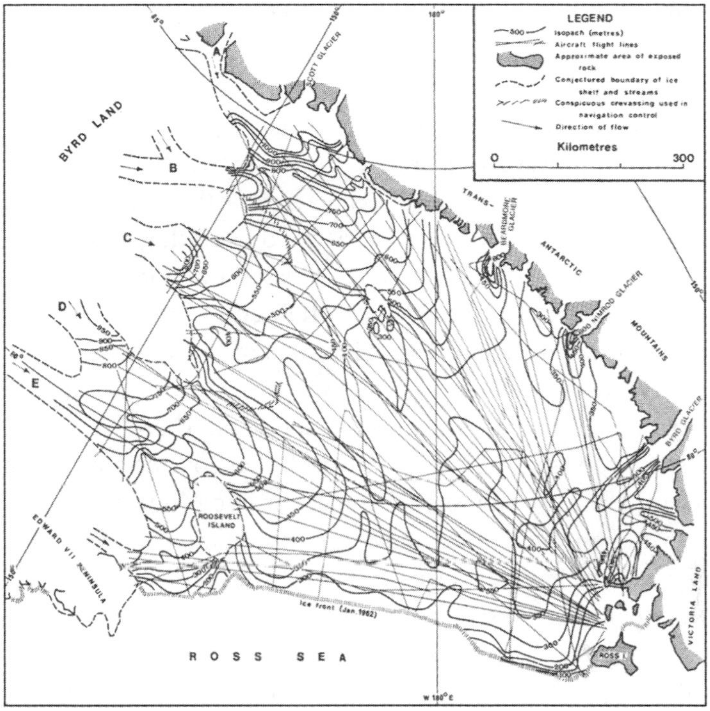

Figure 10.3. Robin's map of the ice thickness of the Ross Ice Shelf, also showing the array of flight lines mostly emanating from McMurdo. The ice streams from Marie Byrd Land are labelled A–E. Crary Ice Rise is in the south-central part of the ice shelf. (Courtesy *Nature*).

full-page maps depicting the ice-thickness distribution, flowlines, and velocity distribution, highlighting many new and interesting features.[220] First and foremost was the general pattern of thickness depicting the ice shelf thinning as it flows outwards across the Ross Sea (Figure 10.3). The inner regions were generally found to be in excess of 500 m, reaching over 1000 m at the southerly head of the ice shelf, at 85°S,150°W. In the central region thicknesses are 400–500 m; the shelf then thins to the order of 100–150 m towards the ice front. In this outer zone there is strong melting at the bottom as ocean currents bring warmer water to the base, and tidal pumping creates turbulent heat. The overall pattern confirmed the notion that ice shelves spread out under their own weight, floating on a frictionless base

[220] Robin, G de Q (1975) Ice shelves and ice flow, *Nature* 253 (5488): 168–72.

(the ocean). Areas of grounding or incipient grounding over seabed peaks, giving rise to concentrations of heavy surface crevassing, were identified, such as at Crary Ice Rise in the south-central portion of the ice shelf, and in one or two locations along the Siple Coast

The second important element was the distinctive bands of thick ice emanating from the mouths of the outlet glaciers of the Transantarctic Mountains, as well as the ice streams of Marie Byrd Land that Rose was studying. Robin's calculations confirmed that the greatest drag on the sides of the ice streams should vary most rapidly near their entry into the ice shelf. Aided by some valuable but sparse determinations of the velocity of the ice surface, he was able to construct some representative flowlines and portray the general pattern of flow rates across the ice shelf. There were striking features. One was the very active Ice Stream B that confined and pushed back the discharge from several of the Transantarctic Mountain outlet glaciers into a narrow zone along the front of the mountains until well north of the Beardmore Glacier. This showed the powerful flow and the volume of ice originating from the West Antarctic Ice Sheet. The satellite image from the MODIS (Moderate Resolution Imaging Spectroradiometer) instrument in Figure 10.4 reveals the integrated nature of the ice sheet, ice streams, ice shelf continuum in West Antarctica.

Preliminary maps of ice thickness, flowlines, and 'balance' velocities were made available at a very early stage to the scientists planning the Ross Ice Shelf project (RISP). The later detailed surface work by the RISP, as well as knowledge gained from drilling through the ice shelf, was highly significant and produced much new research, which concentrated increasingly on the grounding line region in Marie Byrd Land and its response to climate change. These very important studies have somewhat overshadowed Robin's pioneering contribution, without which the planning of the RISP would have been much more uncertain, demanding, and time-consuming.

Chris Neal's research was focused on the same ice shelf but from a different perspective. It has already been noted he was keen to examine the nature of the radio reflections from the ice/water interface to explore whether the radar records could reveal if the ice shelf base was melting or freezing. Such investigations would be important for studies of changes due to ocean warming. In addition, he considered that other features might be detected, such as fissures and fractures (bottom-penetrating crevasses), and possibly debris carried from the grounding line or outlet glaciers. The substantial dataset from the 1974–75 season was still at the early stages of analy-

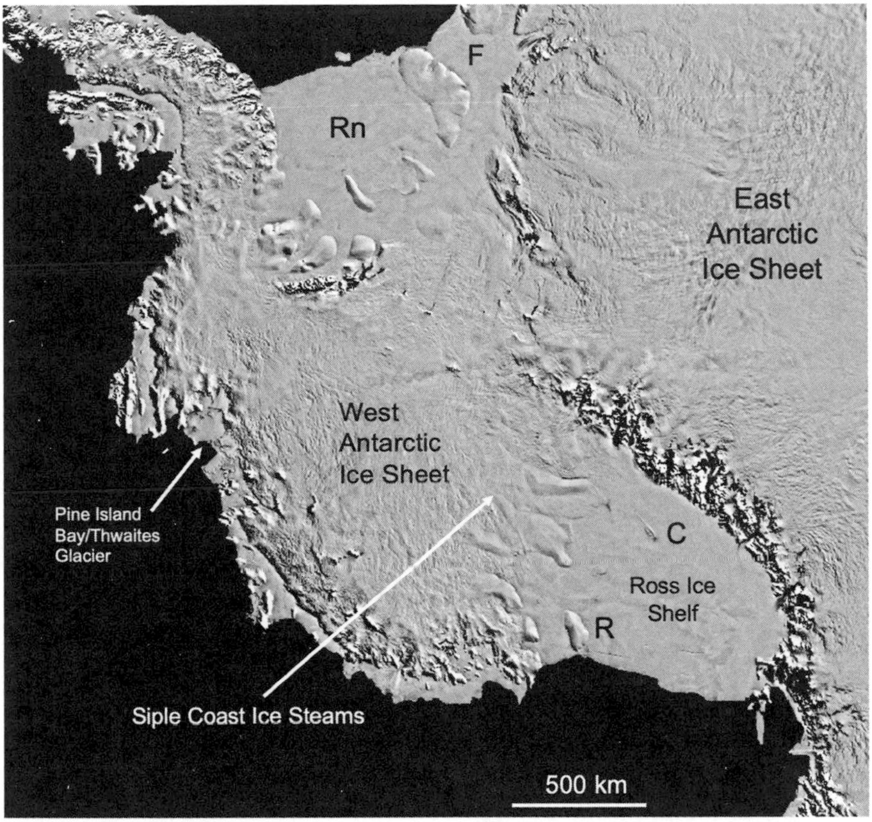

Figure 10.4 This high-definition satellite image demonstrates the highly coupled nature of the ice flow in West Antarctica between the ice sheet, the several distinctive ice streams, and the Ross Ice Shelf. Rn = Ronne Ice Shelf; F = Filchner Ice Shelf; C = Crary Ice Rise, R = Roosevelt Island. (From NASA MODIS instrument, courtesy NASA NSIDC DAAC).

sis. It would be a year or more before he was ready to publish and complete his thesis.

10.3 Some Geological Investigations

The SPRI welcomed visiting scholars from various organisations and universities around the world as their interactions fertilised the intellectual atmosphere of the Institute. Several came to work on data and ideas emanating from the RES programme. These were in addition to those that joined on Antarctic missions. Professor Parker E Calkin, it will be recalled, spent a year's sabbatical in Cambridge to work on RES data. Parker had worked for several seasons in the Dry Valleys and was looking to gain a new insight from the radar records. Lunches and regular swimming schedules with him, along with Hans Weertman, another visiting scholar, from Northwestern University at Evanston, Illinois, proved valuable opportunities to discuss the latest science. Hans' understated, quiet manner but penetrating intellect was hugely stimulating. Parker published a paper in 1974 summarising his work regarding the erosion and evolution of the valleys.[221]

Further studies of the bedrock were incorporated into the author's thesis, completed in 1973 and written up in several papers. I was keen to present ideas differentiating subglacial regions using statistics to describe the 'roughness' of the terrain derived from the RES profiles. I found I was able to cluster similar tracts of the bedrock landscape and indicate possible associations or characteristics related to geological factors.[222]

10.4 Politics Intervene

The 1974–75 season had unquestionably been one of the most extensive and effective. The hours flown in 50 missions had been a formidable accomplishment. The lines of RES were covering increasingly greater areas of Antarctica and, in places, becoming denser. What they were revealing was all new science, especially regarding the ice sheet and its internal properties (such as layering); the nature of the ice/bedrock interface, including melting and presence of water bodies; the process on ice shelves; the identification and disposition of ice streams; and the geological configuration of the land beneath the ice. The team in Cambridge was justifiably proud of its achievements to date. But success can lead to resentment, envy, and tension. The support given by the US NSF to the SPRI was founded on a key principle adopted early on by those nations working in this hostile, remote, and difficult environment: that cooperation, sharing, and support in many forms should be pursued whenever necessary and feasible. As explained in chapter 8, the RES programme at this time had been built on British scientific innovation, Danish technological skill, and American logistic capabilities. So far, this arrangement had worked in exemplary fashion. By the early 1970s

[221] Calkin, P E (1974) Sub-glacial geomorphology surrounding the Dry Valleys of southern Victoria Land, Antarctica, *Journal of Glaciology* 13 (69): 415–30.

[222] Drewry, D J (1975) Terrain units in eastern Antarctica, *Nature* 256 (5514): 194–95.

other research groups were developing their own radar sounding expertise, much encouraged by the SPRI which had made SPRI MkII units commercially available, as well as advice and assistance. Furthermore, Robin had been an architect of Antarctic international collaboration through SCAR and had invited participants to the RES field programmes from early days. The 1973 agreement had been crafted to jointly enhance research in both Antarctica and Greenland.

Criticisms, sotto voce, came from US science groups who perceived the US support for a British team disproportionate and considered their own prospects were being compromised. Furthermore, some complained that they were not gaining access to the Antarctic RES data which the 1973 agreement had envisioned. There was lobbying of the NSF, which was caught on the horns of a dilemma. There was little doubt that the SPRI-TUD-NSF programme was highly successful and was contributing fundamental and extensive knowledge of Antarctica, as well as being a vital component of international projects such as IAGP. Switching to another institution in the US to continue the continental survey would be inefficient, and the prospect of losing momentum and international goodwill were not appealing to the NSF at that juncture. Nevertheless, these stirrings were unlikely to abate, and future plans needed to be made with care in this regard. It has been pointed out there were different philosophies at play regarding data access and distribution between the NSF and the SPRI that became ever more apparent.[223] We shall see in chapter 14 that it was one of the more serious factors that conspired to bring about the eventual termination of the programme.

In Cambridge there was reflection on the international cooperation that had already been facilitated. Texas Technology University and Ian Bird, from ANARE, had participated in 1969–70. Soviet scientists had also been invited onto some of the flights as observers (Sergei Miakov in 1970, and later, Mikhail Grosswald). Belgian geophysicist Hugo Decleir joined the programme for the whole 1971–72 season. US scientists Whillans and Calkin had worked at the Institute on RES data.

As observed, the SPRI programme was an integral part of the IAGP, and early high-level results were duly shared with all the national participants, including the NSF. All SPRI publications acknowledged and thanked the

[223] Dean, K; Naylor, S; Turchetti, S; and Siegert, M (2008) Data in Antarctic science and politics, *Social Studies of Science* 38 (4): 571–604.

Division of Polar Programs for its generous support. In similar fashion, as reported earlier in this chapter, draft charts of the Ross Ice Shelf had been shared at the earliest opportunity with colleagues undertaking the RISP. Looking back, it is quite clear, however, that joint publications with US workers were few and far between. One in six RES papers up to 1984 was co-authored with or solely authored by non-SPRI researchers in other countries but only one in eight, with US scientists.

Seeking to engage better with the NSF as well as with UK funding agencies, principally the NERC, a science plan was produced which set out (i) the background and development of the SPRI RES programme, (ii) principal achievements to that time, (iii) a comprehensive bibliography of SPRI RES publications, (iv) plans over the next 3–4 years, and (v) international cooperation in RES. This plan, written by Robin and Drewry, was completed in November 1975; distributed to the principal organisations involved in collaboration, funding, and logistics; and formed the basis for later dialogues.[224]

10.5 Research Accelerates

With no prospect of an Antarctic field programme in the immediate future as a consequence of the airplane losses incurred by the US, and the construction and modification of the new 'R' model (aircraft #131) proving slower than anticipated, activity concentrated on writing up more material for publication and attending conferences. Planning and discussion meetings with colleagues at TUD continued. These were stimulating and advantageous and led to an opportunity to work with them in Greenland. Gudmandsen communicated that a field season in northern summer 1975 would be based at DYE-3 in south-central Greenland, and the SPRI was invited to send a scientist to join them. The project was to conduct a ground-based RES survey in preparation for drilling an ice core for climate-related studies.

A large-scale programme for coring and related investigations in Greenland was planned by a consortium of Swiss, Danish, and US glaciologists called GISP (Greenland Ice Sheet Project) under the leadership of an impressive triumvirate—Willi Dansgaard (University of Copenhagen), Hans

[224] SPRI (November 1975) *Radio-Echo Sounding of Polar Ice Sheets: A Science Plan*, Cambridge, UK, 34pp.

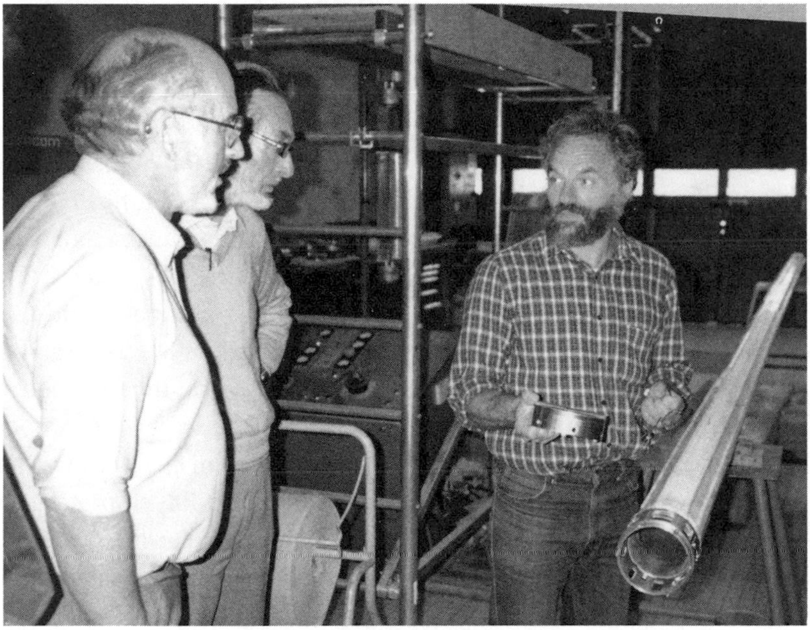

Figure 10.5. Hans Oeschger (left) and Willi Dansgaard (centre) inspect a French ice core drill at the Laboratoire de Glaciologie in Grenoble. Claude Rado provides some details.

Oeschger (Bern University), and Chester (Chet) Langway Jr (US Army Cold Regions Research and Engineering Laboratory (CRREL), Hanover, New Hampshire), who called themselves 'The Three Musketeers'! (Figure 10.5)[225] This plan had started in 1971 with a series of exploratory shallow cores to 372 m using a CRREL thermal drill at DYE-3. The location was one of several DEW Line (Distant Early Warning) stations established during the Cold War to provide surveillance in case of a Soviet attack, so these stations were relatively easy to access by air.

Following the first core, others followed. Several hundred metres of ice were drilled at Milcent and at Crete, which are located more centrally on the ice sheet, both to provide a regional picture and to identify a potential site for drilling to bedrock. In the end, and after considerable scientific debate, DYE-3 appeared to be the only feasible location on logistical and cost

[225] Dansgaard, W (2004) *Frozen Annals: Greenland Ice Cap Research*, Odder, Denmark: Narayana Press, 122pp.

grounds for a deep ice core (over 2000 m), although its glaciological setting was far from ideal.[226]

The 1975 fieldwork campaign organised by TUD involved measuring ice thickness and determining the presence of internal layers and the topography of the bedrock. These would later aid the interpretation of the core analyses. At SPRI it was decided it would be valuable and scientifically appropriate for Neil Hargreaves to join the TUD programme during August; he had participated in the last Antarctic season, and his PhD research fitted into the campaign. His plan was to undertake a series of experiments to investigate whether the orientation of the crystals in the ice sheet and the stress field set up by the ice flow would have an effect on the returned signals from internally reflecting horizons (this was the least understood of the three mechanisms for producing layers described in chapter 14). Theoretical considerations suggested the returns should exhibit elliptical polarisation. Hargreaves used the TUD 300 MHz radar system fitted onto two sledges which were then towed behind a tracked vehicle. Measurements were made at 20 sites. The results indeed showed that internal layers returned radio signals that were elliptically polarised and probably caused by the birefringence, at radio frequencies, of the ice mass, indicating the ice sheet is a layered dielectric medium.[227]

During 1975 other work proceeded back in Cambridge. Michael Gorman, along with Chris Doake from the BAS and Stan Paterson in Ottawa at the Polar Continental Shelf Project, were busy finishing a study on one aspect of the 1973 Devon Island experiments—deducing the flow velocity of the ice cap using the fading patterns of the returned radio pulses. Prior to the widespread use of accurate satellite-determined ice surface velocities, glaciologists were seeking new methods to measure ice flow in places where optical survey from fixed and known base stations was impossible. RES was seen as a possible technique.

Chris Doake had conducted a survey on Fleming Glacier in the Antarctic Peninsula following an earlier study by Mike Walford.[228] The velocities from the fading patterns compared with an optical survey showed significant differences that were interpreted by Doake as the result of the glacier

[226] Dansgaard, W (2004) (footnote 225).

[227] Hargreaves, N D (1977) The polarization of radio signals in the radio-echo sounding of ice sheets, *Journal of Physics* 10:285–304.

[228] Both Walford and Doake used the SPRI MkIV radar sounder. On Devon Island a longer pulse (1 μs) was used that provided more detail in the fading patterns.

sliding over its bed or owing to a moraine layer at the base. It should be recalled that the flow of these large ice masses is far from simple. The surface motion is an expression of internal deformation within the bulk of the ice mass according to reasonably well-established rheological theory, combined with possible sliding at the base if the bottom of the ice reaches the pressure melting point, which in turn is determined by the temperature regime and overburden pressure. The Fleming Glacier was considered by Doake to be at or near the melting point, so sliding was certainly possible. The most likely interpretation, he deduced, was that the radar was tracking a moving layer of moraine material near the base.

The experiments conducted by the SPRI team on Devon Island, separated by a year (June 1973 and June 1974), were less ambiguous, since borehole measurements confirmed that the ice cap is frozen to the bed (–18°C). The calculated speed of 2.58 ± 0.11 m a^{-1} was found to be virtually the same as that obtained by independent methods, a positive verification of the methodology.[229] The RES fading-pattern technique has been rendered superfluous by GPS but can still contribute to an understanding of the nature and processes at or close to a glacier bed. Soviet scientists, for example, have made a number of determinations and studies in Antarctica by similar imaging methods.[230]

10.6 Automated Data Reduction

The phrase 'curate's egg' (as commonly used, meaning good only in parts) applies most aptly to this aspect of the RES programme. It has already been referred to briefly in section 8.6. Overall, it is a tale of high expectations, frustration, mixed outcomes at best, and a significant dose of dismal failure.

The requirements for digitising the 35mm RES film were considered to be amenable to some form of automation whereby the film would be scanned and advanced under program control. The digital output could then be 'interpreted' by the user, and ice thicknesses derived or other features examined depending on resolution. This was a rational plan, but the computing and scanning capacity of systems in the early 1970s to achieve these objectives was limited.

229 Doake, CSM; Gorman, M R; and Paterson, WSB (1976) A further comparison of glacier velocities measured by radio-echo and survey methods, *Journal of Glaciology* 17 (75): 35–38.

230 Bogorodsky, V V; et al. (1985) (footnote 43).

In discussions with Beverley Ewen-Smith, now working for a commercial computer company in Cambridge, it appeared that he and his company might be able to design a bespoke system to meet the SPRI's needs. A design specification followed, and a contract was let to Digico. The system proposed was based on a flying-spot scanning (FSS) technique controlled by a minicomputer (Micro16V). A prototype machine was installed at the SPRI in 1974, and the hardware and software were successively improved in later years.

Nevertheless, the conclusion was that the plan for automated digitisation had mostly failed in speed and reliability. The bulk of the later digitisation was conducted using expanded prints laid out on the D-Mac digitising table. It would be another three decades before new technology would be able to scan the archived 35mm films and render them a vital digital resource for continued investigations of the ice sheet, through an initiative by Dustin (Dusty) Schroeder at Stanford University in collaboration with the SPRI (chapter 16).

10.7 Royal Society Discusses Antarctic Science

The mid-1970s witnessed a rising tempo of curiosity by the scientific community regarding Antarctica and a discussion meeting on Antarctic research was organised at the Royal Society on 19 and 20 May 1976, by Sir Vivian Fuchs and Dick Laws, the latter having taken over in 1973 as the director of BAS from Fuchs. The meeting was held at the Society's headquarters in London and proved a rich meeting ground for the participants and their ideas. The SPRI was keen to participate fully, and a paper authored by Robin, Drewry, and David Meldrum entitled 'International Studies of Ice Sheet and Bedrock' was presented at the meeting by Robin.[231]

The Royal Society event was one of several conferences that featured a contribution from the SPRI—the International Symposium on the Thermal Regime of Glaciers and Ice Sheets at Simon Fraser University, Burnaby, British Columbia, Canada, April 1975 (Robin presented two papers); a geodynamics meeting in Durham in April 1976 (I presented a paper on the Wilkes Subglacial Basin); a council meeting of the IAGP in Madison, Wisconsin, in September 1976. All these were stimulating and did much to ensure con-

[231] Robin, G de Q; et al. (1977) (footnote 208).

tinued engagement by the SPRI group in polar glaciological research, but the missing component was the absence of further fieldwork in Antarctica.

It had now been two years since the team had been preparing for the previous season, and individuals were eager to pursue further airborne campaigns, to extend the coverage of RES across that vast landmass, and to undertake new and stimulating investigations. A degree of restlessness pervaded the upper storey of the Institute, as well as a hunger for the physical and intellectual challenge of that remote and extraordinary continent.

11

Changing Planes

A planning meeting was held in Cambridge on 15 April 1976 between the SPRI and TUD. It was understood the NSF was still anticipating 1976–77 as the next possible Antarctic RES season and a further campaign the following austral summer. Furthermore, there was considerable anticipation that the new C-130 would be available for operations. Much discussion ensued regarding equipment modifications, upgrades, and availability. The 300 MHz system was becoming an important additional element in sounding shallower ice areas with greater detail from its narrower beam. An important outcome was the need for a further meeting with NSF. It was agreed to send a joint letter raising issues on (i) the overall timetable for airborne RES; (ii) the availability of the newly configured LC-130, #131; (iii) progress on funding (by NSF), fabrication, and fitting of the antenna; (iv) progress on fixing equipment in racks and the installation of a magnetometer.[232] The desirability of a workshop was addressed, and early June was proposed for the meeting in Cambridge, to fit Gudmandsen's schedule.

11.1 The NSF Sets Out Its Plans

The Division of Polar Programs[233] at NSF agreed a meeting would be useful. It was called on 8 June, but in Washington. Drewry attended for the SPRI, Gudmandsen for TUD. On the NSF side the programme manager for glaciology, Dick Cameron (a genial and experienced polar scientist), was

[232] For some time it had been recognised that flying a magnetometer concurrently with the RES would allow information to be acquired regarding the magnetic properties of the bedrock underlying the ice. These data could in turn be interpreted and modelled to give broad indications of the types of rocks and structures. Together the two techniques added powerfully to interpreting the broad geology of Antarctica.

[233] The name for this office in the NSF changed over the years: Office of Antarctic Programs (designated 26 May 1961); Office of Polar Programs (designated 19 December 1969), Division of Polar Programs (designated 19 April 1976), Office of Polar Programs (redesignated 16 December 2016).

present alongside Dwayne Anderson (the somewhat gritty chief scientist) and Ken Moulton, a long-standing and easy-going staffer in DPP. Representing the VXE-6 at the meeting were Jerry Pilon, Bill Kosar, and Al Fowler.[234] In addition, NSF had invited Charlie Bentley from Madison, and John Clough was there representing him. This latter invitation was understandable, as Bentley's team was the only US geophysical group at that time already involved in RES, and they had been conducting their own work on the Ross Ice Shelf as part of RIGGS.

Anderson chaired the session and set the scene. He opened by commenting that there had been fundamental changes during the previous year in responsibilities and roles at NSF, resulting mainly from the switchover in Antarctic operations from the US Navy to the NSF. There were, he recounted, enormous pressures (maybe of a magnitude that those outside the US did not fully comprehend) from 'the electorate' and other bodies (maybe he was hinting at the Senate) to change the role of US science in Antarctica. He did not elaborate, but the subtext was uncertainty over the status and range of international collaborations.

There followed discussion on the practicalities of future RES seasons—radar and other geophysical systems, data logging equipment, and specific operations in Greenland and Antarctica. It was clear that the loss of two aircraft and the demands for the rebuilding of South Pole Station were having a considerable impact. Cameron outlined these concerns, and Pilon explained that for the next season logistics operations would be down to four, not five, airplanes and that all future remote-sensing work would be concentrated on the specially configured C-130R designated #03 (previously referred to as #131).

The SPRI priority was stated as working within the objectives of the IAGP and that given the constraints as outlined, it appeared impracticable to operate RES in the 1976–77 season. Gudmandsen described TUD work on improvements to the 300 MHz system, but their manpower was limited, and they looked to undertake some tests in Greenland in 1977 or not until an Antarctic season with SPRI, perhaps in 1977–78. He reported that earlier

[234] Captain Alfred N Fowler was a highly respected naval aviator and an adroit administrator but above all a warm and generous human being. He commanded the US military task force supporting the US Antarctic Program and after retirement from the Navy was deputy division director of the DPP at the NSF. Subsequently, he was the first executive secretary of the Council of Managers of National Antarctic Programs. It was in this last capacity that I worked with Al for three years as the founding chair of COMNAP. His diplomatic skills, eagerness for work, and dedication to Antarctica and international Antarctic science ensured the council was placed on the firmest of footings.

in 1976 his group had trialled equipment in a Danish C-130 and operated around the Hans Tausen Ice Cap and near Thule. Separately, they were investigating Soviet interests in using side-looking airborne radar (SLAR) for studies of accumulation on the ice sheet. Clough commented that their group would continue with the Ross Ice Shelf and ground-based survey work in support of the Greenland Ice Sheet Project (GISP). I provided a résumé of the work SPRI had been conducting. This covered large-scale mapping, glaciological studies (ground investigations of polarisation, layering, ice deformation, flowlines, Ross Ice Shelf), geophysical interpretations of sub-ice geology, and terrain roughness.

Kosar discussed aircraft developments and indicated NSF hoped for two new aircraft coming into service in 1977, bringing the number to six for the 1977–78 season. He focused on the work to equip #03 with instruments. A group at the US Naval Weapons Center (NWC) at China Lake in California was occupied on hardware for the data logging system. They were inserting apertures for air sampling and providing a vertical camera bay. Importantly, for SPRI interests, he said they had decided to install a magnetometer. This was to prove of considerable scientific benefit.

The dialogue moved from instruments and operations to cooperative strategy. Andersen emphasised that we should discuss this matter in a frank and honest fashion; 'our feelings and greatest fears' were to be solicited. I stated that the SPRI policy was to move to further, more open cooperation with US agencies that indicated their interest in the RES work. We had already benefitted from a number of collaborations, and this process was progressing.

Gudmandsen's response was much more evasive. I had noted this on other occasions, and it probably stemmed from the differences between the political status of Greenland as an integral part of the Kingdom of Denmark (home rule was not established until 1 May 1979) and the circumstances in Antarctica under the Treaty that actively fostered collaboration, and personnel and data exchanges. Gudmandsen was also working closely with the European Space Agency and in that context was readily familiar with multiparty arrangements in instrument construction, testing, missions, and data dissemination. The discussions in Washington spent some time over these matters and of data exchange, in particular the following:

1. Making RES film available to outside investigators, as well as processing and duplicating the records;

2. Safeguarding the scientific interests of groups and individuals, such as by ascribing one year for the principal investigators to take a first look at the data and thereafter making them available to all interested parties; and
3. Publishing a statement that RES data will be available to all accredited parties under a list of conditions.

Clough recommended having a facility in the US for processing the RES film; I was lukewarm on this proposal. Anderson proposed that NSF draft a document setting out this emerging policy to be passed to SPRI and TUD during the next six months, and SPRI should draft a similar document. Anderson commented that in no way should SPRI or TUD feel that NSF wished to phase out their contribution or role, but that the agency was under pressure to increase US activity in this area. It would support RES very strongly during the next few years. In my note of the meeting, I concluded: 'In my opinion SPRI (i.e. the RES programme) must diversify . . . in order to survive.' This was a somewhat dramatic statement, but research activities in Cambridge did broaden in the years ahead, both in scientific ambitions and geographical interest, and were highly productive (some of these developments are elaborated in chapters 16 and 17). Nevertheless, the issues surfaced at this meeting had ignited a slow-burning fuse that would eventually reach the powder keg in 1979; chapter 14 explores the outcomes.

11.2 IAGP, September 1976

It was only a few months after the 'conference' at NSF that the next planning meeting for the IAGP was held in Madison, Wisconsin, between 7 and 9 September. I attended on behalf of SPRI and gave a report on the work over the 1975–76 period. I was able to provide further detail of the results from the Dome C and Dome B areas (additional fieldwork demonstrated the latter is really a long ridge). These results were important for Soviet and French colleagues, who were planning surface-based operations in the sector. Other topics discussed were aspects of large-scale ice sheet flow and internal reflecting horizons, including the identification of the *basal echo-free zone*. This, we had discovered, was a region above the bed where, despite adequate system sensitivity, it was not possible to discern internal layers. We ascribed this phenomenon to the disruption of bottom layering by ice flow over rough terrain, and it had important implications for the interpretation of the lowermost sections of ice cores.

The US made it clear their continuing aircraft problems would impact various aspects of support for IAGP science and, in particular, confirmed there would be no RES field activity in 1976–77. SPRI would continue with data reduction and interpretation and preparations for a field season in 1977–78.

It was during this meeting that I had a second chance to meet Bert Crary. I held him, along with Gordon Robin, in the highest regard as one of the great pioneers of Antarctic geophysical exploration. Charles Bentley and his outgoing and gregarious wife, Marybelle, generously invited the IAGP group for dinner to their home on the shore of Lake Mendota, and Crary was there, too. It was a privilege to discuss our work to date and plans with him, seek his suggestions, and simply enjoy his engaging company.

A major Antarctic geology and geophysics conference under the auspices of SCAR, the International Union of Geological Sciences, and the Inter-Union Commission on Geodynamics was organised by Cam Craddock, a veteran Antarctic geologist based in the Department of Geology at the University of Wisconsin, for August 1977, so a return to Madison was inevitable. The conference was attended by all the main players from around the world in this field of science and was an important opportunity to present results of radio-echo soundings, particularly from studies of 1971–72 and 1974–75 data. During the previous year we had spent some time preparing and writing for this meeting. From the SPRI Keith Rose, Hugh Steed (a new PhD research student) and I attended.

Rose gave a paper on the subglacial topography of Marie Byrd Land, and Steed in his paper similarly discussed the substantial area covered by the grid in Wilkes Land in East Antarctica. I presented a paper on the detailed soundings inland of the McMurdo Dry Valleys, where we had very detailed subglacial definition from the 300 MHz radar. The conference was a success, but the very late publication of the volume of papers was a considerable disappointment; it did not appear until 1982!

There were some interesting social vignettes at the meeting. One of these involved an interaction between the head of the Soviet delegation (Professor Michael Ravich) and Professor Alton Wade. It will be recalled that Wade had accompanied the SPRI team to Antarctica in 1969. Wade was one of the pioneer Antarctic scientists, having overwintered with the Second and Third Byrd Expeditions (1933–35 and 1939–41, respectively). In the latter he was appointed chief scientist. He had married Jane, who was some years younger and retained considerable style. Both Ravich and Wade were about the same age, in their early 70s. At the first evening reception Wade was standing alongside his wife with a glass of wine when Ravich entered, followed by his team, which unusually included several younger members of the Soviet geological fraternity. These latter mingled with us, also young scientists from the West. Ravich, however, spied Wade and his wife. He went over and in halting English said how 'ravishing' Mrs Wade was and how he would like to 'kiss the feet of Wade's beautiful wife', upon which he knelt down and did just that! Well, those nearby were startled by this display, and I could see the reaction of a couple of young Soviets in our small group—if their head of deputation could act like this in the US, then they were now free of a leash!

11.3 Antarctica 1977–78—A Change of Planes

The agreement by the DPP to continue the SPRI-TUD RES programme in the 1977–78 season set out an initial allocation of 450 hours of flying in the newly fabricated C-130 #03 with its specially configured systems for remote sensing. NSF had let a contract to the Applied Physics Laboratory (APL at Johns Hopkins University to design and install the magnetometer assembly (comprising both scalar and vector units) that added significantly to the aircraft's geophysical capability.

Flight plans were drawn up on this basis and forwarded to DPP. They comprised the following:

1. Extension of the 100 km grid within the IAGP sector of East Antarctica
2. Extension of the 50 km grid within West Antarctica inland of Byrd Station
3. Several flight lines between East and West Antarctica to investigate the geological/geophysical nature of this boundary
4. Flights specific to certain research programmes, following requests by IAGP and other investigators:
   a. 100 km rosettes for sub-ice terrain analysis
   b. Flowline studies—upstream of Vostok station; Casey-Vostok flowline (Bill Budd at ANARE); downstream of Byrd Station; selected flowlines over Dome C (Ian Whillans, Institute of Polar Studies, The Ohio State University)
   c. Ross Ice Shelf in support of RISP (Clough, RISP)
   d. Lines around volcanic centres in Marie Byrd Land (Wes LeMasurier, Colorado University; and Whillans, IPS)

e. Detailed lines around Dome C drill hole (IAGP; and Lorius, Grenoble)
f. Sub-ice lakes north of Vostok Station

The new radio-echo sounding units, the magnetometer installation, the interface with the RES equipment, and the new data recording system would all be carefully evaluated.

In late September 1977 David Meldrum visited the NWC at China Lake for a preliminary test of equipment and the new installation of antennas. He was accompanied by Hugh Steed and Chris Hereward; the latter had joined the group as replacement technician for David Mackie. Afterwards there was a trial of the magnetometer package at Naval Air Station Patuxent River, Maryland. The NWC was responsible for the newly configured Airborne Research Data System (ARDS).

This data logging facility was to interrogate various sensors and navigational instruments throughout the aircraft; these had been evaluated and specified as a result of the meeting Zwally and I had attended at Kaman Aerospace in 1973. The ARDS could record up to 100 channels of six-digit data on magnetic and analogue tapes at a sampling rate of 3.3 ms per channel, and a printout was also available. It was the main system for providing the data to reconstruct the flight tracks and reduce the altimetry to give absolute heights of the ice sheet and bedrock surfaces.

By October the team was assembled for the season. Meldrum would oversee all the RES technical details, supported by Mogens Pallisgaard from TUD, who had been with us in 1974–75, and Hereward. Research students also would be part of the team. Hugh Steed had come to SPRI in 1976 to work on geophysical and geological aspects of the RES data we had collected in East Antarctica and was anxious to extend any flying into the coastal regions of Terre Adélie. Earlier in the year Ed Jankowski had joined from Durham University, where he had read geophysics; it was planned that he would work on a combination of the RES and the simultaneous aeromagnetic data, principally in West Antarctica. Here we were to extend the grid towards the Ellsworth Mountains, the highest in Antarctica. We needed additional support for general tasks supporting the aircraft operations; back-to-back flights increased the pressure considerably. At Meldrum's suggestion we employed, on a temporary basis, a young student, Andrew Brimelow, who proved both a capable and companionable addition. As part of our new policy and arrangements with the NSF to further involve US (and other national scientists) we welcomed Larry Irons from the University of

Nebraska and Colin Brown, a geophysicist from Victoria University of Wellington, New Zealand. This total team of nine would travel in two parties, an advance group in mid-December and the second team to leave once the schedule for deploying #03 to Christchurch was confirmed.

In early December, close to departure, unwelcome news emanated from the NSF. It became clear that the original allocation of flight hours would have to be reduced owing to logistic constraints placed upon C-130 hours by the resupply and rebuilding of the US station, Siple, at the southern end of the Antarctic Peninsula; 250 hours was considered to be the new target. It was a situation to which we had become accustomed, but 250 hours still represented a very worthwhile season.

11.4 Christchurch, New Zealand—The Programme Hangs in the Balance

On 18 December, Meldrum, Pallisgaard, Colin Brown, and Drewry arrived in Christchurch by commercial air. We were met by Margaret Lanyon from the NSF/USARP office and taken to Warner's Hotel in Cathedral Square. This was a change from a B&B in the suburbs, and although the accommodation was fairly basic, we had better access to shops and other facilities. The bar of the hotel, furthermore, was a congenial rendezvous for scientists and support personnel heading to the 'Ice', and we met several colleagues there for convivial evenings. Later, on 23 December, we moved to the West End Motel, closer to the airport. We also obtained bicycles and used these to travel between the motel and the airport and for other visits. Once again, the handy, inexpensive, and quality facilities of the US Navy base were made available. Over the following days, we set about checking the equipment that had been shipped variously from UK, China Lake, and Denmark prior to its installation in #03. On 21 December our colleagues from APL arrived—Abe Finkel and Phil Von Gunten, who would be responsible for the operation of the magnetometer.

There was regular liaison with the NSF Office. Walt Seelig, our longstanding and hospitable colleague from DPP, was the USARP representative. Margaret Lanyon, his assistant who permanently maintained the office, provided daily updates. It became clear that our aircraft, currently in McMurdo, would not fly north to Christchurch until after Christmas—indeed, probably after the New Year—as the NSF was keen to stage our installation work concurrently with its next scheduled (phase) maintenance.

This meant we would have some further time to spend in Christchurch. We enjoyed drinks on Christmas Eve in the city centre and a fine dinner on Christmas Day at the US base with our two colleagues from APL. A former SPRI PhD student, John Walker, was resident in Christchurch at Canterbury University, and he and his wife generously took our party out to the Port Hills on Boxing Day. Everybody appreciated the change of scene.

The following week considerable activity was detected in the USARP offices. I recorded the following:

> December 30th . . . we suspected that there was some behind-the-scenes business going on at NSF, Washington, and today we heard the full story. Apparently, because this year is such a big USARP season, the DDP looks as if it has bitten off more than it can chew—too many projects and tasks and too few airplanes and flying hours to accomplish them (seriously over-programmed). The main culprit has been the rebuilding of Siple Station—all flights are going there to carry cargo and fuel. Because they are so far behind in their schedule, NSF held a critical meeting. Ken Moulton in McMurdo and Commander Claude ("Lefty") Nordhill (the Naval Air Support Force Commander in McMurdo) had spent some time calculating flight hours, and yesterday decided to cut the RES programme! Price Lewis, one of the senior staffers in Washington, rang Gordon Robin with this news. Robin immediately called Dwayne Anderson (Chief Scientist) and Director of DPP (Ed Todd) and made his anger felt—whereupon they recanted and as of now the situation is that we will go down as scheduled about 10–12 January and fly as much as we can, but if the plane is required for cargo hauling our programme will be terminated!

Our plans had been shredded. The original allocation of 450 hours had been whittled down to 250 hours during the summer as we learned of the demands of Siple Station and was now probably half of that again; we were facing a very uncertain logistics environment. NSF planning warranted much criticism. According to the DPP schedule the RES programme was to cease on 9 February. In Christchurch, Meldrum and I, as the leaders of the group, discussed the situation and various possibilities. We proposed (i) a short window of operations in which RES would be given high priority; (ii) simultaneous RES if cargo hauling was required, with the ARDS pallet being removed to provide additional cargo pallet positions; the most useful flights would be to South Pole and adjacent areas; and (iii) a period during which cuts might be made at any time. Seelig agreed to put these possibilities to Moulton in McMurdo.[235] While we awaited a response we undertook a number of calculations on the likely amount of fuel on the aircraft for RES flying. We estimated that with RES equipment onboard (i.e. option ii) only 4000 lbs of cargo could be carried—pretty small. Would it be worthwhile?

From these and similar considerations we concluded we could obtain 300 km of RES track sounding should we have to take cargo to Siple, and about 1800 km on a fuel flight to Byrd Station. It was time to replan the season's priorities should we be able to achieve a minimum of 10 exclusive RES flights. It also seemed essential given all the uncertainties that we should have a presence in McMurdo as early as possible. This would be to maintain pressure on the schedule and undertake the necessary liaisons with the senior staff in VXE-6, aircrews, logistics support personnel under contract to Holmes and Narver, as well, of course, as the NSF officials. On 2 January I wrote to Walt Seelig stressing the necessity for this advance party to deploy to McMurdo and suggested he schedule me on 4 January, when we knew there would be two aircraft rotations from McMurdo to Christchurch. The next day Seelig received a response from Ken Moulton indicating that I should not go to the ice prior to talking with David Srite (VXE-6 operations officer), who was due to head north on 6 January. There was the possibility of my going to McMurdo on 8 January. We learned independently that operations on the ice were behind schedule, with 60 hours of flying planned for #03 before it headed to Christchurch for phase maintenance (PM) and the installation of the RES equipment. We calculated that the plane would arrive on 8 or 9 January at the earliest. The PM would take four days, so the first opportunity for launching South would be 12 or 13 January.

Meldrum and I ruminated on these various signals, reading into them a concern for our programme overall. The two of us held a meeting with Seelig on 3 January and summarised our position: we had experienced two years of delays since our last field season in 1974–75 despite promises of earlier operations. Should we be delayed another season, it would prove difficult to hold our group, with its considerable experience, together. The RES was an integral part of the IAGP and was providing data to a number of other

[235] Drewry, D J; and Meldrum, D T (undated but probably December 1977) A revised proposal for Radio-echo Sounding DF-78 (unpublished), 4pp.

Figure 11.1. Assembling the palletised racks in the VXE-6 hanger at Christchurch airport. Left: ARDS pallet. Right: the RES equipment, primarily TUD 60 MHz and 300 MHz radars. Pallisgaard is inspecting paperwork.

projects, such as drilling at Dome C; these would be compromised if our planned activity did not go ahead. I hoped these sentiments would be transmitted to Moulton in McMurdo.

On 5 January the remainder of the SPRI team—Hugh Steed, Ed Jankowski, Chris Hereward, and Andrew Brimelow—arrived in Christchurch on a MAC flight, and Mogens Pallisgaard a little later. Two personnel from China Lake arrived on the same flight, Gary Ahr and Jim Klever; they would be operating the ARDs equipment. The SPRI group were eager to learn the details of our situation, having been at the other end of a rather tenuous information chain.

We were all settled into the motel, where it was rather cosy, as we had to use four additional camp beds, but we hoped it would not be for very long. The weather was mixed, but to relieve some of the tension we enjoyed a barbecue and swimming at one of the local outdoor pools. The next day all were stuck into equipment preparations at the airport in anticipation of the arrival of the aircraft (Figure 11.1).

Cmdr David Srite landed in Christchurch from the ice, and we arranged for a discussion on 6 January. This was a helpful meeting, and we found Srite anxious to assist. Most of our discussions related to the operational parameters of the aircraft. Following this Meldrum and I worked on various scenarios, as well as on developing new checklists for the RES flights. These were immensely useful and mirrored the checklists used by the aircrew. We developed a 24 hrs Pre-flight Check, McMurdo Pre-flight Check, Strip Pre-Flight, Taxi and Take-off Check, In-Flight Check, and Post-flight Check.

News came in from McMurdo in response to my request to go to the ice in advance of our main deployment—and I was scheduled for 8 January. It was vital that Meldrum remain in CHCH to oversee the technical aspects. This decision was a great relief, as it augured more positively for the eventual commencement of the programme than hitherto. I decided that I would be accompanied by Jankowski, as he had spent a good deal of time on flight planning back in the SPRI and would be able to assist on practical details. We made our preparations and were at the airport with the rest of the group to welcome #03 on a rapid rotation.

Jankowski and I left at 1330 in a crowded hold with 60 people aboard, and eight hours later arrived at Williams Field in McMurdo Sound. It was a huge relief to have got this far. We were lodged in the 'Mammoth Mountain Inn', a 'hotel'-style facility recent erected and with comfortable double rooms. The next few days were spent in making contacts with various key individuals on the base—the NSF representative (Ken Moulton had just been replaced by Dave Bresnahan); the Holmes and Narver representative looking after field equipment and certain aspects of transportation; the main air operations personnel in VXE-6, including the commanding officer, Cmdr Jim Jaeger; and the meteorological staff, with whom we would liaise on a daily basis.

Besides locating some of our cargo that had already been shipped to McMurdo, a vital element was to secure laboratory and office space for our equipment and for planning. I arranged a meeting with Jaeger, which was extremely cordial and encouraging. He indicated he would give us as much support as possible and added there would be an extra 10% flight hours to cover the rotations out of Christchurch to the ice. He commented, very candidly, that he had advised cancellation of our programme at an earlier time because of the commitments of the squadron to Siple Station rebuild. However, he said, it was NSF that made the decisions; he undertook the flying. He offered space to be provided for us in the same area in the VXE-6 Squadron building as before, which was an excellent arrangement for which

I thanked him greatly. This would again enable us to maintain close interaction with the aircrew of #03.

Jankowski and I uncovered, however, another problem. The primary recording medium for the RES was 35mm black-and-white RAR photographic film. In the early summer our photographic requirements had been sent to the DPP. Despite assurances received when SPRI personnel were in China Lake in September and later in Christchurch in December the NSF photographic laboratory had failed to obtain this film for our programme! It was not there awaiting us in McMurdo. This situation could have been disastrous, possibly resulting in cancellation at worst or a delay by some weeks at best while alternative supplies could be sourced in New Zealand. As it turned out by a quirk of fortune, a batch of film that had arrived *late* for the 1974–75 season was still in cold storage and, although out-of-date and not the requested specification, was good enough for us to use. There were several side-band calls to Christchurch to discuss this finding with Meldrum; his verbal reactions are unprintable.

On 9 January #03 deployed north for its PM and fitting out of the RES, ARDS, and magnetometer equipment. Once more we breathed a sigh of relief as this next piece of the jigsaw seemed to slip into place. The early preparations undertaken by Meldrum, the keen and efficient working of the SPRI and TUD members, the helpfulness of the VXE-6 ground team at the airport, and the diligence of the ARDS and the APL technicians all paid off in completing the PM and installing our equipment in record time, so that by 12 January #03 was prepared for a test flight (Figure 11.2). The RES antennas had been affixed to the hard points beneath the wings, and the magnetometer boom, or *stinger*, mounted on the tail of the airplane. The various pallets of equipment were rolled into the fuselage. A test flight took place over the Pacific Ocean east of Christchurch and included important compensation and calibration runs for the magnetometer. As almost all the programme-critical elements were in working order and the PM had been completed expeditiously, #03 would launch to the ice the next day.

11.5 Antarctica at Last!

The whole team of SPRI, ARDS, and APL assembled in McMurdo on 13 January—we were in business! Ed and I had worked hard to ensure the arrangements in McMurdo would be ready for the rest of team and therefore prepared, aircraft dependent, to launch on our first missions immediately.

Figure 11.2. Part of the ARDS installation. The magnetic tape recorder is on the extreme left (see Figure 11.8). (Courtesy SPRI).

Bresnahan had confirmed 150 hours would be available to us. Any fuel we took at South Pole and Byrd Stations would be prorated and subtracted from our airtime allocation; we had calculated those elements accordingly. He also informed us that he would be taking a low profile regarding our programme, but that Lt Bill Wheat had been designated to act as liaison between VXE-6 and us. This was an interesting development. Meldrum and I had anticipated we would, as previously, be establishing our own rapport with the airplane commanders and flight crew, so another link in the chain seemed unnecessary and cumbersome. Transportation to and from 'The Hill' (i.e. McMurdo Station) to Williams Field Skiway was another significant issue. In the past we had managed to obtain the exclusive use of a truck that would take items of equipment, as well as personnel. Bresnahan advised we would have to take the 'shuttle bus' that was available or use the helicopters, some from the icebreaker that was now stationed off Hut Point (Figure 11.3). We agreed we would make the necessary arrangements with the aircrew (they had access to a truck!)

The flight plans were ready and waiting. To aid flexibility we had, back in Cambridge, worked on simplifying and speeding up the flight planning

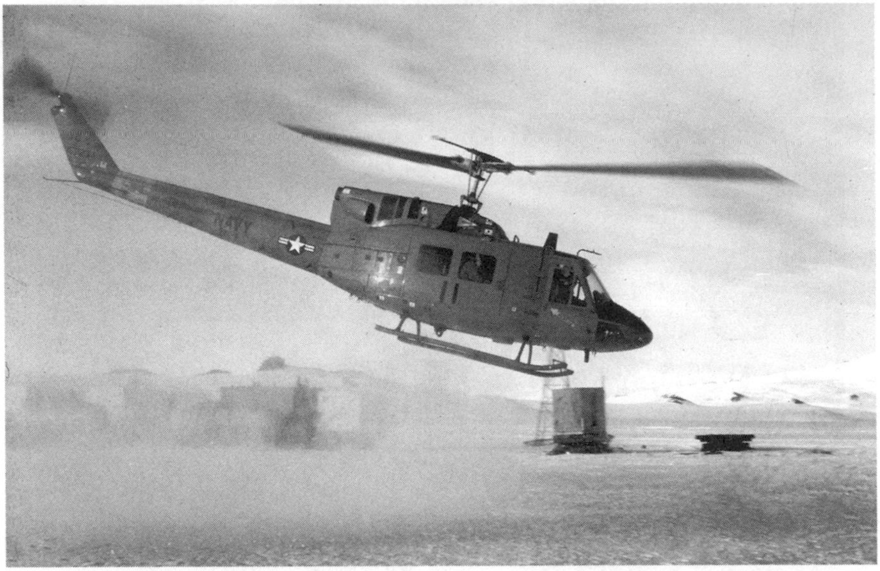

Figure 11.3. On occasion, transfers between McMurdo (The Hill) and Williams Field were by helicopter (a 5- to 10-minute ride!).

process. We had constructed flight lines on the 50 km square grid pattern we had earlier computed and plotted on Air/Jet Navigation Charts as the base maps. In East Antarctica the grid was oriented parallel to the 0°–180° meridian, in West Antarctica parallel to the 135° meridian. The geographic and grid coordinates of all the intersections of the matrix of lines had been computed by Jankowski and were printed and then bound into a flip-top booklet, 'The Flight Planning Directory'. Using this made it relatively easy to pick the coordinates of turning points for a sounding mission and prepare the flight plan (Figure 11.4). The coordinates would later be input into the aircraft's INS. Furthermore, it allowed for maximum flexibility; in-flight changes could be made quickly and precisely should bad weather, aircraft problems, or other deviations be necessary. From our past experiences we had realised it is rarely possible to arrive in Antarctica with a predetermined set of flight plans and hope to achieve them as arranged.

The initial list for the reduced season of 150 hours consisted of four flights in West Antarctica, including two with refuelling (one each at Byrd and South Pole Stations). These were to provide data for Jankowski's research. Similarly, there would be three flights in East Antarctica concentrated on the Sabrina and Banzare Coasts for Hugh Steed, with two others near the

Figure 11.4. Ed Jankowski working on flight plans in the SPRI office, VXE-6 Squadron building at McMurdo.

South Pole to fill in a major gap. We also confirmed an earlier priority of a flight along the flowline from Vostok to Dome A. Chris Neal had requested several specific lines on the Ross Ice Shelf to complete his analyses, and there was a request from RISP which we could probably incorporate.

11.6 An 'Operational Day'

Over the previous seasons we had honed our activities in regard to flight planning, operation of the RES equipment, and in-flight procedures. We had learned a great deal from our Navy aircrews, in particular the preparation of checklists for the RES operators and for the glaciologist. We considered we were a pretty well-oiled machine. Described next is a typical day for a RES flight in the 1977–78 and 1978–79 seasons.

Twelve hours before a flight we would visit the Meteorological (Met.) Center to discuss the latest synoptic situation with the forecaster and examine the most recently downloaded satellite images (Figure 11.5). In the early days these had low resolution and poor detail. In addition to data from US satellites such as Tiros, ESSA, and Nimbus, Soviet Meteor imagery was available. Normally, conditions needed to be free of low cloud (icing of the antennas could occur if we flew too long in low clouds) and generally clear

Figure 11.5. Lt Dell (forecaster) in the meteorological center at McMurdo.

weather in the area of RES. On the basis of the forecast, we would select the appropriate flight and inform the operations duty officer (ODO) of the flight line.

The next day or five hours pre-flight we would check again with the Met. Center. We would confirm the proposed flight or, should the circumstances have changed, select another to avoid poor flying conditions. The flight plan would then be finalised and issued. Three hours pre-flight we would return to the Met. Center to rendezvous with the aircraft commander (CO) and navigator and would give the latter the relevant Air/Jet Navigation Chart with the accurately plotted way points. Once the aircraft CO and navigator were satisfied with the plan, we would arrange to meet the aircrew at the strip (Williams Field Skiway). We had an important note to check that there were in-flight rations for the aircrew!

Before departing McMurdo Station, we would collect all the required maps and charts, as well as instruments and calculators, logbooks, and the logging sheets for the INS. Among the many backup practices we had developed over years was that the navigator routinely filled in the INS logs with latitude and longitude at regular intervals.[236] Similarly, the logbooks

236 In his article Robert Nyden refers to this additional activity for the navigator: '[B]ecause the INS technology was new to our squadron and the invested project time and money was so great, the SPRI

were a record of many of the aircraft avionics outputs, as well as a host of other observations. These were insurance in case our digital recording systems failed. If it was possible, the previous or disembarking crew would be asked about any equipment malfunction—there would always be a note on a door in our accommodation or on the aircraft once we boarded. It was important not to forget to collect cold-weather survival gear, typically in a large orange USARP bag containing clothing and other essentials in case of emergency evacuation of the aircraft in the field or at some other station. Often, there were odd tins of some edible delicacy secreted in them. The RES team clothing was a comfortable grey-green aircrew flight suit. Once we had satisfied ourselves and ticked off all the items on the Pre-Flight Check List we would, we hoped, take a truck from the USARP Chalet for Williams Field. The so-called shuttle or a helicopter was, respectively, seriously uncomfortable or required arrangements too far in advance for practicality.

On our arrival Strip Maintenance would provide a pre-heater on the aircraft. If the plane had been stood out on the ramp for several hours, it would become 'cold soaked', and we needed the hot air to warm up the masses of electronics on the pallets in the cargo bay, as well as making life comfortable for ourselves. The RES antennas would be examined for any cracked plates, mountings, and missing bolts (Figures 11.6 and 11.7). Various other checks would take place including with the photo man onboard that the vertical camera system was loaded and functioning and with sufficient film (#03 had not been fitted with the three-camera trimetrogon system). The team members operating the RES and the ARDS and magnetometer would go through their pre-flight checks, importantly, ensuring that all the RES and aircraft systems were time-synchronised to GMT. The most critical were the INS alignments (at high latitudes this process can take some time), and the altimeter readout on the ARDS was set to the strip (i.e. Willams Field) atmospheric pressure reading. David Meldrum obtained a spare altimeter, which he linked by static line and mounted on the RES rack, which gave a handy readout for the operators in the hold.

All ready and secure, the aircraft would taxi out to the ice runway or skiway. The glaciologist would be seated on a bench at the rear of the flight deck, the RES and ARDS operators in their seats by their consoles, and any

---

scientists didn't want to take any chances. So for the first flights, . . . *we were* . . . recording the data from both INS units every five minutes.' (footnote 146).

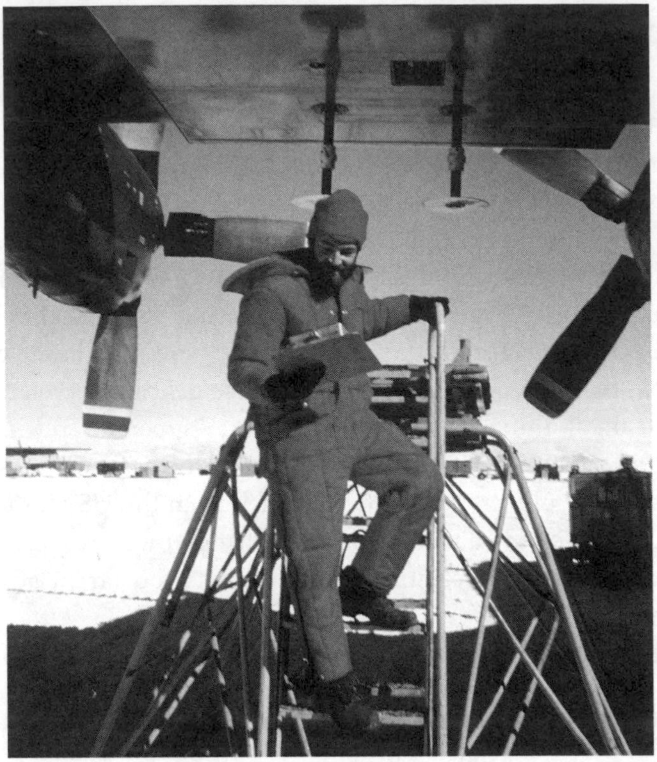

Figure 11.6. Niels Skou inspecting the 300 MHz antenna (1974–75 season).

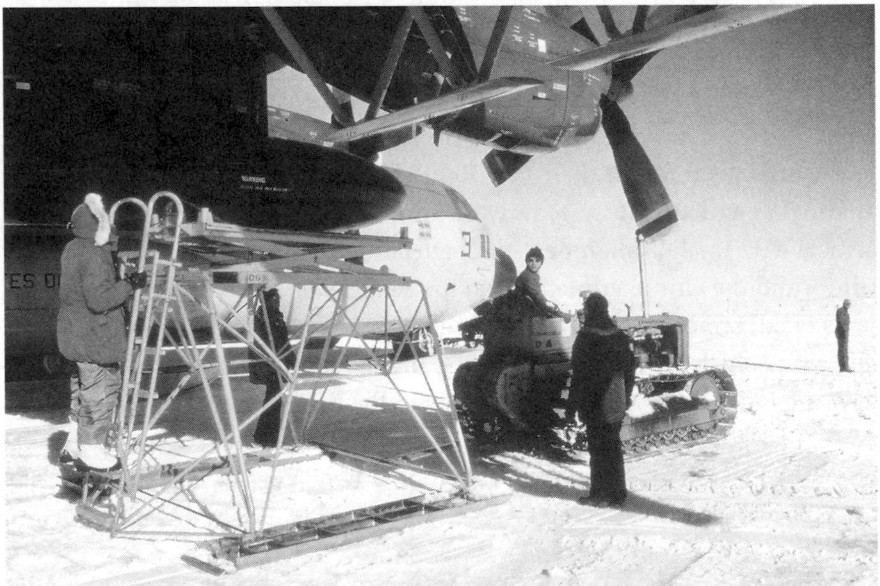

Figure 11.7. Checking the 60 MHz aerials.

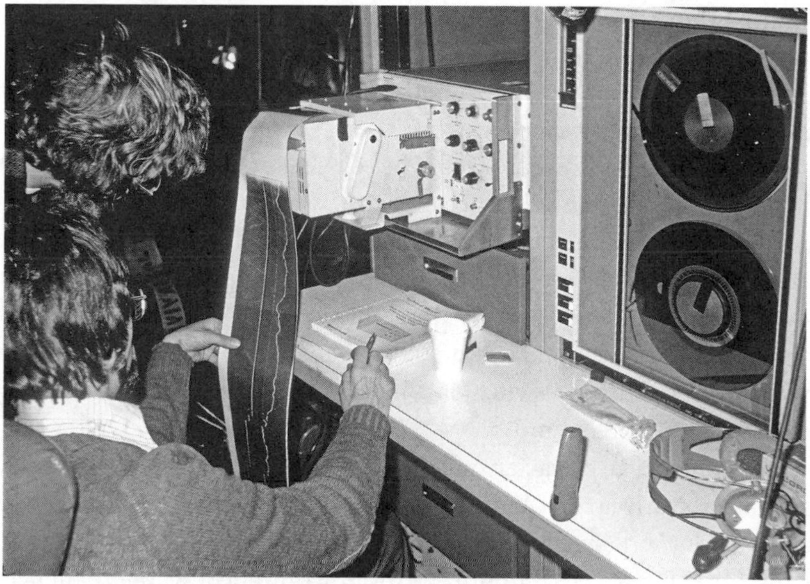

Figure 11.8. Honeywell chart recorder with a strip profile emerging. To the right is the high-speed magnetic tape recorder for the ARDS (navigational and magnetics data). (Courtesy SPRI).

other individuals strapped into various webbing seats in the cargo hold. Take-off would be straightforward, often revealing views of Mount Erebus or Mount Discovery and the impressive Royal Society Range as we climbed to altitude. Quickly we would move into operational mode. Even on high-level outward and return sectors to a more distant target area the RES would be operating; the ARDS of course would be logging navigational data continuously, beginning to end. As we approached the start of a flight pattern the aircraft would descend, and the glaciologist would liaise with the pilot to determine the precise flight level or terrain clearance. It was then a matter of following the flight plan and noting any particular or unusual features both on the ice sheet surface (such as mountains, nunataks, and surface crevassing) and from the RES as the Honeywell chart steadily churned a strip of 'Z' output (Figure 11.8).[237] The immediate availability of these profiles gave an invaluable boost to the enthusiasm of both aircrew and scientists and provided data of the highest quality.

[237] One of the most important developments in 1977–78 was the installation of a Honeywell 1856A fibre-optic recording oscillograph that yielded continuous, high-resolution profiles from the 60 MHz radar, with a delay of only a few minutes for the heat-processing of the recording paper.

Quite often something of interest would be revealed—very deep ice, a range of subglacial mountains, a possible sub-ice lake, very-well defined internal layering—or maybe something completely puzzling! (Gordon Oswald has given an account from the RES operator's perspective in section 7.3). Such observations would be relayed to all onboard via the internal communications system, since everyone was on a headset. The headset was an important item, as it also acted to damp out the continuous drone from the four turboprop engines. Hercules transports are noisy birds, and ear plugs were frequently used in addition. The glaciologist would log turns and major track and altitude changes, and any significant feature could be marked using a switch on a box on the flight deck. This would be identified on the RES film.

On long flights an extra person would join the team who would take over duties, particularly at the RES console, so the first operator could take a rest, even "rack out" on one of the webbing stretchers mounted on the inside of the fuselage (Figure 11.9). Routinely there would be a whole sequence of checks—that the ARDS was running and recording, line printer was on and running legibly, the magnetometer functioning, cameras advancing.

We carried food onboard for the flight, and the Navy IFRs (in-flight rations) were always available. These consisted of a cardboard box about 8 in. square containing a variety of canned items. There would be meat, fruit, and cookies. Some of the tins could be heated in the oven in the galley area on the plane. On one occasion during a flight in 1970 there was an alarming dull explosion onboard. The door of the oven, located halfway up steps to the flight deck, flew open, and a hail of spaghetti and tomato sauce burst out and into the forward part of the cargo bay, just at the rear of the RES racks. The navigator, Lt Springate, having put his IFR tin in the oven had failed to pierce the top, as instructed! We spent many days afterwards extracting strands of pasta from the already spaghetti-like masses of our wiring.

A long flight over the 'featureless' ice sheet could be quite draining. I was impressed that the airplane commander would routinely undertake some crew training on the high-altitude sectors. With one of the flight manuals opened on his knee, he would ask around what procedures were required by a certain indicator light or popped bus on a control panel. At other times we all would discuss how to achieve maximum range or endurance, tuning engines to fly with given flow of fuel, changing the angle of attack, and even shutting down engines. I recall on several missions over the Ross Ice Shelf

Figure 11.9. The author 'racked-out' (taking a short nap) during a RES sortie in the 1978–79 season. (Courtesy SPRI).

we shut down the two inboard engines for several hours until we commenced our return to McMurdo.

On some occasions, circumstances could get quite tense. I recall vividly flying over the Byrd Subglacial Basin in West Antarctica, where there is very thick (3000–4000 m) and 'warm' ice. This combination leads to greater absorption of the radio signals and can result in loss of any returned echoes from the bed. On this particular day the flying conditions were perfect—the atmosphere was brittle, the sky brilliantly clear. The surface of the ice stretched out before us, scored with lines and corrugations of hard-packed snow glittering as with a patina of tiny frozen diamonds. But we were picking up signals from the bottom only intermittently. I asked the pilot if he would descend from our height of about 5000 ft. He obliged and switched off the autopilot to handle the descent himself (Figure 11.10). I added, 'Come down to whatever height you feel comfortable'. As the altimeter wound lower the aircraft engineer, seated behind the pilots, steadily arched forwards staring at the dial on the pilot's instrument panel and the diminishing clearance between the airplane and the frigid surface. Suddenly he snapped back into his seat. I turned to him and asked, 'OK, now Bones?', his nickname. 'Yeah', he drawled in a slight Southern accent, 'at this height if anything happens,

Figure 11.10. Low-level flying over the West Antarctic Ice Sheet (<500 ft above the ice surface).

we'll know nothing about it!'. The pilot was now flying this great airplane at less than 25 ft (8 m) above the ice; we were in ground effect (increased lifting and lowered drag)! At 250 knots the ice sheet streaked by at lightning speed, its details a blur of white and grey. It was unreal, and we watched mesmerised. The concentration was palpable; it filled the flight deck with a dynamic tension. After several hundred exhausting miles, at the end of the line the pilot pulled back on the yoke, and the C-130 climbed back to a safer level. The tension eased, the tight jaw of the engineer went slack, and we all breathed more easily. 'Well, I hope you got the results you need', the pilot said. 'Thanks', I replied—not just for him but the rest of the aircrew. I didn't say at the time that we still hadn't recorded very much of a signal for all that dangerous effort! Science can yield startling results, it can break new frontiers and help us understand the complex workings of our natural world, but at other times it can be fickle and exasperating.

Despite all the planning, Antarctic weather is capricious; on our return to McMurdo after several hours of flying we could experience dramatically changed and difficult conditions. Sometimes we could divert and land at the South Pole or Byrd Stations. In one case a flight landed at Vostok to see out a storm in the Ross Sea. The most anxious, and thankfully rarest, moments are in a 'whiteout' landing. With heavy falling snow, the ground loses definition, and the horizon disappears. In the 1970s, automated landings even at busy city airports were only just beginning, and in Antarctica all the air support systems were fairly basic. The US Navy had landing radars that enabled the pilot to be talked down by ground control, once they had locked onto the TACAN[238], with a glide path to descend to the runway and adjust flaps and speed accordingly. If this was not possible, the plane would fly to the Ross Ice Shelf beyond Ross Island, an extensive and relatively featureless stretch of ice and snow, where the pilot would commence a very carefully controlled rate of descent with skis down until the radio altimeter indicated the plane was near to the surface, everyone on the flight deck braced and peering out to spot any sign of the surface, all other crew tightly strapped in. There was then a tense wait until the skis hit the surface, and all aboard prayed the impact was not severe. The aircraft would then slow quickly to a stop before making the long, bumpy taxi over the ice to Williams Field, with all crew relieved it had been a successful landing.

Chris Neal recalled one such very rare incident during the 1974–75 season. The good weather at McMurdo deteriorated rapidly on the aircraft's return, and the plane commander, Art Herr, informed all onboard of the decision to make a whiteout landing. Gordon Robin was on the flight deck, and Neal and other SPRI crew were operating the RES equipment in the cargo bay. The procedure outlined was duly performed, and the aircraft returned safely to McMurdo. Once the plane was parked, Art Herr came back into the cargo bay to talk to everybody. He mentioned that the very first pilot to complete such a landing received a military medal for the courageous act, and although many such landings had been made subsequently, it was still highly dangerous. He said how impressed he had been by the calm and responsible performance of the SPRI team during this hazardous manoeuvre.

At the end of a flight, it was important to ensure that nothing was turned off until the final INS and strip atmospheric pressure readings had been logged. After a long flight we would be pretty ravenous, our IFRs seeming a long time past. The mess hours at McMurdo might be some way off if we landed in the wee hours, but there was always food at the strip mess, and we would pile into the steamy Jamesway and join the crew and maintenance staff for dinner (Figure 11.11). Before we were able to turn in, there were still post-flight duties to perform: rectifying any faults in equipment, reloading film cassettes, photocopying logbooks, and checking and annotating the Honeywell output.

[238] Tactical Air Navigation system, used by military aircraft. It provides the user with bearing and distance to a ground or ship-borne station

Figure 11.11. A late dinner in the mess at Williams Field.

11.7 1977–78 Operations

The first RES flight took place the day after the team's arrival, 14 January, to the inner zone of the Ross Ice Shelf to undertake cross profiles at the mouths of two of the major outlet glaciers cutting through the Transantarctic Mountains (Scott and Liv Glaciers). These were to fill in some gaps at the request of Charles Swithinbank at BAS, who, it will be recalled from chapter 4, had worked on these glaciers and estimated their discharge into the Ross Ice Shelf. It was quite exciting low-level flying over the crevassed ice surface and the adjacent ridges of spectacular mountains. The remainder of the flight was out towards the inner grounding zone of the Filchner Ice Shelf on the other, the 'Atlantic', side of the West Antarctic Ice Sheet. This initial mission demonstrated all the equipment was working more or less satisfactorily, and we were in a position to fly as frequently as the aircraft, crew, and weather permitted.

Following this first sortie, the pressure was on, and the team was flying more or less continuously with mostly two flights a day for the next 10 days. While the RES personnel divided into two groups to operate on a rotational basis, the ARDS staff—Ahr and Klever, who had no relief personnel—were put under exceptional strain. We had completed 13 flights by Sunday, 23 January, with a cumulative total of 141 hours operational flying, achieving approximately 80 hours and 24,000 km of on-station soundings. We refuelled

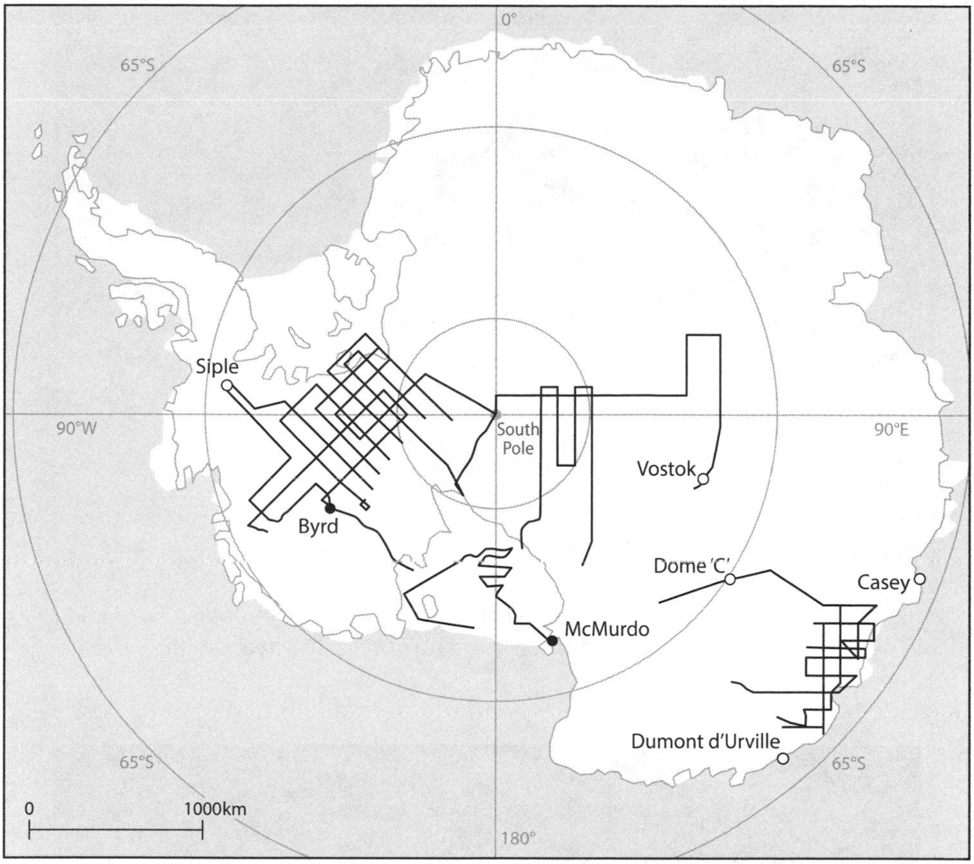

Figure 11.12. Flight lines conducted in the 1977–78 season. Only on-station track is shown. Refuelling was undertaken at Byrd and South Pole Stations.

on two occasions each at South Pole and Byrd Stations (Figure 11.12). The aircraft remained in excellent condition throughout the period. It was a remarkable achievement given the vicissitudes already experienced and the dire prospects of cancellation only a few weeks previously.

But there was little time to congratulate ourselves. Three hours following our return from the last flight over Marie Byrd Land, #03 was scheduled to return to Christchurch! We had expected sending some of our personnel north at the first opportunity, but this was precipitous. Four went with the plane—Hereward, Brown, Irons, and Pallisgaard. On 25 January, Meldrum and I had a debriefing session with Srite and Wheat, covering a range of issues—navigation, aircraft systems, charts, spares, film stock, and

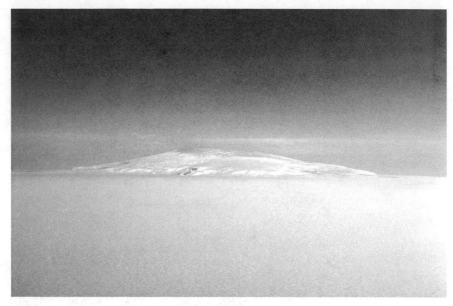

Figure 11.13. Mount Takahe, a large shield volcano in northern Marie Byrd Land (3,460 m; 11,350 ft) in height above sea level and rising 2,100 m above the ice level).

Figure 11.14. The Ellsworth Mountains looking east. The highest peak in Antarctica, Vinson Massif at 16,080 ft (4902 m), is distinguishable as the first block of mountains on the left. Note the stacked lee-wave clouds over the mountains—bumpy flying if you went there!

maintenance. It proved a positive meeting following our effective but short period of activity.

We once more suggested that for the next season it would be preferable to undertake the RES earlier, prior to the closure of the sea-ice runway, which was typically about 20 December. Such timing would provide an additional 8 klbs of fuel, equivalent to 1½ hours flying. We also discussed flying techniques to maximise the endurance of the aircraft. In the report of the season[239] we set out in some technical detail and following discussions with the pilots, that all flying should be at *specific range*, that is, air or ground miles per pound of fuel. This was better than 'long-range cruise', which most pilots would adopt. I also reasoned that we needed to be able to provide the flight crews a thorough pre-ops briefing for RES flights. The pace of operations in this season would not have been sustainable should we have had more hours available; we recommended that a window of 14–21 days would be preferable with, ideally, the scheduling of one flight per day. We both expressed our very sincere gratitude to the VXE-6 staff and crew, who worked exceptionally hard on our behalf.

We had been fortunate with the weather. It was excellent at McMurdo for the whole of our working window, and conditions over the sounding areas were also very good. We had completed the designated six flights over East and six over West Antarctica plus one over the Ross Ice Shelf.

I noted in my diary: 'I spent most of my flights in West Antarctica. We saw some really spectacular scenery—immense shield volcanoes in northern Byrd Land—up to 10,000–12,000 feet high (Figure 11.13). West Antarctica, in places, is just a mass of small nunataks rising up through the ice sheet every few hundred miles—making the flying fascinating. We also flew at low altitude along the west side of the Ellsworth Mountains—the highest and most impressive chain in Antarctica—peaks rising to 15,000–16,000 feet high.' (Figure 11.14)

11.8 Dry Valleys

In the 1974–75 season I had commenced a project examining the flow of ice from the East Antarctic Ice Sheet into the Dry Valleys of southern Victoria Land (section 9.5). I had planned a further investigation this season

[239] Drewry, D J; and Meldrum, D T (November 1978) *Preliminary Report: Radio-Echo Sounding of the Antarctic Ice Sheet DF-78, Event S-168*, SPRI, 32pp.

Figure 11.15. Ed Jankowski with SIPRE auger and three shallow horizontal core holes at the base of Taylor Glacier snout.

and had organised through the ever-helpful Dave Bresnahan, the USARP representative, to have helicopter support to take ice samples at Taylor and Wright Upper Glaciers and on the small, newly identified dome (that I called 'Taylor Dome') behind the exposed mountains. This project was to continue into the following 1978–79 season; the scientific rationale is elaborated in section 12.4. A first foray had taken place during the week prior to the commencement of the RES activity, when Jankowski and I had drilled short ice cores from the exposed cliffs of the Taylor Glacier snout (Figure 11.15).

Some of these Jankowski and I had analysed in the Bio-Lab for sediment content; the melted ice samples were for isotopic analysis. On 27 January all five of the SPRI party remaining at McMurdo took a helicopter to Taylor Valley to gather further samples. It was the last helicopter flight of the season, and the day was calm and warm by Antarctic standards! We spent several hours around the snout of the glacier, privileged with outstanding views, and realised we were now the only people along the whole length of the Transantarctic Mountains, for 2000 km; all the other field parties had been brought back to McMurdo. There was such a feeling of tranquillity as we savoured the majesty of nature and the enormous scale of this extraordinary continent.

The helicopter returned for us, and we flew back to McMurdo and to a small reception by the helo crews, complete with smoke bombs and much cheering. The last mission was complete, and now all the pilots and ground crew would be leaving to return to the US. That evening we held a small party for all the people who had assisted our programme. Three days later, on 1 February, Meldrum, Steed, Jankowski, and Brimelow departed for Christchurch, leaving me for another two weeks in McMurdo. I busied myself with writing the end-of-season report and in discussions with some of the remaining scientists. I visited the New Zealanders at Scott Base and spent a day with Howard Brady, a glacial geologist from Macquarie University in Australia, examining glacial features along Hut Point Peninsula. I departed on 16 February, one of the last scientists to leave McMurdo that season.

11.9 Retrospective

Despite all the stress and uncertainties that had continuously plagued the programme, we had eventually pulled a modest rabbit out of the battered hat. The teams, aircrews, and other support personnel had been outstanding. The new 'R' model aircraft (#03) had performed exceedingly well, as it had greater range and more sophisticated systems than its 'F' model predecessor (the faithful #320). The 60 MHz and 300 MHz radar sounders designed and built by TUD in collaboration with SPRI operated effectively. Unlike in 1974–75, all the equipment was mounted on pallets that could be driven into the cargo bay of the C-130. The two antenna systems performed effectively, with some minor failures, but there was no loss of performance. Equally, the ARDS delivered its promise, although SPRI personnel considered it grossly over-specified. The inertial navigation system on #03 demonstrated its high level of capability, and closures at the end of flights of 4000 km were typically of the order of 3 km. The superior avionics were beneficial, and the static pressure transducer essential for tracking the aircraft altitude and hence the height of the ice sheet surface had a maximum error of 22 Pa and root mean square of 16 Pa, which is equivalent to 2.5 m. Some initial noise problems with the magnetometer associated with the aircraft power supply were resolved, and thereafter the system functioned well. When operating in cloud, noise amplitudes could exceed 100 nT. The compensation flights were conducted in New Zealand and Antarctica with calibrations over observatories in Otago and at Scott Base and Vostok Station. We recommended that a continuously recording magnetometer should be operated close to Williams Field during the next season to assist in calibration and removal of diurnal variations.

The preliminary scientific results were also very satisfying but tantalising in respect to what might have been acquired had the original allocation of flight hours materialised. The data along the coastal area of East Antarctica to be used by Steed looked very promising and identified the valleys and deep troughs of some of the important outlet glaciers. A flight over Dome C camp in support of the French deep drilling (part of the IAGP) was undertaken to detect layers in the top 1000 m that might be correlated with ice core parameters. The bedrock surface confirmed an irregular relief of 200–300 m from previous RES. In a similar fashion a flowline from Ridge B to Vostok had been sounded to assist in the study of the Soviet deep drilling (it had reached 980 m depth at that time). The terrain upstream was found to be very rugged owing to the presence of the Vostok Subglacial Highlands part of the Gamburtsev Mountains. Such roughness could generate complications for modelling and interpreting the lower portion of ice column record.

In West Antarctica the grid of data, which would be used by Jankowski, was of good quality and would assist in interpreting the flow towards the Ronne and Filchner Ice Shelves—a very poorly understood region. The grounding line of the latter was shown to be located 50–100 km further inland than previously suggested, and a new ice stream was identified that we named after the SPRI—'Institute Ice Stream'.[240] The data from this season would mesh with the 1974–75 lines that Keith Rose worked on. An interesting observation was made of several sub-ice glacial valleys draining westward away from the exposed ranges of the Ellsworth Mountains (a direction opposite to the current ice flow), signifying that these mountains formed a major growth centre of the early West Antarctic Ice Sheet. A set of tracks had been flown on one mission over the Ross Ice Shelf at the request of the RISP. It also supplied additional information of bottom characteristics for Chris Neal in delimiting areas of melting and freezing.

[240] This was seen as appropriate, since the adjacent large ice stream to the east was named after the NSF—'Foundation Ice Stream'.

The first large sub-ice lake to be discovered in West Antarctica was detected along the western flank of the Ellsworth Mountains between the Sentinel and the Heritage Ranges. It was determined to be at least 13 km in length occupying a bedrock trough beneath about 3500 m of ice.

All personnel were back in Cambridge by the beginning of March, and the task ahead was to commence the reduction of these data. The meetings that had been held in McMurdo were strongly suggestive of continued support into 1978–79 with an allocation of 150 hours and an early season start. On 14 April Gordon Robin made a visit to the DPP at the NSF and met with Director Ed Todd to discuss the future operations of the NSF-SPRI-TUD project. The indications were positive. On 12 May Todd wrote to Robin confirming another season in 1978–79. The game was still on! In his letter Todd had set out a number of 'understandings' that had resulted from Robin's visit, namely, that (i) the RES missions would be flown off the sea-ice runway, (ii) 150 hours would be scheduled, (iii) fuel could be taken at Siple and South Pole Station and replaced using flight hours deducted from the RES share, (iv) 30–40 hours would be flown in the Dufek Massif collecting ice-thickness and magnetometer data for John Behrendt of the USGS, (v) the Dufek data would be made jointly available to each investigator, and (iv) magnetometer data were to be collected on all RES missions. This was an excellent outcome, shaded perhaps by the limitation of 150 flight hours—pretty much the same as in the previous season. The team would, of course, have wished for a much greater allocation. The principal matter of concern, however, was that the next season was only nine months away.

12

The Final Season, 1978–79

12.1 Magnetic Moves

Work continued over the summer of 1978 on several fronts. There was a progress report to write regarding the NERC research grant covering the period 1976–1979; and papers by several of the RES team had been accepted for a significant glaciological conference to be held in Ottawa in August, and the manuscripts had to be completed. Importantly, preparations had to be pursued for the next Antarctic season. This was the first time that one season had followed directly after the other, with all the consequential pressures.

Meldrum, assisted by Neal and Hereward, worked on the technical front, pursuing regular liaison with colleagues in TUD. Neal had completed his thesis in 1977, and it had been decided to offer him a research assistant contract to continue his research, support the RES programme, and participate in the next Antarctic season. A new doctoral student had joined the group, David Millar. His general aim was to study the internal reflecting horizons that are depicted extensively on the RES records (Figure 12.1). We also had a request from Professor David Sugden at Edinburgh University enquiring whether it would be possible for one of his research students to work on RES data from a geomorphological perspective. We were keen wherever possible to encourage other research groups in the UK to work with us and share data. I knew David Sugden well, a convivial and perceptive geomorphologist wedded to the polar regions; he had worked in South Georgia, as well as in Greenland, and was developing a research project with George Denton at University of Maine. His proposal seemed eminently sensible. Furthermore, we needed an additional person for the operational side of the next season, so we agreed to take his student, David Perkins, as a member of the team.

Figure 12.1. Part of the RES team on the terrace of the top floor of the SPRI, circa summer 1977. Left to right: the author, Chris Neal, Hugh Steed, Keith Rose.

Earlier in the year John Behrendt at the US Geological Survey, Branch of Regional Geophysics, based to the west of Denver, Colorado, in Golden (famous also for the brewing of Coors beer) had approached us to collaborate on a project to conduct a pattern of closely spaced flight lines over the Dufek Massif in the Pensacola Mountains. The proposal was to trace the extent and determine the characteristics of this igneous intrusive body beyond the rock exposures. Behrendt had become aware of the combined aeromagnetic and RES system installed on #03 from his contacts in the NSF, who had encouraged his involvement. I knew Behrendt from his participation in recent international conferences on Antarctic earth sciences and had read his earlier work on seismic ice-depth sounding. He was an old Antarctic hand, having participated in the American team that established Ellsworth Station on the Filchner Ice Shelf in 1956–57 as part of the IGY. Behrendt wintered over as assistant seismologist involved in the oversnow traverse that explored the ice shelf and its surroundings, including the first visit to the impressive Dufek Massif. His experiences are colourfully and

Figure 12.2. Walker Peak in the Dufek Massif. It is possible to distinguish some of the igneous layering in the upper part of the spires.

skillfully recounted in his book *Innocents on the Ice*.[241] Subsequently, he directed a geophysical programme in Liberia, West Africa, and conducted substantial airborne geophysical work, especially magnetic sounding along the Atlantic seaboard of the US.

Behrendt was now turning his interests back to Antarctica at a time when the mineral potential of the continent and its offshore areas was actively being debated. He was brought in by the State Department to advise on the US position within the Antarctic Treaty during the period when there was the possibility of agreeing a Convention on Regulation of Antarctic Mineral Resource Activities (CRAMRA). This measure was stillborn, being superseded by the Madrid Protocol banning minerals exploration and exploitation for 50 years, but that remarkable development was still a decade into the future.[242] Along with his USGS colleague Art Ford from Menlo Park, California, Behrendt was part of a project to undertake further geological and geophysical work over the Dufek Massif (Figure 12.2).

[241] Behrendt, J C (1998) (footnote 23).

[242] The Madrid Protocol was signed on 4 October 1991 and entered into force in 1998.

This range is considered to be one of a few and distinctive geological structures in the world rich in an assemblage of ore minerals, several of which are of high commercial value, such as platinum. The layered mafic igneous intrusion of the Dufek exhibits striking similarities with the Proterozoic Bushveld Complex in South Africa and the Neo-Archean Stillwater Complex in Montana. It is estimated to be between 8 to 9 km thick, dipping gently east. Above the ice 3.5 km is exposed in the Forrestal Range and Dufek Massif, with the remainder most likely lying beneath the ice-covered areas. The sequence has been dated to between 169 and 179 Ma BP, and its emplacement is associated with the rifting between Africa and Antarctica, possibly the location of a triple junction. I was keen that we should work with Behrendt on the survey he proposed of the region. Behrendt visited SPRI on 23 May 1978, during which we discussed in detail his ideas and some of the technical requirements. These were subsequently worked into our plans. Behrendt's considerable experience in the interpretation of airborne magnetic data would prove to be invaluable more generally, and we arranged for Jankowski to visit the USGS in Golden to work with him following the field season.

12.2 Dynamics of Large Ice Masses

Organised by the International Glaciological Society, the conference on the dynamics of large ice masses at Carleton University in Ottawa, 21–25 August, proved a further opportunity to report and showcase some of the highlights of the RES programme. Gordon Robin, Chris Neal, Keith Rose, and I attended, and it provided an excellent forum to engage with colleagues from around the world to discuss Antarctic matters in a stimulating environment.

Robin gave a wide-ranging review of ice shelves, with an emphasis on the melting and freezing processes in relation to salinity, temperature, and pressure of the water circulating beneath them.[243] Much of his discussion was based on the work we had conducted on the Ross Ice Shelf. Keith Rose presented a study on the characteristic of ice flow in Marie Byrd Land, a significant part of the West Antarctic Ice Sheet and feeding ice into the Ross Ice Shelf.[244] His findings have been discussed in section 10.1.

[243] Robin, G de Q (1979) Formation, flow and disintegration of ice shelves, *Journal of Glaciology* 24 (90): 259–72.

[244] Rose, K E (1979) (footnote 218).

Figure 12.3. The bottom of the J9 ice core showing the large ice crystals formed from freezing on of saline sea water to the underside of the Ross Ice Shelf at a depth of approximately 416 m (scale is in inches). (Courtesy NSF).

Chris Neal gave a fascinating talk about his investigations, also on the Ross Ice Shelf.[245] He was able to produce a new map of the ice thickness at a 50 m contour interval from the detailed 1974–75 data. A distinctive feature on the records was the disappearance of the bottom echo and the presence of an internal reflecting layer. This Neal attributed to the intrusion of brine associated with upward-propagating bottom crevasses. These brine percolation layers, he considered, were generated near the grounding line and could be tracked to the ice front, thus providing a means of following and mapping the flowlines in the ice shelf. His study was in close agreement with Robin's 1975 paper. Neal also looked at the fading pattern of the RES signals using his echo-strength measurements from the ice/water interface to locate zones of basal melting (with strong echoes) and areas of possible freezing (accretion) to the bottom of the shelf. The latter characterised the region around J-9 where drilling through the ice shelf as part of RISP had recovered a frozen saline layer 6 m thick at the base of the 416 m core,[246] thus confirming his findings (Figure 12.3). Recent studies have extended these early investigations. Between 2015 and 2017, RES surveys of the Ross Ice Shelf were conducted by the ROSETTA-Ice project of Lamont-Doherty Earth Observatory at Columbia University, using radars at 2 GHz and 188 MHz, yielding 61,000 km of profiling.[247]

[245] Neal, C S (1979) Dynamics of the Ross Ice Shelf as revealed by radio-echo sounding, *Journal of Glaciology* 24 (90): 295–308.

[246] Zotikov, I A; Zagorodnov, V S; and Raikovsky, Ju V (1980) Core drilling through the Ross Ice Shelf (Antarctica) confirmed basal freezing, *Science* 207 (4438), 1463–65.

[247] These data have enabled melt rates at the base to be calculated. These are close to zero over much of the shelf, with higher rates (0.5–2.0 m a^{-1}) at several 'hot spots' towards the ice front. Additionally, RES data combined with satellite imagery and altimetry have revealed an extensive network of

I contributed a paper on a theme different from the radio-echo sounding work, proposing a reconstruction of the Ross Sea area during the last major glaciation in Late Wisconsin times.[248] This was quite a contentious offering and distinct from the 'accepted' view. The group led by George Denton had posited that there would be a large ice sheet filling the Ross Sea embayment and extending to the edge of the continental shelf at this time (~18 ka BP). I was not entirely convinced by the extrapolation of glacial geological studies in the Dry Valleys of southern Victoria Land to the whole of the Ross Sea region and was willing to propose an alternative model to stimulate, perhaps even to provoke, discussion. I had been fortunate as an undergraduate to be introduced to the concept of 'multiple working hypotheses' elaborated by the geologist T C Chamberlin at the end of the nineteenth century. His dictum was to seek out all possible explanations and causes of an observed phenomenon, work through them systematically, and avoid becoming wedded to one particular hypothesis. Chamberlin wrote in 1890: 'The effort is to bring up into view every rational explanation of new phenomena and to develop every tenable hypothesis respecting their cause and history'.[249] My alternative concept, stimulated by the pattern of ice flow in Marie Byrd Land that had been revealed by our RES work and described by Keith Rose, was one of an expanded ice shelf regime with ice streams and enlarged areas of grounded domes and ridges. I backed up my ideas with marine sedimentary records and estimates of past sea levels. I had specific ideas about what was happening in the western Ross Sea area that affected the Dry Valley area. Needless to say, this paper attracted considerable discussion and led to an occasionally tense relationship with Denton in the following few years. Later research showed that the Last Glacial Maximum in the Ross Sea was characterized by a large ice sheet extending to the edge of the continental shelf, and my model was incorrect but perhaps

---

channels in the base of the ice shelf associated with zones of weakness and crevassing. 'Warm' ocean water flowing in these channels is melting into and eroding the ice base, with implications for climate-induced thinning and possible shelf loss of mass. (Das, I; et al. (2020) Multidecadal basal melt rates and structure of the Ross Ice Shelf, Antarctica, using airborne ice penetrating radar, *Journal of Geophysical Research* 125, no. 3, https://doi.org/10.1029/2019JF005241; Alley, K E; Scambos, T A; Siegfried, M R; and Fricker, H A (2016) Impacts of warm water on Antarctic ice shelf stability through basal channel formation, *Nature Geoscience* 9:290–93, https://doi.org/10.1038/ngeo2675).

[248] Drewry, D J (1979) Late Wisconsin reconstruction for the Ross Sea Region, Antarctica, *Journal of Glaciology* 24 (90): 231–44.

[249] Chamberlin, T C (1890) The method of multiple working hypotheses, *Science* 15:92–96; reprinted (1965) *Science* 148:754–59.

represented an intermediate stage during Holocene retreat. These issues are explored further in section 12.4.

The conference on the dynamics of large ice masses was preceded by another symposium, in Ottawa, on 'Glacier Beds: The Ice-Rock Interface', that Robin and I also attended. At this meeting I presented a paper on estimating the basal heat flux over ice-covered areas from RES, which was a worthwhile development, as it could provide a means of approximately determining this quantity over large areas of the ice sheet devoid of boreholes to bed or exposed rock.[250]

12.3 The Final Season Advances

By September, preparations were in full swing; there were only three months before we would be back in McMurdo. Our experiences of previous seasons had spawned several practical and useful internal planning documents, prepared by various of the SPRI team. They reflected how the programme was maturing and becoming increasingly rigorous:

1. NSF-SPRI-TUD 1978–79, Flight Planning Guide and Glaciologist Checklist
2. Inertial Navigation Systems for Antarctic C-130 Geophysical Operations (by K E Rose), October 1978
3. Altimetry Reduction and Elevation Control for Radio-echo Sounding Studies of Antarctica (by K E Rose), October 1978
4. NSF-SPRI-TUD Radio Echo Sounding Programme 1977–78, Guide to Film Interpretation (by E J Jankowski), SPRI, 1978

In the middle of the month Preben Gudmandsen wrote to inform us that owing to lack of funds and time pressures it was unlikely TUD would be able to send one of their staff to work in Antarctica. This was concerning, as an experienced electrical engineer who had been involved intimately in construction of the radar sounders was invaluable. Furthermore, a very satisfactory rapport had developed with TUD personnel in previous seasons, and their participation was greatly welcomed. Without such support much more work would be placed on Meldrum's shoulders, along with Neal's and Hereward's. I had several telephone calls with Gudmandsen, and by 18 Sep-

[250] Drewry, D J (1979) Estimation of basal heat flux over ice-covered areas from radio-echo sounding (abstract), *Journal of Glaciology* 23 (89): 405–6.

tember he had come around to entertaining the possibility of someone joining us. The next day Nils Skou rang and explained that while he could not go south, TUD was willing to send Finn Søndergaard to complete the team, but they were still short of funds. I proposed some SPRI support in the form of Chris Neal; he would go over to TUD to assist for a few days during the first week in October—it was a deal. By mid-October our cargo and the TUD equipment had to be ready for shipping via the MAC to Christchurch. It left on 2 November. Another important task was to nail down the vital film supplies that had caused so much anxiety the previous year. This involved exchanges with the NSF and the Photo Laboratory in McMurdo. We also had to keep our liaison officer at DPP, Bill Wheat, informed of our progress. Later, we discussed with him in some detail our plans for the copying of the RES material after completion of the season. This was an important outcome of the discussions with the NSF to make RES data more widely available to US investigators. In addition, Meldrum needed to liaise with the ARDS group in China Lake and APL at Johns Hopkins. Commercial flights for our team had to be booked as well as our accommodation in New Zealand. The arrangements were that Meldrum, Neal, and I from SPRI and Søndergaard from TUD would arrive in Christchurch on 30 November; Hereward, Jankowski, Millar, and Perkins would follow on 2 December. Larry Irons from Nebraska would join again as part of the team. Our colleague from the US Geological Survey, John Behrendt, had also scheduled his arrival on the 2nd.

The advance party flew into Auckland on 26 November and after a 72-hour stopover continued to Christchurch on the 30th. On 1 December everybody moved into the Ilam Motel, where two large units accommodated the whole team.

A meeting with Walt Seelig at the USARP office followed when we requested information on the status of the sea-ice runway and the latest traffic schedules on #03, and queried the film order. Seelig revealed that there were likely to be delays to the start of our flying campaign, so there was nothing new to that report! Delay is one of the recurring elements of working in Antarctica. There will inevitably be holdups, changes of plans, deferments, and frustration. You have to stay calm and be resigned to waiting!

A message on 5 December from Dick Cameron in Washington advised that #03 would return to Christchurch on 7 December for its phase maintenance and our equipment installation, which would mean departing for the ice at the earliest on 11 December. I requested that Seelig organise for me and one other to depart two to three days earlier to make the necessary connections and arrangements in McMurdo. This happened very quickly, and Jankowski and I left at 1300 on 6 December on a New Zealand Air Force C-130 for a wheeled landing on the sea-ice runway at McMurdo. David Meldrum would oversee the detailed installation and testing of the RES in Christchurch.

We had expected #03 to be heading south on 11 December with the remainder of the group, but a structural problem necessitated airframe repairs at Harewood in the Air New Zealand hanger. This gave us time to further consider flight plans, and Jankowski and I worked on these for five full days in McMurdo—attempting to squeeze as much out of our allotted hours as possible and allowing for a rapid start to sounding once #03 arrived. Detailed talks with the aircrew revealed that the aircraft weighed 2000 lbs heavier than in the previous season, reducing our fuel payload and hence range, which we had to factor in. We calculated that 10 or 11 flights, almost all of which would be long, at least 13 hours each, would probably complete the season.

Before leaving Christchurch, we had received a copy of a long and tetchy message addressed to the NSF representative in Antarctica from Charlie Bentley—who was at Dome C with Ian Whillans from The Ohio State University undertaking a ground survey—requesting urgently a number of flight lines over that area. He stated that the survey was vital for planning of the geophysical and glaciological programmes at Dome C in the next season and to assist future deep core drilling locations. He also asked for a flight along the first 100 km of the Soviet Pioneerskaya oversnow traverse route. His request was for a 50 km square with 10 km line spacings in both directions. The phrasing of his communication was rather unfortunate: 'The IAGP Group has recommended this local coverage, and waited for it, for several years. It can no longer be put off without seriously interfering with an active, current ground program. I can see no such urgency for expending all available flight time (after the aeromagnetic flights over the Dufek Massif have been completed) in West Antarctica.' Bentley referred to the IAGP Council meeting that had taken place in Chamonix in May 1978 in which these requests had been affirmed as important and were to coincide with Bentley's programme at Dome C.

The response, on 27 November, from the NSF representative in McMurdo was fairly uncompromising: '1) Radio-echo sounding flight plans presented by Drewry accepted as submitted. 2) As overall flight hours had to be limited and had to include Dufek Massif for magnetometry and ice sounding NSF did not consider it reasonable to require special flights this season.'

The fact was that Bentley had never chased up his request with SPRI during the preparations for the 1977–78 season. We did not know whether his programme had been approved by the NSF and, if so, what the timing may have been for his presence at Dome C. In any case we would have required, well in advance, the precise location and coordinates of the 50 km box he was requesting. The comments regarding the siting of the drill hole were also overblown, as we had in the previous season made several reconnaissance passes over the top of the dome for this very reason and were fully cognizant of the IAGP priorities. Being approached by Bentley much earlier would have placed us in a far better position to consider the inclusion of his plan in our programme. Furthermore, owing to the increasing concern of the emerging climate change community regarding the putative instability of the West Antarctic Ice Sheet, accumulating further RES lines there was becoming a priority. Despite this unreasonableness, I spent some time with Jankowski to examine whether we could incorporate Bentley's request into our already very tight schedule, and on 12 December I held a good-humoured sideband radio discussion with him at Dome C to obtain the necessary geographical coordinates for a flight to undertake the 50 km box whenever possible.

A number of scientists pass through McMurdo, affording a rich and interesting opportunity for discussion. One was Terry Hughes from the University of Maine working on determining the velocity across Byrd Glacier. Hughes had written to me earlier seeking any possibility of our obtaining ice-thickness data for Byrd Glacier (Figures 12.4 and 12.5). This is one of the principal outlet glaciers through the Transantarctic Mountains and contributes substantially to the discharge of ice into the western Ross Ice Shelf (see footnote 87). I had responded that we would do our best to fit a flight line down the glacier. Hughes and his team were already in the field undertaking survey work to establish flow characteristics, and he had asked Charles Swithinbank to join them based on his experience back in the 1960s. Swithinbank would later accompany us on one of our RES flights. Another long-standing colleague in McMurdo was Tony Gow, from CRREL, and I spent an engaging if busy evening with him in the laboratory assisting in the preparation of thin sections of ice cores he had collected on the McMurdo Ice Shelf.

Figure 12.4. The mighty Byrd Glacier—cutting through the Transantarctic Mountains and flowing into the Ross Ice Shelf (bottom of image); the glacier is 30 km wide at this point. (Landsat image courtesy of the US Geological Survey).

Figure 12.5 Byrd Glacier showing the intensely crevassed surface on and across which Hughes and Swithinbank established velocity markers (by helicopter!).

The USARP representative in McMurdo was once again the helpful Dave Bresnahan, and an early meeting with him allowed me to go through the list of our requirements—accommodation (which was entirely satisfactory), laboratory space in the VXE-6 building once more, flight hours allocation, and refuelling at South Pole and Byrd Stations. The film we ordered was available, thankfully, but there were other photo requirements needing attention.

All seemed to be under control, and on 9 December I had a long and detailed sideband radio conversation with David Meldrum in Christchurch. The aircraft repairs there were going well, and they expected to fly down on 13 December. I suggested they fly via the Rennick Glacier in northern Victoria Land as a test; it was being investigated collaboratively with Paul Mayewski, a glacial geologist at the University of New Hampshire.[251] Meldrum reported several glitches and breakdowns with equipment, but the main task of installing the antennas had been completed (Figure 12.6). We also agreed the initial flight teams. Team A would comprise Finn Søndergaard, Chris Neal, David Millar, and me; Team B was to be David Meldrum, Ed Jankowski, David Perkins, and Larry Irons. Hereward would float between groups, and Behrendt would join all the Dufek Massif flights.

Following the airlift to McMurdo the team settled into its accommodation. The first flight was scheduled on 15 December to the Dufek Massif to commence Behrendt's flight lines, followed quickly by another the day after. Both were long flights of between 13 and 13½ hours and involved refuelling at South Pole Station.

Team A took the first flight to the Dufek Massif and was joined by Charles Swithinbank (Figure 12.7). He recounts in his book the considerable advances made in the RES technology since his 1967 missions.[252] The flight departed over the Ross Ice Shelf and then, at the head of the ice shelf, over the Wisconsin Range, more or less the southern end of the continuous swathe of the Transantarctic Mountains chain. The mission continued south of the Thiel Mountains and on past other scattered nunataks marking the boundary between East and West Antarctica to arrive at the Pensacola Mountains on the edge of the Filchner Ice Shelf (Figure P.1).

[251] Mayewski, P A; Attig Jr, J W; and Drewry D J (1979) Pattern of ice surface lowering for the Rennick Glacier, northern Victoria Land, *Journal of Glaciology* 22 (86): 53–65.

[252] Swithinbank, CWM (1997) (footnote 81).

Figure 12.6. Operating interior of #03. All the equipment is mounted on pallets. Above: Hugh Steed at the ARDS console, beyond the RES systems. Below: In front are various individual survival bags. The small pallet behind these supports the magnetometer system. Beyond, Finn Søndergaard sits at the RES console. Webbing seats along the right (starboard) side of the fuselage are clearly visible, as are two stretchers above. These were used for short rests between operator duty activity (racking-out).

Figure 12.7. Part of the SPRI team at McMurdo. Left to right: Gisela Dreschhoff (NSF Aircraft projects manager); Charles Swithinbank (BAS with Hughes' Byrd Glacier Project); Finn Søndergaard (TUD); Ed Jankowski, Ben Millar, David Meldrum, John Behrendt (USGS).

It was a long journey even before the sounding work commenced. Behrendt had plotted parallel lines at variable spacing but mostly between 10 and 25 km apart that extended from the Schmidt Hills north across the grounding line of the Filchner Ice Shelf and beyond for a further 200 km (see Figures 12.12 and 14.1). The exposed area of the mountains was spectacular. We flew very low over the terrain—jagged spires and peaks rose out of the ice fields on a crystal-clear and perfect day (Figures 12.2 and 12.8). Back and forth we flew, drinking in the beauty and grandeur of this place. Meanwhile we were obtaining excellent ice-thickness and magnetic measurements. Behrendt was working hard on the flight deck making adjustments with the pilot to ensure the track connected with the planned positions in the mountains, triggering aerial photographs to ensure we could locate the aircraft above specific and known features. We completed 1600 km on-station before heading for South Pole Station to refuel and return to McMurdo. It had been an exhilarating and successful first mission. There was not much time to regroup, and the next day Behrendt was again flying with Team B back to the Dufek for another full schedule of flight lines.

Although these Dufek flights presaged a good start, the aircraft was experiencing continual mechanical breakdowns—hydraulic system leaks, a generator failure, ski damage, radar malfunction, and minor maintenance issues. Unlike in the previous season, when operations had been more or less trouble-free, these problems were frustrating. There were some snags with the RES equipment too. The inboard antenna element of the 60 MHz system had to be removed after three missions owing to a sheared bracket,

Figure 12.8. Approaching the Dufek Massif, viewed from the flight deck of #03.

and a 300 MHz plate had cracked and been replaced. The former, fortunately, did not significantly reduce the overall performance.

Then, after four flights a serious split was discovered in one of the aircraft wheels, but wheel changes could not be made at McMurdo, and the alternative of flying back to Christchurch would count against our hours and total time and was clearly not practicable. This failure resulted in our discontinuing operations from the rapidly deteriorating sea-ice runway, moving to the skiway, and subsequently losing 8 klbs of onboard fuel.

More bad news followed. After our eighth flight, on 22 December, the starboard external (pylon) fuel tank was damaged during removal of a check stand, preventing the aircraft's further use and resulting in an additional loss of fuel capacity by 9 klbs! The total fuel at take-off was now at 46 klbs, down from 60 klbs at the beginning of the season. All these problems triggered delays, and no flight left on schedule; a pattern emerged of a 13-hour flight followed by 12 hours of downtime. Nevertheless, we were sanguine about these matters and appreciated the hard work being undertaken by the aircrews and ground support staff.

We were also experiencing other frustrations. These arose from the measurable increase in the level of bureaucratic annoyance. I noted in a letter I sent to Robin, summarising our views at the end of the season:

> The burgeoning administration adds an additional bad taste to working here. The problem stems . . . from the simple fact that there are too

> many people trying to get in on the act. The APL group (3 people) who run the ARDS equipment consider themselves the vital element in any flight, so much so that many of the missions have been dubbed 'ARDS Missions' not Radio-echo or Ice Sensing! Although we would not say that there was friction between our groups the APL attitude made the situation difficult. The Senior APL man was constantly assuming the role of the 'expert' on aircraft operations, electronics and dealing with NSF!!!

There was an additional factor. The NSF had appointed an 'Aircraft Projects Manager' in the communications and organisational chain, in the person of Dr Gisela Dreschhoff. She was required to act as the channel between ourselves and almost anybody else with whom we needed to communicate. Frankly, her job was unenviable, since she inevitably wished to do a good job, yet we felt that her functions were entirely superfluous. I commented: 'Naturally, confusion reigns with too many people trying to tell each other what needs to be done; as a result our carefully nurtured relationship with VXE-6 is in jeopardy and we lose some credibility.' I had known Dreschhoff for some years. She had been a regular visitor to McMurdo from the University of Kansas undertaking radiometric measurements by helicopter over areas of the Transantarctic Mountains with her co-principal investigator, Ed Zeller. Our personal relations then and later as fellow *scientists* were very cordial. We met up in Boulder, Colorado, in 2018 for a dinner with other Antarctic colleagues and enjoyed looking back on those days in McMurdo.

In spite of these vexations—mechanical, electronic, and human—we pushed on with the programme with a steady launch of flights, mostly to West Antarctica, one per day until 23 December. On that day we conducted a flight that tracked down the centreline of Thwaites Glacier (Figure 12.9[253]). Even at this time, towards the end of the 1970s, it was becoming clear the fast-moving outlet glaciers unprotected by fringing ice shelves were critical features in evaluating the long-term stability of the West Antarctic Ice Sheet.

Terry Hughes had expressed these views following the earlier suggestions by John Mercer. Bob Thomas, a somewhat peripatetic glaciologist, attached

[253] Haran, T; Bohlander, J; Scambos, T; Painter, T; and Fahnestock, M (2014) MODIS Mosaic of Antarctica 2008–2009 (MOA2009) Image Map. Boulder, Colorado: National Snow and Ice Data Center, http://dx.doi.org/10.7265/N5KP8037.

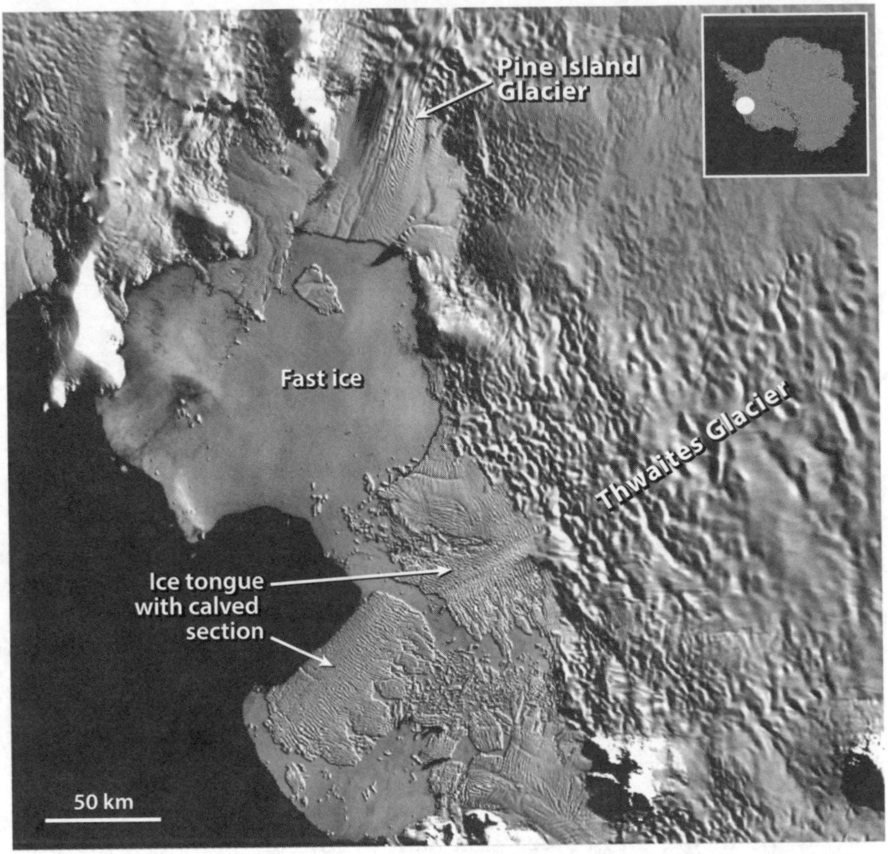

Figure 12.9. Image of inner Pine Island Bay showing the Thwaites and Pine Island Glaciers. (From NASA MODIS instrument, courtesy NASA NSIDC DAAC).

at that time to the University of Maine and who had spent four months at the SPRI (March–July 1978), was working on West Antarctic Ice Sheet dynamics and was very interested in the data that Keith Rose had been analysing in Marie Byrd Land. Together with Tim Sanderson at the BAS, Thomas and Rose prepared a paper for publication in *Nature* on the effects of climate warming on the West Antarctic Ice Sheet.[254] These scientists speculated that Thwaites and its near neighbour, Pine Island Glacier, could be in a precarious state with possible 'collapse' unless their flows were 'protected' by high bedrock sills. They proposed that the stability of such ice

[254] Thomas, R H; Sanderson, T J O; and Rose, K E (1979) The effects of climate warming on the West Antarctic Ice Sheet, *Nature* 277 (5695): 355–58.

streams, in the absence of ice shelves, is crucially dependent upon the depth of such bed thresholds and suggested an elevation no lower than 400 m below sea level to provide sufficient back pressure. The RES profile, although only from a single line down the centre of Thwaites Glacier, indicated a threshold about 17 km inland of the then grounding line at 700 m below sea level. This was well below that suggested by Thomas, Rose, and Sanderson as critical for stability. The data also suggested the presence of a deep trough. The 1978–79 RES sounding line was later remastered and compared with considerable and more recent radar thickness measurements to determine any changes over a 30-year period, which suggested a thinning by 115 m.[255] In 2019 a major international research programme commenced to gather detailed data on this outlet glacier to investigate its past and future behaviour. These included, among other investigations, surface glaciology on ice dynamics, airborne RES, drilling through the floating glacier tongue to investigate ice/ocean processes at and beyond the grounding line, and seabed scanning to investigate former ice fluctuations. The glacier and its drainage basin are viewed as a critical component in the stability of the ice sheet and hence to predicting future global sea levels.[256]

Following a break at Christmas and atrocious weather conditions that prevented flying, the programme recommenced on 27 December, to complete the remaining four flights. On that day Charlie Bentley and Ian Whillans arrived at McMurdo from Dome C, and I was able to inform them that we still intended to undertake a flight to acquire the data they had requested. Indeed, we flew to East Antarctica shortly thereafter and completed that project. I felt vindicated in that we demonstrated our willingness to collaborate and share resources; we had pulled out all the stops to assist Bentley and meet his request (Figure 12.10). On the return to McMurdo, we flew down Taylor Glacier in the Dry Valleys, which was the object of research I was conducting on the present and past dynamics of the edge of the ice sheet. It was beautiful weather, and the impressive mountains framing the valley were a continued reminder of the glory of Antarctica. It was the last mission.

No sooner had we landed than we discovered we were scheduled to depart immediately to conduct a magnetic compensation flight for a couple

[255] Schroeder, D M; et al. (2019) Multi-decadal observations of the Antarctic Ice Sheet from restored analog radar records, *PNAS* 116 (38): 18867–73.

[256] International Thwaites Glacier Collaboration (thwaitesglacier.org).

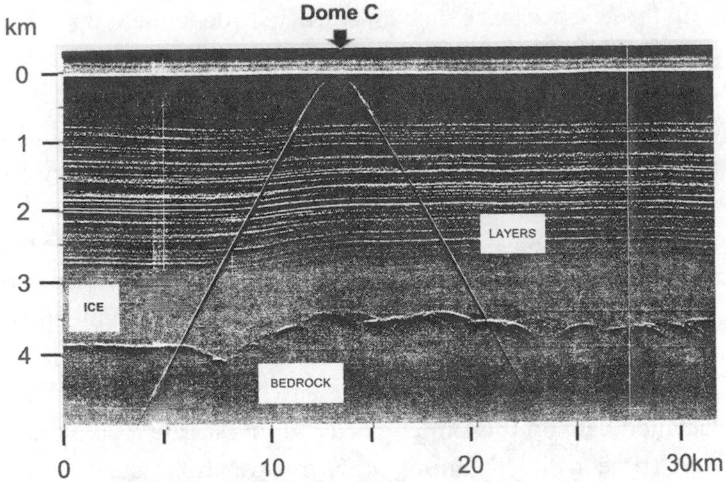

Figure 12.10. A 60 MHz record over Dome C (copy from the Honeywell chart). The slanting return is from the surface buildings, and 200–300 m of relief is evident beneath the intended drilling site.

of hours over a relatively magnetically 'quiet' area of the central Ross Sea. This is an important procedure used to calculate and remove any effects of the aircraft and its magnetic signature on the magnetic field measurements. The compensation involves flying all four cardinal and four ordinal directions of the compass in both directions and along each line with a sequence of pitches, rolls, and yaws. There were a couple of mechanical problems with #03 before we commenced this flight, which departed about seven hours late—on top of the eight-hour flight we had just completed. The navigator came onboard with a bundle of large black plastic sacks, which he handed out saying this was going to be a really rough flight. And so it proved. The aerobatics were so bad almost all the crew were sick with the stomach-churning g-forces during the sinusoidal pitches, including me. I wrote at the time, 'What a way to end the season!' (Figure 12.11).

When we eventually got to bed that night, we had been up for 48 hours. But that was not the end to flying for several of the team, as #03 refuelled directly upon our return to Williams Field and then departed north for Christchurch with Neal, Millar, Jankowski, Hereward, Søndergaard, and Irons. The season had really ended abruptly for them.

The RES programme had been a success despite the myriad aircraft problems that had bedevilled us, a number of snags with our own radar equip-

Figure 12.11. Last group photo of the SPRI RES team in operational mode at Williams Field, shortly after the gruelling magnetics compensation flight, with #03 in the background. Left to right: Chris Neal, the author, David Meldrum, David Perkins, Finn Søndergaard, Ed Jankowski, Larry Irons, and Chris Hereward. David Millar took the photograph.

ment, and tedious bureaucracy. The aircrews had been outstanding in their support, flying our missions with great dedication and skill, especially the challenging tightly spaced lines in the Dufek Massif; these demanded high levels of professional expertise, which they were proud and determined to demonstrate. The SPRI team, too, had worked very well in the field, having flown 130 hours with 61 hours on-station and having covered 26,000 km of track profiling (Figure 12.12). Luckily, there was no charge for the refuelling hours at outlying stations, which represented a considerable bonus. We were well satisfied given the initial modest allocation of flight time.

From the concentration on West Antarctica, it had been possible to extend the grid towards the Ellsworth Mountains and across the important and complex geological/tectonic boundary between East and West Antarctica—the West Antarctic Rift System, as we now know it (Figure 12.13). At this stage we were obtaining tantalising glimpses of how the combined RES and magnetic data could assist interpretation. The transition was illustrated dramatically in the area north of the Horlick Mountains. On the West Antarctic side, we observed an abrupt change with bedrock lying 1500 m below the ice sheet surface (Figure 12.14). This transition

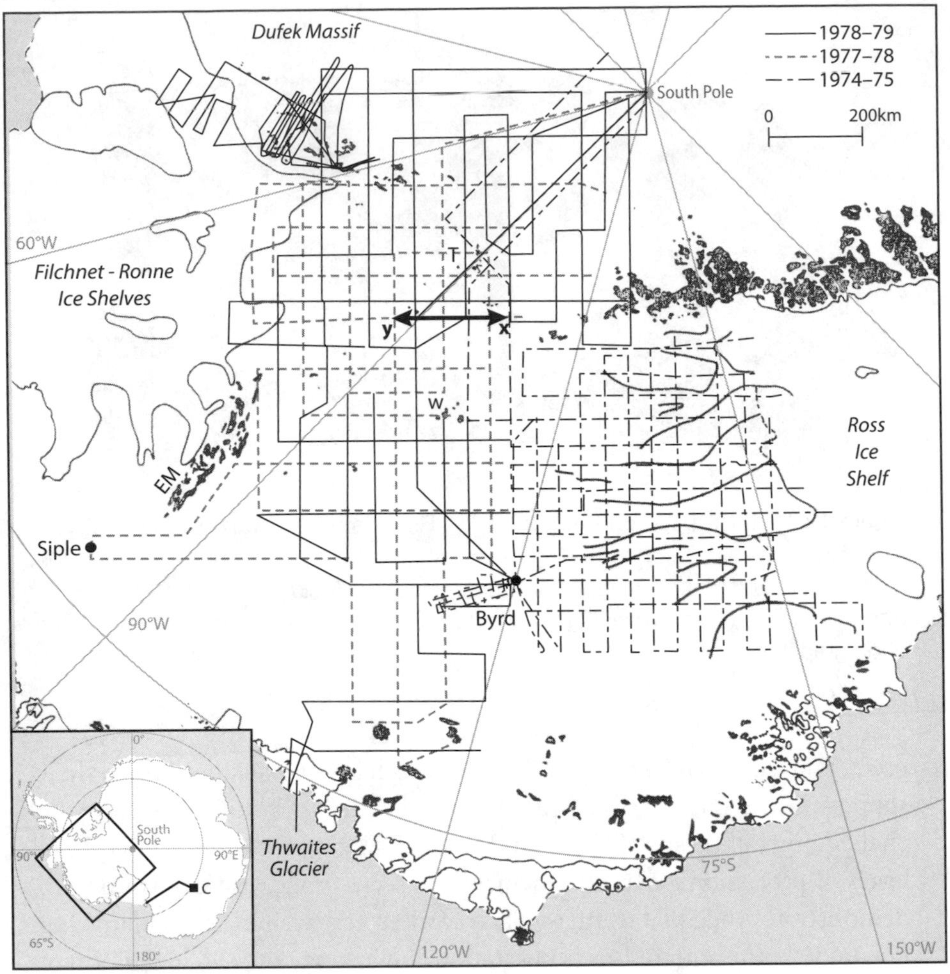

Figure 12.12. Radio-echo and magnetic sounding flight lines in West Antarctica in 1978–79. Missions accomplished during the 1977–78 and 1974–75 seasons are also shown (the latter without simultaneous magnetics). EM = Ellsworth Mountains; T = Thiel Mountains; W = Whitmore Mountains. One 1978–79 mission to Dome C is shown in the inset, where a 50 km box pattern was flown at a 10 km line spacing in each direction in support of the IAGP surface glaciological activity.

appeared to be associated with a positive magnetic anomaly of a few hundred nanotesla across the southern flank of the rift.[257]

[257] Jankowski, E J; Drewry, D J; and Behrendt, J C (1983) 'Magnetic Studies of Upper Crustal Structure in West Antarctica and the Boundary with East Antarctica', in Oliver, R L; James, P R; and Jago, J B (eds.) *Antarctic Earth Sciences*, Canberra: Australian Academy of Sciences, 197–203.

Figure 12.13. The escarpment of the Thiel Mountains (south is to the left). This topographically subdued block is an extension of the Transantarctic Mountains and lies on the East Antarctic margin of the West Antarctic Rift.

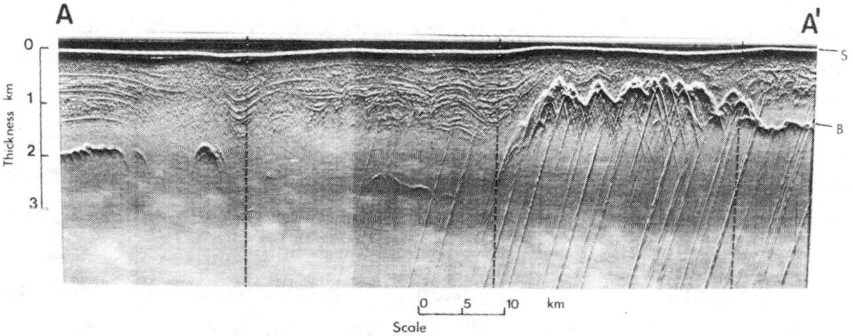

Figure 12.14. RES profile across the boundary between East and West Antarctica in the vicinity of the Horlick Mountains (see Figure P.4) (right side, A′); (S = ice surface; B = bedrock surface).

The power of the simultaneous magnetic and RES records became apparent. A completely new feature, a sinuous ridge, was identified within the extensive low-lying Byrd Subglacial Basin (up to 1 km above the basin floor) that possesses a pronounced magnetic anomaly. This was interpreted as of volcanic origin and the basin constituting a series of interbedded sedimentary

Figure 12.15. Erebus ice tongue. It is about 11–12 km long and surrounded by shore fast (sea) ice. In the background, part of Mount Terror, Ross Island.

and volcanic strata, reflecting the opening of this rift. Later geophysical surveys and geological work has built a picture of this region as comprising one of the most extensive zones of stretched continental crust with substantial volcanism.[258] It is comparable in size and setting to the East African Rift System and the Basin and Range Province of the western US.[259]

There had been one or two other supplements to the programme. Two missions were flown across the McMurdo Ice Shelf where bottom freezing had long been detected, and one down the skinny and serrated-edged Erebus Ice Tongue on Ross Island, the focus of some research into ice/ocean interactions (Figure 12.15). This is further discussed in section 13.4. Nor did we forget the request by Terry Hughes for ice-thickness data on Byrd Glacier. We had tried twice before to obtain depths (1969–70 and 1977–78) but had failed to penetrate the thickest ice, because bottom reflections were obscured by scattering from extensive surface crevassing. We flew the line at an elevation of 150 m above the surface of this enormous, spectacular, and fast-flowing outlet glacier in an attempt to reduce the clutter echoes and were rewarded with an almost continuous profile. We identified the grounding line from a significant change in the character of the bottom echo and where the depth was 1300 m. Further up-glacier by about 25 km the ice had thickened astonishingly, to between 3100 and 3300 m.

[258] Blankenship, D D; Bell, R E; Hodge, S M; Brozena, J M; Behrendt, J C; and Finn, C A (1993) Active volcanism beneath the West Antarctic ice sheet and implications for ice-sheet stability, *Nature* 361:526–29.

[259] van Wyk de Vries, M; Bingham, R G; and Hein, A S (2017) *A New Volcanic Province: An Inventory of Subglacial Volcanoes in West Antarctica, Geological Society, London, Special Publications* 461: 231–24, https://doi.org/10.1144/SP461.7.

Those of us who were left commenced packing to retrograde our equipment and in particular the vital records to Denmark and the UK. There was quite a volume of the latter, and all had to be carefully documented. We were anticipating leaving a trunk in McMurdo for the following season (1979–80). Meldrum discussed with Hickerson modifications to the ARDS to improve its performance based on the season's experience.

12.4 Taylor Glacier Project

The three personnel remaining in McMurdo (Meldrum, Perkins, and I) now participated in a short programme of ice sampling in the Dry Valley area in and around Taylor Glacier (Figure 12.16). It will be recalled that in the previous season I instigated the collection of ice cores accompanied by Jankowski. I had submitted a detailed plan to NSF, and they had agreed to provide helicopter support for three days. I continued to be amazed at their capacity that enabled a relatively ad hoc activity to be quickly and effectively expedited.

The project had come about as a result of RES undertaken in 1974–75, recounted in section 10.5. George Denton had acquired an impressive body of knowledge regarding the glacial history of this Ross Sea sector. His work was becoming increasingly important in understanding the role of the changing extent and volume of the Antarctic Ice Sheet in modulating global atmospheric and oceanographic processes and with consequent implications for world sea levels.

One of Denton's assumptions was that fluctuations of Taylor Glacier reflected the changes in the size and flow of the East Antarctic Ice Sheet. I was not entirely convinced of this large-scale extrapolation. I had suggested that a local dome inland of the Dry Valleys could be the source of Taylor Glacier ice, thus restricting the glacier's wider sensitivity (Figure 9.10). To pursue these ideas further and more rigorously I had devised some experiments. By taking ice samples from the lowest and hence oldest layers at the front of Taylor Glacier and comparing their chemical signatures with those of ice taken from the top of the local dome it would be possible to determine whether the ice had been derived locally or from deep within East Antarctica. In addition, a number of samples were taken from glaciers in the region for comparative purposes.

Our preparations entailed a good deal of work collecting the necessary equipment (ice augers, snow saws, shovels, ladder, ice picks, a spring balance,

Figure 12.16. Snout of Taylor Glacier and frozen Lake Bonney in the foreground.

cylinders, altimeters, theodolite and tripods, and tape measures, as well as numerous poly bags and similar items for samples, including cooler boxes). None of these had we brought with us, but I was confident we could be supplied by the expert outfitters at the Berg Field Center.

Our first foray was on 2 January to the Wilson Piedmont, which lies 120 km (75 mi) to the west, at the mouths of the Dry Valleys (Figure 12.17[260]). We were landed by helicopter on the crest, and here we excavated a pit to a depth of 2 m and then cored further down for snow samples (Figure 12.18). We made density measurements using the cylinder and balance. It was a glorious day. Blue sky from horizon to horizon and, importantly, not a breath of wind! We looked out across the sound to Ross Island and the volcanic edifices of Mounts Erebus and Terror and then to the east into Wright Valley and the peaks and ridges of southern Victoria Land disappearing into the distance and beyond the ice sheet. Digging the pit was hard work, and we stripped off our shirts, working bare-chested, a rare occurrence in Antarctica. Having completed and collected our samples we were returned to McMurdo.

260 Discussion with other scientists at McMurdo who were prepared to assist enabled collections from several other locations. I am particularly grateful to Ed Zeller for samples in Wright Valley and the late Terry Hughes from Byrd Glacier.

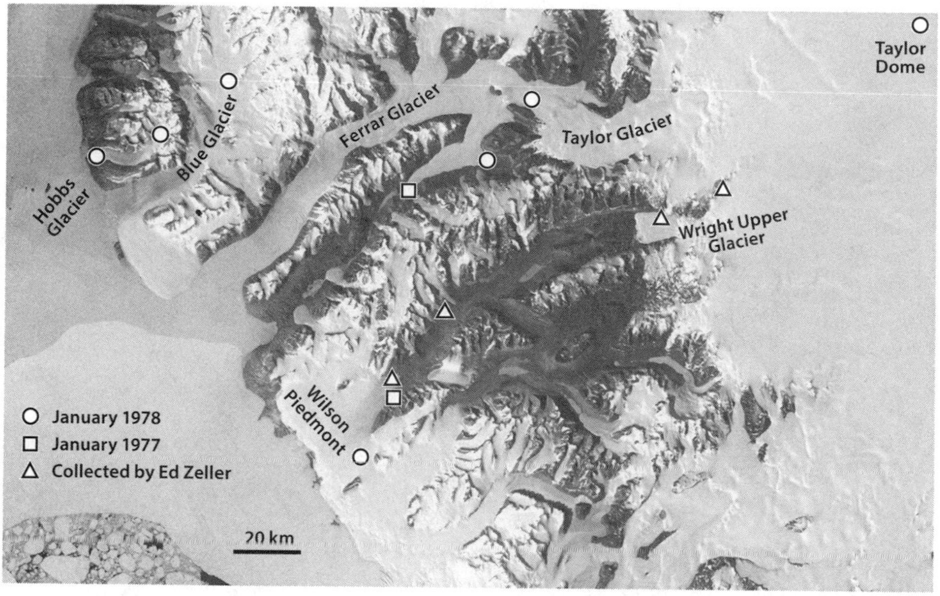

Figure 12.17. Satellite image of the Dry Valleys, southern Victoria Land, showing the ice sampling locations. (Landsat image courtesy US Geological Survey).

Figure 12.18. The author (left) on the crest of Wilson Piedmont Glacier; Mount Bird on Ross Island is in the background. David Perkins is valiantly commencing excavation of a snow pit. It was an extraordinarily sunny and calm day—hence the attire whilst working energetically!

Figure 12.19. The glazed ice surface of Taylor Glacier (exposing Holocene ice) opposite Solitary Rocks, looking up-glacier.

The next day commenced with some ice fog, which delayed departure to the Taylor Glacier. It cleared to reveal another bright and sunny day in which it was a privilege to be working in these breathtaking mountainous surroundings. From here we could access the snout of Taylor Glacier, where we extracted two horizontal cores from the lowest exposed basal layers. These were added to the samples Jankowski and I had already collected.

When the helicopter returned, we flew further up the glacier to an area of bare ice opposite Solitary Rocks and close by the impressive Pandora Spires. Here the down-glacier winds continuously scour the surface, blowing away any fallen snow and eroding some of the surface layers so that deeper and older 'blue' ice is revealed that would have come from further inland. The ice surface was very slick, and the helicopter skidded around as it landed. We had to wear crampons to move over the highly 'polished' surface and give us some traction when we started coring (Figure 12.19). We returned by flying down the adjacent, parallel valley filled with the Ferrar Glacier.

On 4 January we undertook our third and final day of exploring this region. In my notes I recalled the following:

> Today proved the 'pièce de résistance'. Only DTM (Meldrum) and I went, as David Perkins was scheduled to fly back to CHCH. We set

> off at 0930 am and headed out for the edge of the ice sheet about 1½ hours away (and via Victoria Upper Glacier and Webb Valley). Once past the mountains we flew on for another 30 miles to the top of the small ice dome. This would usually lie beyond the area the helicopters operate. We landed to take a 4 m core. It was cold on the plateau edge at 8000 feet (approximately 2450 m) with a light wind. The physical effort of coring proved the only way to keep warm! The snow was so soft, the helo's skids sank in well and truly and all of us had to jump up and down to release them. Coming back, we lost our way a little (fortunately for us as we had some of the most stupendous views of mountains you could wish for). We called in for lunch at the summer camp run by the Kiwis in Wright Valley—Vanda Station. There we had soup, tea and hot buttered scones! Then we pressed on with our work, to the bottom of Wright Valley to refuel and thereafter to the head of Blue Glacier which lies in an awe-inspiring place at the very foot of the Royal Society Range. More coring (in the névé area) and then to Hobbs Glacier where we took a core in the upper (névé) part and afterwards dropped down to the snout to be left for a couple of hours in the warm sun (Figure 12.20). We took more cores at the foot of the spectacular ice cliff and made a short reconnaissance before it was time to be picked up and go back to McMurdo. What a day. With such fabulous weather it was truly an epic of unimaginable scenery. A fitting end to our exploits.

The outcome of this expedition-within-an-expedition over three seasons was most satisfactory. The ice samples (see Figure 12.17 for locations) were sent off to laboratories in Grenoble, France (Laboratoire de Glaciologie), where my good colleague Claude Lorius had agreed to undertake the deuterium assays, and to the British Geological Survey in London for the oxygen isotopes. The isotopic signatures demonstrated that the basal ice of Taylor Glacier could not have had its origins deep in central East Antarctica but, rather, in the local Taylor Dome. Furthermore, the data suggested strongly that the glacier has been in near isotopic equilibrium for the lifetime of the oldest ice sampled, which would be several thousand years.

I also drew on information from the meteorites collected in the Allan Hills area on the inland edge of the Transantarctic Mountains. These meteorites would have been flushed into adjacent outlet glaciers should the ice sheet have thickened significantly, as there is only a low topographical

Figure 12.20. David Meldrum and the author taking a snack beneath the front of Hobbs Glacier. All the baggage is emergency survival equipment should there be delays with the helicopter pickup.

barrier in this locality. I calculated that the time to accumulate the approximately 50 'parent' meteorites in the vicinity of the nunataks was between 17 and 33 ka, and thus the ice sheet could not have thickened substantially during this period. These findings were published in *Nature*[261] and fit with later studies that suggest the glacier extended almost to the eastern end of the valley between 70 and 100 ka ago, whence it retreated. Recent analyses of ice along Taylor Glacier surface suggest a period of stability during the Holocene.[262] I emphasized it was important to separate out the interaction of the East Antarctic Ice sheet with the local Taylor Dome and Taylor Glacier from the ice fluctuations of the Ross Ice Shelf. Later investigations have amply demonstrated that the Ross Sea was filled with an enlarged ice sheet

[261] Drewry, D J (1980) Pleistocene bimodal response of Antarctic ice, *Nature* 287 (5779): 214–16.

[262] Higgins, S M; Hendy, C H; and Denton, G H (2000) Geochronology of Bonney Drift, Taylor Valley, Antarctica: Evidence for interglacial expansions of Taylor Glacier, *Geografiska Annaler Series A, Physical Geography* (82):391–409, https://doi.org/10.1111/j.0435-3676.2000.00130.x; The RAISED Consortium, Bentley, M J; et al. (2014) A community-based geological reconstruction of Antarctic Ice Sheet deglaciations since the Last Glacial Maximum, *Quaternary Science Reviews* 100:1–9.

reaching out to the edge of the continental shelf at the Last Glacial Maximum. I have already mentioned that my own reconstruction of that region was incorrect but perhaps illustrated a stage during the rapid retreat and collapse of this ice sheet. The Taylor Dome proved an interesting site and a core to a depth of 554 m was drilled there in 1994. Its interpretation has proved somewhat complex owing to a strong accumulation gradient across the dome.[263]

The completion of this project, investigating questions arising from the airborne campaigns, ended the 1978–79 season. To the personnel returning to Cambridge that northern spring, all the signals suggested one further season of RES would be undertaken in the following year.

[263] Grootes, P M; Steig, E J, Stuiver, M; Waddington, E D; Morse, D L (1994) A new ice core record from Taylor Dome, Antarctica. *EOS Transactions* 75:225; Baggenstos, D (2015) 'Taylor Glacier as an Archive of Ancient Ice for Large-Volume Samples: Chronology, Gases, Dust, and Climate', PhD dissertation, Scripps Institute of Oceanography, University of California San Diego.

13

The Axe Falls

13.1 A Telegram Arrives

In late February 1979 Gordon Robin received a telegram from the NSF indicating that the agency was withdrawing its logistics support for the SPRI-TUD programme. This was a severe blow. The reasons for its decision were only partially explained and referred to possible radio-echo sounding from the space shuttle.[264]

For some time, it had been recognised that the SPRI programme was unlikely to extend beyond one more season, 1979–80, for several reasons. Any research project has a finite life, and new ventures will need to be funded and supported. Second, we were coming towards the end of what could be accomplished in terms of geographical coverage using the range capability of the C-130 out of McMurdo. Unless we could operate from a more distant research station, we would be unable to extend our grid by very much. This plan was unlikely to be supported due to the logistics implications, cost of fuel, and aircraft maintenance coincident with the squeeze on the NSF budget. Third, other countries were developing RES capabilities and would be able to fill in the regions from their bases the SPRI could not cover. We welcomed this evolution, as it would stimulate the science and lead to the eventual full coverage of the continent by RES. Fourth, as we were soon to discover, pressures within the DPP/NSF were growing to move away from the existing regime led by the SPRI and TUD towards a greater US involve-

264 This suggestion was considered fanciful and irrelevant. The SPRI was engaged in exploring the potential of using space-borne radar systems for surface mapping of ice sheets and was commencing a collaboration with other groups in the UK. It had already been concluded that they were unlikely to replace airborne sounding for measuring ice thicknesses (due to radio frequency, power, and geometry), or for flowline and individual glacier investigations. This remains true today.

ment. Besides the strong sentiments being expressed by some American researchers, we had already picked up that NSF was keen to develop its own RES facility. As in any scientific field, over time new techniques will be developed further. We learned that an 'informal' all-US meeting funded by the DPP had been convened in April of 1978 at the University of New Hampshire with some 40 scientists participating. The present state of the art of RES systems was reviewed, and the report of the meeting noted: 'After discussions on the requirements of the users, the workshop concluded with recommendations on the possible specifications of future systems'.[265] Fifth, and linked to the last point, there had been disagreements over access to and sharing of the RES data. This matter had come to a head in 1978 when there was a round of correspondence between Gordon Robin and Dwayne Anderson at NSF concerning availability of the magnetic data collected in the 1977–78 season. Robin argued that unless the magnetics data were released to SPRI for evaluation prior to the imminent next season (1978–79), SPRI would be unable to supply RES data to other investigators. The fact was that the two datasets were inextricably linked; the magnetics data required the ice-thickness and elevation data from the radar soundings to be of any value. From the SPRI perspective the programme in 1977–78 had been planned on the basis that the combined datasets would be available to SPRI scientists for interpretation. The flights had been designed by SPRI to utilise the combined geophysical package and flown by SPRI personnel in cooperation with the APL technicians (who were there to collect but not interpret the data). We have seen that for the subsequent 1978–79 season, Director Ed Todd had agreed to this access (section 11.9). NSF, meanwhile, had come under increasing pressure to ensure ready and easy availability of the RES data. SPRI had committed itself earlier to this principle, albeit stating that in the first instance researchers should visit the SPRI and undertake part or all of their initial analysis in Cambridge, where reduction facilities and expertise were on hand to assist. Regardless of these provisos, copies of the Honeywell radio-echo sounding traces were microfilmed in Christchurch at the end of season, with one set despatched to the NSF for its use.

[265] Sivaprasad, K (ed.) (1978) *Report of Radio-echo Sounding Workshop*, Department of Electrical and Computer Engineering, University of New Hampshire, Durham, 15pp.

13.1.1 Data, Access, and Political Myopia

Data access issues affecting the relations between SPRI and NSF were discussed in some detail in a paper by Dean and others in 2008.[266] They pointed to fundamental and philosophical differences of approach between the NSF, which saw itself as the distributor of RES data to interested parties, and the SPRI, that fostered collaboration on data but undertaken at or through the Institute. They commented: 'The data-access models put forward by both SPRI and NSF can thus be understood as strategic responses on the part of institutions and scientists to international data networks. It seems that in each case the approach to data taken by the NSF and SPRI can be situated in broader governmental contexts.'

On reflection it is quite clear that the earlier agreement between the NSF and the SPRI was underpinned by different but unconscious perceptions of each party. NSF, lying within the ambit of and accountable to government, was a funder and coordinator of programmes and supplier of data; the SPRI, as an active research centre within the independent university sector, maintained an approach predicated upon the acquisition and interpretation of data (where appropriate in collaboration with any third party), followed by publication. The SPRI had agreed fully to the principle of sharing and or supplying data, albeit after the usual period for principal investigators to undertake their initial analyses. It was not, however, well set up organisationally, logistically, or temperamentally to supply quantities of RES data to meet requests. The institution lacked the handling facilities to duplicate and despatch substantial amounts of material. These would require photographic replication with the addition of considerable explanatory texts detailing the RES parameters and information of the codings on the films. Furthermore, the data needed to be linked to the navigation files with comparable explanations. Ice-thickness data already digitised were easier to provide but still necessitated extraction from a non-too-robust database and with corresponding notes. There were no spare staff to undertake these preparatory tasks, and such 'chores' would fall to research students or research assistants to accomplish in between other work. In some cases, requests, particularly for material of lesser interest to SPRI's own principal investigators, were seen as a distraction to the primary focus to push forward its own research.

[266] Dean, K; et al. 2008 (footnote 223).

Cambridge was and remains a fiercely competitive intellectual environment, and individual and institutional goals were pursued single-mindedly and with tenacity. As a very small 'department' within a Cambridge faculty comprising several 'big hitters', such as engineering, chemistry, and the Cavendish Laboratory (physics), the Institute was highly focussed on its research, obtaining grant funding, and publications.[267] In respect of the agreement with NSF, the SPRI approach was uneven and perfunctory at best. The lack of attention to the important consequences of these shortcomings was an issue that led the NSF to cease supporting the programme—a case of institutional myopia. Dean and her colleagues argue that the SPRI's behaviour and responses were based on an 'imperial' legacy. It is my view that the high-octane and competitive university environment and the drive for research leadership (as prevalent in other countries as in Britain), along with the practical constraints described above, were the primary causes.

It should be recorded that requests for data were fulfilled—Terry Hughes at Maine received the ice thicknesses for Byrd Glacier, Ian Whillans was provided with the data along the custom-flown Byrd Station Strain Network, Charles Bentley had data for both the Ross Ice Shelf and Dome C, Hal Borns at Maine got the profiles of the Wilson Piedmont Glacier, and Claude Lorius in Grenoble was sent the RES results along the traverse from Dumont D'Urville to Dome C. It should not be overlooked that there were also several requests that resulted in collaboration and joint publications between SPRI and US scientists (e.g. studies of the Filchner-Ronne Ice Shelves, Lake Vostok, Dufek Massif, Rennick Glacier, West Antarctic and its ice streams, and the Erebus Ice Tongue). On reflection, however, the numbers were small, the cases selective, and the overall pattern inconsistent.

The various issues and developments that conspired to end the Antarctic RES activities occupied our minds. Nevertheless, the inevitable was accepted when it arrived, although perhaps not the method by which the message was delivered. Ian Whillans, at The Ohio State University and one of the leading glaciologists in the US at that time wrote to me upon learning of the situation: 'I am very sorry to hear about the termination of the R/E program. . . . This represents the further dismemberment of the US field effort on polar ice sheets. What remains now are investigations of 100 m ice cores, studies in the French Dome C core, and some snow pit work. The

[267] Remarkably, the SPRI performed extremely well on metrics of research funding per staff member.

Table 13.1. Summary of NSF-SPRI-TUD Antarctic RES Missions

Antarctic Season	Missions	Flight Time (hrs)	Flight Distance (km)
1967–68	10	94	30,000
1969–70	29	332	120,000
1971–72	16	162	60,000
1974–75	50	320	135,000
1977–78	13	141.5	25,000*
1978–79	11	130	26,000*
Total	**129**	**1052**	**396,000**

*On-station distance

announcement was made on May 17; I haven't found a single scientist who was consulted on the decision. . . . The decision is, of course, ruinous to our Dome C plans. I haven't anything good to say, I just wish to pass on sympathy'.[268] His comments suggest that there were other factors in play that affected the wider support being given to glaciology in the NSF-run Antarctic programme.

At the SPRI attention turned to ways to capitalise on the enormous scientific value of what had already been collected. Over the six seasons there had been almost 130 RES missions totalling in excess of 1000 hours flown across the continent and acquiring 400,000 line-kilometres of valuable data (Table 13.1).

An early plan had been to synthesise much of the resulting information in map and diagram form. We were sufficiently seized by this concept to publish a short note explaining our intentions in the *Polar Record* and the *Antarctic Journal of the United States*.[269] We reported that material was being amalgamated for a comprehensive Antarctic Glaciological and Geophysical Folio, with an expected publication date in 1982. Ambitiously we proposed that the compilation would be presented in a large-scale atlas format, containing upwards of 24 four-colour sheets depicting parameters derived directly or interpreted principally from measured RES and magnetic data.[270]

268 Letter to D J Drewry from I Whillans, 22 June 1979.

269 Drewry, D J; and Jordan, S R (1980) Compilation of an Antarctic glaciological and geophysical folio, *Polar Record* 20 (126): 288; Drewry, D J; and Jordan, S R (1980) Compilation of Antarctic glaciological and geophysical folio, *Antarctic Journal of the United States* 15 (5): 224–25.

270 Drewry, D J; and Jordan, S R (1980) (footnote 269).

Map scales would vary depending on the quantity of data, and sheets would include tabulated and graphical information and cross profiles. There would be explanatory text of approximately 2000 words and references accompanying each map. It was an important statement of our intended direction of travel, but, as we shall see, there were many twists and turns before the project was completed, and not in its entirety. That 24 sheets could be produced was soon discovered to be exceedingly optimistic, and a more rational and achievable goal was collapsed into half the number.

At more or less the same time in February as the NSF telegram was received, Robin and I were sent invitations to attend a meeting in Washington, 28 and 29 March, convened by the NSF to discuss long-range plans for airborne magnetometry in Antarctica and related oceanographic research. This seemed ironic, but since our American friends 'don't do irony', we considered it was a means of obtaining our input and experience for their future plans. It was obvious that ice-depth sounding was inextricably linked to any continental magnetometry campaigns and would need to be discussed at the meeting. It was decided that only one of us should attend—I was there for the SPRI and determined to extract a more detailed explanation of NSF's actions regarding the termination of the RES programme and its possible replacement.

I arrived in Washington the day before and arranged to meet with Dick Cameron and Gisella Dreschhoff for dinner. Over our meal I learned that our potential 1979–80 Antarctic season had been cancelled owing to a very heavy commitment on C-130 hours and heard the same story that NSF was exploring undertaking radar sounding from the space shuttle. The next day I learned more about what was developing in DPP when I met for a side conversation with Dick Cameron and Director Ed Todd. The circumstances regarding the RES programme had, however, to be sweated out of Cameron with the admission that NSF was seeking to develop a US RES capability—possibly a fully digital system—and that $2000 had been allocated to APL to undertake a 2- to 3-year feasibility study. The notion was to use the existing C-130 antenna configuration (that NSF had paid for). Cameron could not confirm whether any TUD equipment might be incorporated. What was left in the balance was whether there would be any role for SPRI and TUD in the future. My own view was that the writing was clearly written on the wall—time to move on!

13.2 NSF Magnetics Meeting

The consultation on magnetics kicked off in the morning of Wednesday, 28 March. A large number of DPP staff, including Todd, Cameron, Mort Turner (lead for geology), were present, as was Gisela Dreschhoff. VXE-6 air operations staff and a cohort of academic representatives from the US Antarctic geosciences and oceanographic communities also were in attendance. These included John Behrendt from the USGS and John LeBreque and Jim Heirtzler from Lamont-Doherty Earth Observatory of Columbia University and Woods Hole Oceanographic Institution, respectively. Charles Swithinbank attended for the BAS and was the only other UK attendee. Denmark had sent Leif Thorning from the Greenland Geological Survey. Invited but absent were Preben Gudmandsen (TUD) and Charles Bentley (University of Wisconsin).

It emerged that there were two camps vying to control any future magnetics project; this was not a great surprise, as it fell along science lines—the one led by John Behrendt, and the marine group led by Heirtzler and LeBreque. A working group was proposed but despite calling the meeting, DPP prevaricated; no commitments could be made for at least 2–3 years. This left a number of the participants frustrated. The whole matter of the fate of or future developments in RES (essential for any over-ice campaigns) was never debated.

The day after these desultory discussions I had a full diary of engagements and contacts to chase up in Washington and by telephone. Terry Hughes rang in the morning to invite me to Maine, as he did Charles Swithinbank, to discuss Byrd Glacier, but I had to decline owing to my pre-existing schedule. I needed to confer with Bill Kosco of the USGS in Reston, Virginia, on the aerial photos taken during our last season. I was also keen to talk to Behrendt about data. I left later that day to fly back to the UK and Cambridge to consider the situation with Robin. Much had occurred.

13.3 New Initiatives, New Opportunities

The frontier of science is propelled ever forward; to stand still is to fall behind. Where one avenue closes off, other pathways emerge. So, it was to be with the SPRI radio-echo programme. We had gained and were continuing to acquire unprecedented insights into the characteristics and behaviour of the great ice sheets in Antarctica. Each new foray south had stimulated original thinking, often challenging earlier ideas. Canny thoughts had created innovative techniques to unlock fresh perspectives, and new puzzles appeared as we pored over the results. We had harnessed the methodologies developed in other research fields to our particular ends. Nevertheless, we could now discern technological opportunities to extend and deepen our study of the polar ice masses.

As the Antarctic door closed for the medium term we were keen to exploit our expertise by seeking interesting scientific questions in other regions and at different scales. Thus it was that three important and exciting developments emerged during the course of 1980—first was constructing new radar sounders, for mostly ground-based activity, to reveal the fine detail of layering in the upper region of the ice; second was to engage actively in the embryonic area of satellite remote sensing of ice masses; and third was the opportunity to study smaller glaciers and ice caps in the Arctic that exhibited distinctive, surging behaviour—perhaps as a model for the future unstable response of West Antarctica to climate warming.

13.3.1 Satellite Studies of Polar Ice

Scientists at the Rutherford Appleton Laboratory (RAL), then located in Slough, to the immediate west of London, approached us to seek assistance in a study examining the potential contribution of satellite radar altimetry to climate, oceanographic, and glaciological research. The SPRI capability in RES was seen as relevant, as well as its expertise in the study of sea ice in one of our other groups, led by Peter Wadhams. Andrew Cowan, a member of that team who was engaged in remote-sensing aspects of sea ice, joined me to work with RAL on this initial assessment. It was the beginning of a new research frontier for SPRI; it is explored further in chapter 16.

13.3.2 Svalbard

Olav Orheim at the Norsk Polarinstitutt (NP), based at that time in Oslo, Norway, telephoned early in 1979 to enquire whether it would be possible to use our RES equipment to investigate glaciers in Svalbard. I had known Olav for some years; he was a glaciologist and had worked in Antarctica for a long period. We had met on several occasions and enjoyed each other's company. I considered his proposal very seriously and arranged that we should get together to discuss this matter later in the year and fixed a date in June in Oslo. Here was another door opening. I reflected on some notes

I had made prior to departing for the 1978–79 Antarctic season, which looked into the future. I had written: 'I think we should try to expand our fieldwork base and develop some independence of NSF; e.g. work in the North Polar areas—is Spitsbergen a possibility?'

In June 1979 I travelled to Oslo to meet with Orheim and review the putative project in Spitsbergen. I recall vividly celebrating Midsummer's Day at Orheim's home in the suburbs of Oslo prior to the meeting. How we had the strength or sobriety the next day to undertake the task was a puzzle, but we did. Orheim suggested we could collaborate and plan for a season in May 1980 using helicopters. The NP and other university groups in Norway were studying several glaciers. Some of these were surge type, and it was considered important to detect the presence of basal water. Some were tidewater glaciers, and the location of the grounding line would be a key boundary to be detected.

We scrutinised aerial photos and discussed sounding possibilities for several of these ice masses. We relied on the expertise of Tore Siggerud from the logistics side of the NP in considering the likely operational capabilities of the helicopter (Bell 206 JetRanger) and navigational requirements. Orheim thought that 30 hours of flying might be available. The project would be fully collaborative. SPRI would fund the RES activity, as well as travel and subsistence. The NP would organise the helicopter hire, accommodation in Longyearbyen through its links with the coal mining company SNSK that ran the bulk of the facilities in the small mining town, and liaison with the Norwegian governor on Svalbard (Sysselman). This was an exciting opportunity to keep our RES team together, maintain field operations, and switch to a northern area of investigation with new challenges.

By early 1980 we were in the thick of preparations for the Svalbard field programme. Chris Neal had made excellent headway with the technical preparations. We would be using the SPRI MkIV 60 MHz system; it would fit inside the Bell 206 JetRanger along with one monitoring and two recording oscilloscopes. A simple dipole antenna would be fitted externally on and parallel to the starboard float (Figure 13.1). By using the short-pulse option (250 ns), we would be better able to sound the thinner and warmer ice of the Spitsbergen glaciers. We reckoned the system sensitivity would be about 191dB with a long pulse (1000 ns) but somewhat lower with the shorter pulse.[271] In any case this was to be an exploratory season commencing in

[271] Nominal system sensitivity (170 dB), antenna gain (+24 dB), transmit/receive switch loss (−1 dB), cable loss (−2 dB) = total 190 dB.

Figure 13.1. Bell 206 JetRanger at Longyearbyen Airport, Svalbard. The 60 MHz RES antenna is located offset from the float.

Figure 13.2. Discussion at Barentsburg, the Soviet coal-mining village. Facing the camera, left to right: Olav Orheim, Olav Liestøl, LS Troitsky, and Yevgeny Zinger (the latter two from the Institute for Geography in Moscow). Bjorn Wold (nearest right) has his back to the camera.

late April. The trip would involve only Neal and me from SPRI, along with Orheim and his colleague Olav Liestøl from the NP (Figure 13.2).

We completed 10 missions, totalling 20½ hours, out of Longyearbyen with some refuelling at Ny-Ålesund and at the Polish Research Station at Hornsund (Figure 13.3), and sounded 45 glaciers. The SPRI system worked

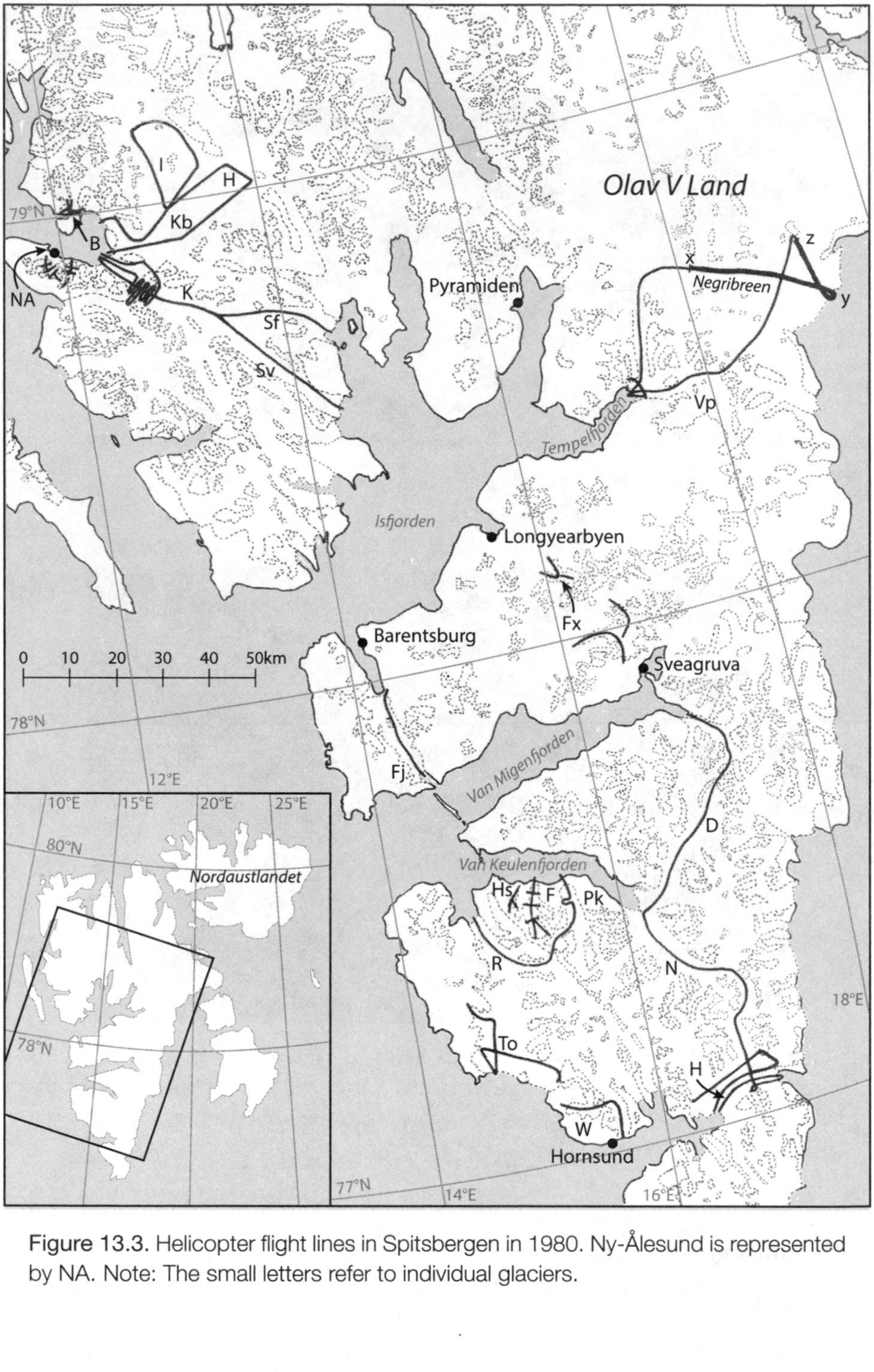

Figure 13.3. Helicopter flight lines in Spitsbergen in 1980. Ny-Ålesund is represented by NA. Note: The small letters refer to individual glaciers.

effectively from the helicopter, and our Norwegian colleagues were well pleased with the results. These were reported in detail in a series of papers and formed part of the material to be worked on by two new research students who joined the SPRI during the following three years—Julian Dowdeswell (1981) and Jonathan Bamber (1983).[272] Neal and I had learned a great deal regarding the glaciology of Svalbard and identified some potential future opportunities. We had, on an entirely aesthetic level, been captivated by the stark mountainous scenery of Spitsbergen. We would be back.

13.3.3 Short-Pulse Radar

Chris Neal was interested in developing an impulse radar to investigate the relationship between RES layers and the stratigraphy in the upper part of ice sheets, particularly as revealed in ice cores where there were physical (in particular, density information) and chemical studies available to explore the linkages. In late 1979 the overall concept had been scoped; it would operate at 1 GHz with a resolution of between 100 and 200 mm. Materials were being ordered, and Neal and Michael Gorman (who had returned to work at the Institute in late 1980) would commence to design the radar unit and antenna. It was hoped to test the system in the European Alps before deploying it to a suitable Arctic location.

The Royal Society in London was contacted, as it was party to an agreement to make use of the Swiss Research Station located at the Jungfraujoch, at the head of the largest glacier in the Alps, the Aletsch Gletscher.[273] It was readily agreed that we could avail ourselves of their facilities. Chris Neal and Michael Gorman completed the construction of the sounder with the assistance of Chris Hereward in February 1981. Alongside other equipment it was loaded into a hired van and driven via the Dover ferry and across France to Switzerland on 14 March. At Kleine Scheidegg the funicular took us and the equipment up to the adjacent research station, a massive stone construction at 3454 m above sea level with the spectacular backdrop of the mountains of the Bernese Oberland—Jungfrau, Eiger and Mönch. The living accommodation

[272] Drewry, D J; et al. (1980) Airborne radio-echo sounding of glaciers in Svalbard. *Polar Record* 20 (126): 261–75; Dowdeswell, J A; et al. (1984) *Airborne Radio Echo Sounding of Sub-Polar Glaciers in Spitsbergen*, Oslo: Norsk Polarinstitutt, Skrifter Nr. 182, 41pp; Bamber, J L (1987) Internal reflecting horizons in Spitsbergen glaciers. *Annals of Glaciology* 9:1–6.

[273] Hochalpine Forschungsstation Jungfraujoch. We also received assistance from Eidgenössische Technische Hochschule (ETH) in Zürich.

Figure 13.4. The impulse sounder operating on the Jungfraufirn, Switzerland. It is being winched across the snow surface by the author. The cable (right) leads to the control unit and recorder.

and research facilities were extremely comfortable, and there was laboratory space for us to undertake any preparation and modification work. A door led out of the rear of the labs onto the high firn basin of the Upper Aletsch, where we could undertake our tests—Jungfraufirn (Figure 13.4). The snowfall here was relatively high (approximately 2 m a^{-1}), with some summer melting, which we believed would produce a number of distinguishable layers in the upper few tens of metres.

We also needed to sound through denser ice and were given permission to run our sledge along the corridors of the 'Ice Palace'. This is a public area for tourists and comprises a series of corridors and chambers carved into the solid ice of the glacier. An interesting feature was a layer of ash in the ice resulting from the destruction by fire of the hotel that occupied part of the site in October 1972. We hoped this would be observable on the radar records.

We spent our few days undertaking the experiments, but we could use the Ice Palace only after the public had left and the last funicular had departed the Jungfraujoch. This meant we had to work through the night in the deserted and slightly eerie tunnels. On one occasion, well after midnight, I had to return to the laboratory to collect additional equipment, and I thought I heard a noise but recalled that only the three of us were in resi-

Figure 13.5. Paul Cooper demonstrating the 300 MHz backpack RES developed at the SPRI. The antenna is on a harness on his back; the electronics and readout are in front.

dence. I walked on through the dimly lit ice corridor, when suddenly a man jumped out in front of me from a cross passage! To say I was startled was an understatement. The man was a Japanese tourist who had become disoriented and had missed the last funicular back to Kleine Scheidegg, now many hours previously. He was cold, frightened, and confused. We later took him back to the research station, found him some food and a bed, and returned him to the funicular the next morning.

The trip proved worthwhile, and we were able to distinguish what we believed were annual layers and the ash horizon. This was sufficient for Neal to travel to Canada later in May, accompanied by Neil McIntyre and Julian Paren, for similar experiments on the Agassiz Ice Cap.[274] McIntyre had joined the RES group as a research student in October 1980 to investigate details of the surface morphology of the Antarctic Ice Sheet to deduce basal conditions and the characteristics of ice flow.

Another initiative was to construct a basic radio-echo sounding device that could be used for small-scale surveys in relatively thin ice and carried by an individual. The 'backpack' sounder was prototyped and is shown in Figure 13.5. It operated at 300 MHz, and was known as IDIOT

[274] Neal, C S (1981) High resolution radio-echo sounding on Ellesmere Island, Northwest Territories, *Polar Record* 21 (130): 61–64.

(Ice Depth Instrument, Operator Transported). The ice thickness could be read directly by using a rotating knob to move a marker on the oscilloscope screen to line up with the return from the bedrock. Although developed at the SPRI, the backpack sounder was for use by Hugo Decleir, based in Ghent, Belgium, who had accompanied the RES team in Antarctica in 1971–72.

13.4 North American Engagements

Early in the New Year of 1980 I made a series of visits to organisations in the US and Canada with two aims, first, to continue a liaison with US colleagues on our RES work and, second, to use the opportunity to seek funding for the emerging Antarctic Folio. My first stop was Washington, DC, and the NSF, where I met with Cameron, Todd, and Frank Williamson (head of science). We had not completely given up on continuing some form of RES activity with them. I discussed a proposal to study ice sheet layering using the high-frequency system being developed by Chris Neal. We were seeking support in January 1981 for ground-based soundings at Byrd and South Pole Stations.[275] In addition, I was able to outline the Folio project and mentioned the work in Svalbard. It was clear that funding for the US Antarctic program was being slashed. At a later meeting in Orono, Maine, in April I heard directly from Williamson that the cut in the overall NSF budget was $125M due to military pay raises and increased fuel costs elsewhere in government, resulting in a reduction of between $6M and $8M in the DPP funds for the fiscal year 1981. As a consequence, there was a proposal to close Siple Station at the end of the following season, lay up the research trawler *Hero*, and reduce other ship time, particularly icebreaker support. I did not give our quite modest proposal much chance and was proved correct.

The next day (5 February), I flew to Denver to liaise with Behrendt on the Dufek Massif work. I stayed six days, using my time to learn more of Behrendt's experience with airborne magnetics and meet some of his colleagues. I also took time to visit the Institute for Arctic and Alpine Research in Boulder.

[275] Drewry, D J; and Neal, C S (1979) Proposal to NSF (Division of Polar Programs), "Ground-based radar studies of layering in the Antarctic Ice Sheet", SPRI, 20pp. In the event this opportunity did not materialize.

I left for Tulsa, Oklahoma, where I was picked up at the airport and taken to Bartlesville, the home of Phillips Petroleum. Here I was involved in discussions on Antarctic geology. In the late 1970s and into the early 1980s, there was increasing interest in the mineral potential of Antarctica. We shall return to discussions with Phillips in support of the Antarctic Folio in the next chapter (14).

Cam Craddock, head of the Department of Geology and Geophysics at the University of Wisconsin, and his students had amassed a comprehensive collection rock samples from West Antarctica. Many were from the scattered nunataks we had observed during our flights over the area in the 1977–78 and 1978–79 seasons. In mid-February I was flying north to Madison, where I was able to make magnetic susceptibility measurements on these rock samples. These would assist Jankowski in modelling the magnetic data along the flight lines to speculate on the sub-ice geology and structures.[276] I also took the opportunity to meet with Charlie Bentley in the Geophysical and Polar Research Center. We were going to provide him and his group with our data from the Ross Ice Shelf. This included new maps of flight lines, data we had upstream of the Crary Ice Rise, as well as any data from 1971–72. Neal's ESM records were to be copied, but their use had to involve Neal, and it was suggested that Bentley send over to Cambridge one of his colleagues. The bottom line was that we were pleased to cooperate and provide these data following our own synthesis.

My final visit of this extended North American tour was to Ottawa. I had meetings planned with the Polar Continental Shelf Programme (PCSP), Environment Canada, and the Canadian Hydrographic Service (CHS)—a division of the science branch of Fisheries and Oceans. Robin and I had been designated coordinators for the topographic detail of the Antarctic sheet of the General Bathymetric Charts of the Oceans (GEBCO) This was a series of definitive charts of all the oceans, and the fifth edition was being produced by the Geoscience Mapping Section of the CHS under David Monahan's direction.[277] On 18 February I met with him to discuss the incorporation of

[276] Many years later in a memorial conference for Craddock, who had died in 2006, Jankowski and I published these susceptibility measurements: Drewry, D J; and Jankowski, E J (2007) 'Magnetic Susceptibility of West Antarctic Rocks', in Cooper, A; Raymond, C; and the 10th ISAES Editorial Team (eds.) *Antarctica; A Keystone in a Changing World*, Online Proceedings for the 10th International Symposium on Antarctic Earth Sciences, Santa Barbara, California, August 6–September 7, 2007; U.S. Geological Survey Open-File Report 2007–1047, http://pubs.usgs.gov/of/2007/1047.

[277] Leonard Johnson of the US Naval Oceanographic office (ONR), and well-known in the SPRI, was scientific coordinator of the Antarctic Sheet.

our data, including a plot of the Antarctic coastline, as the existing GEBCO outline was a complete hodgepodge, with very outdated and inaccurate material. The plan was for this sheet to be finalised and printed later in the year or early 1981.[278]

Shortly afterwards, a meeting had been arranged with Gerry Holdsworth, a glaciologist at Environment Canada. He was working with Stan Jacobs, a US oceanographer from Lamont-Doherty Earth Observatory at Columbia University, and a colleague from Cambridge, Herbert Huppert.[279] They were investigating the Erebus Ice Tongue that flows off the ice-clad slopes of Ross Island and into McMurdo Sound, and their studies were focused on examining the thermodynamics of this and similar extended floating ice features. During the previous season we had gathered RES data for Holdsworth. This was a chance to review progress.

The team had identified step-like thermohaline structures in the water column adjacent to the ice tongue from a US Coast Guard icebreaker operating in McMurdo Sound. It was assumed that these structures were the result of melting at the base of the ice tongue, with the water spreading out laterally in layers with an approximate depth of 17 m. The RES profile also showed two distinct levels and a strong bottom return, suggesting melting. Combining these records, it was possible to investigate the processes in the water column and compare the observed data with theory and laboratory experiments that Huppert had conducted in Cambridge. This study resulted in an interesting paper demonstrating that a very small section of radio-echo sounding can be scientifically extremely beneficial.[280]

At the PCSP I linked up with Stan Paterson, Fritz Koerner, and David Fisher. Paterson had been instrumental in organising the Devon Island field experiments in 1973. We discussed the possibility of RES work in the Canadian Arctic using Neal's new impulse radar and explored a suggestion of sounding on the Agassiz Ice Cap located on the eastern side of Ellesmere Island. The Canadians had very good surface glaciological data and comprehensive physico-chemical studies of the ice core. The plan was set for May 1981.

[278] Drewry, D J; and Robin, G de Q (special coordinators) (1980) *General Bathymetric Chart of the Oceans* (GEBCO), Sheet 5–18, 'Antarctica', scale 1:6M.

[279] Herbert Huppert worked at the Department of Applied Mathematics and Theoretical Physics (DAMPT)

[280] Jacobs, S S; Huppert, H E; Holdsworth, G; and Drewry D J (1981) Thermohaline steps induced by melting beneath the Erebus Ice Tongue, *Journal of Geophysical Research* 86 (C7): 6547–55.

This had been a pretty gruelling visit, but it had been hugely productive. We had maintained good contact with the NSF, moved forward our collaboration with Behrendt and the USGS, obtained valuable magnetic susceptibility data on the rocks in West Antarctica, and had set up the RES fieldwork in Arctic Canada. Importantly, we had opened a dialogue with Phillips Petroleum for financial support of our Antarctic Folio. It was time for this work to commence!

14

The Antarctic Folio

The maps published in 1974 following the 1971–72 RES season had been positively received by the scientific community and provided important information for those working on projects under the banner of the IAGP (see Figure 7.8). In addition to the many papers we had written, the maps had conveyed the power of the RES technique when flown in a regular survey-style pattern. They had revealed a new landscape beneath the ice and a configuration of the ice sheet, including the startling presence of many scores of internal reflecting horizons, which enabled better understanding of its flow and associated dynamics. The following seasons had added massive quantities of data that assisted our colleagues in other institutions around the world engaged in projects such as deep core drilling and ice sheet modelling.

We were, as well, exploring the RES technique itself, learning more about the propagation of radio waves in and the effects of natural ice masses. The ice shelves were revealing their secrets, particularly the processes of melting and freezing at the ice/water interface, and vast lakes had been discovered at the bed deep in the interior of the East Antarctic Ice Sheet. It became apparent that assembling the ice thickness, surface elevation, and bedrock data into map form was a logical and indeed necessary process. But was there more we could accomplish to summarise and present the very large body of information we had obtained? We discussed this amongst the group in SPRI following our return from the 1978–79 season.

From time to time we referred to the modelling study undertaken by our colleagues in Melbourne under the leadership of Bill Budd entitled 'Derived Physical Characteristics of the Antarctic Ice Sheet'.[281] Although it was several years old, it depicted many glaciological parameters in map form on

[281] Budd, W F; Jenssen, D; and Radok, U (1970) *Derived Physical Characteristics of the Antarctic Ice Sheet*, ANARE Interim Reports 120, Antarctic Division, Hobart, 178pp.

separate sheets, but only at A4 size. A similar set of maps could be produced, we argued, but rather than being *derived* they would be the *measured* physical characteristics, and we could publish them at much larger scale commensurate with the information we had obtained. Another analogue for a possible format was the American Geographical Society *Antarctic Map Folio Series*, published between 1964 and 1975. It had several folios devoted to glaciology, geology, and geophysics, although the contouring and synthetic profiles they depicted were based on sparse data. We also cast our eyes over the elderly but impressive Soviet *Atlas Antarktiki*. This was a substantial volume, dating from 1966, that covered an enormous span of geographical and scientific topics, and the maps were of high quality. However, the maps were at small scale and bound into the atlas, which allowed their perusal but effectively precluded their use as research material. In the end there was little doubt that a large-scale loose-leaf format would most likely be the final choice for our own production.

Alongside these plans for a future folio the SPRI was a hive of activity. Hargreaves had already submitted his PhD thesis in 1977 and had left. Rose had submitted in October 1978. Jankowski was busy with the reduction of the West Antarctic RES and magnetic data. John Behrendt made a welcome visit to the SPRI while routing back from business in Saudi Arabia. We held a brief but productive meeting on 22 January that set the course for our collaboration on both the Dufek Massif data and other West Antarctic results during the next year and was a prelude to Jankowski visiting the USGS to work with Behrendt in the US. The worked-up aeromagnetic and RES data for the Dufek Massif area are shown in Figure 14.1[282]; they gave a compelling picture of the much greater extent of the igneous body extending beneath the inner Filchner Ice Shelf. Millar was commencing his studies in earnest on layering. Chris Neal was completing his thesis, as well as writing papers on the Ross Ice Shelf. Robin was now occupied with organising a major workshop on the climate record in polar ice sheets to be held at the SPRI and editing the volume of proceedings. I was working actively on the folio concept, as well as completing several papers—a review of the radio-echo sounding technique and its high-level results, contributing to the sub-ice topographic details on the GEBCO sheet, and engaged in determining

[282] Behrendt, J C; Drewry, D J; Jankowski, E J; and Grim, MS (1980) Aeromagnetic and radio-echo ice-sounding measurements show much greater area of the Dufek Intrusion, Antarctica, *Science* 209 (4460): 1014–17.

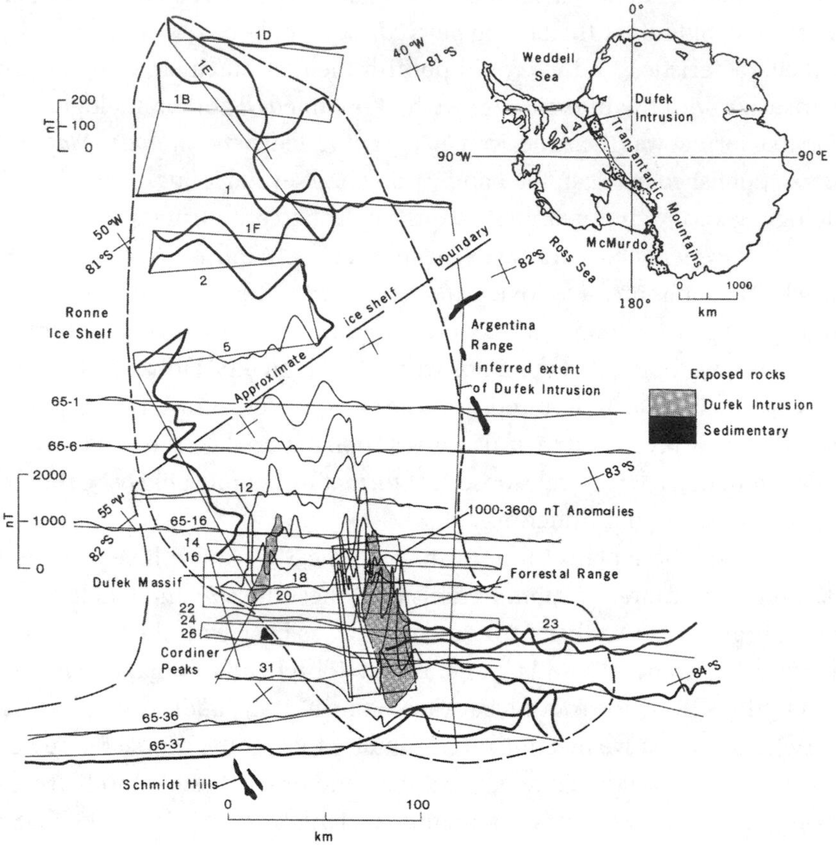

Figure 14.1. Aeromagnetic profiles across the Dufek Massif and region. The anomalies associated with the Dufek intrusion continue into the Filchner-Ronne Ice Shelf. (From Behrendt at al. (1980); see footnote 282). (Reprinted with permission from AAAS).

the glaciological dynamics of the edge of the East Antarctic Ice Sheet adjacent to the Dry Valleys from the ice cores and radar soundings we had taken at the Taylor and adjacent glaciers.

14.1 Developing the Portfolio

SPRI scientific activity, particularly the development of RES and our fruitful campaigns in Antarctica, was funded by a substantial research grant from the Natural Environment Research Council (NERC) that had been running

Figure 14.2. Sue Jordan working in a 'corner' of the SPRI Map Room. (Courtesy S R Jordan).

with several renewals for a dozen years.[283] The current grant period would expire at the end of July 1979. We had seen the need to apply for additional funds to continue our work, and Robin and I had expended substantial effort in preparing a new and scientifically compelling application. It stressed the need for resources to execute the reduction and publishing of the results. Happily, we were successful, and we gained additional resources that provided for up to five staff—a research associate (Chris Neal), a research assistant/coordinator, a project computer programmer to work on the data reduction and database management, and a technician (Hereward).[284] The two new positions were vital for our folio plans, as well as for the long-term archiving of data. Regarding the first of these, we were delighted to offer the position to Sue Jordan, a Liverpool University graduate in geology. She joined the team in October 1979 after gaining experience in seismic data processing at Geophysical Services International (GSI) and seismic data interpretation as a geophysicist at Texaco UK. Sue, an infectiously positive and zealous researcher, was to play a pivotal role in the compilation of material for the folio, high-level drafting, liaison with the cartographers, and its final production (Figure 14.2).

It was not long afterwards that Paul Cooper arrived to swell the team as data manager, with a particular responsibility for the database and data presentation for the folio (Figure 13.5). Paul had studied geology at Cambridge,

[283] The research grants from NERC had commenced in August 1967 as follows: 1967–1970 (£30,600); 1970–1973 (£35,223); 1973–1976 (£54,439); and 1976–1979 (£97,534).

[284] This extension was £100,221, running from 1979 to 1983.

and at Churchill College knew Ian Holyer, who was already working at the SPRI. Paul, on leaving Cambridge, had gone to undertake data analysis at geophysical companies and joined the SPRI from Kestrel Data Services in 1979. He was to prove exceptionally dedicated and effective and continued in the RES team after the folio was completed, working in the Svalbard campaigns. Later, both he and Sue Jordan moved to the BAS Mapping and Geographic Information Centre (MAGIC). Cooper and Jordan commenced getting acquainted with the RES programme, and the existing 'database' had to be handed over from Ian Holyer. Digitisation was a high priority to incorporate the latest RES records. The remaining work was undertaken by Hugh Steed, Ed Jankowski, and David Millar.

By the end of the year (1979) activity on data analysis had moved on considerably, as had the concept and possible format of the glaciological and geophysical compilation; I had opened a new notebook entitled 'Antarctic Map Folio' and was holding regular discussions with Cooper and Jordan, primarily to scope the project. A host of practical considerations began to emerge—understandable, as we were to produce the latest and the most detailed large-scale mapping for an entire continent. First was securing funding for any map-making, as this was not covered in the NERC grant. Second was choosing a cartographic company to undertake the layouts and line drawing, and to oversee the production and printing of the maps. Third was deciding the size or scale of maps to be adopted and the number of sheets. Would the maps include text and diagrams? The question of presentation still had to be finalised—in a bound book, loose sheets, in a binder? These questions surfaced relentlessly. It was clear the general layout of a folio and the contents had to be resolved as quickly as possible if we were to engage organisations with clarity and confidence with the aim of raising funds for its production. We needed further professional advice.

Two well-established cartographic companies in the UK were John Bartholomew and Sons and David L Fryer & Co. I held initial discussions with both of these. The first task was to establish whether David Fryer was interested in this new and much larger venture. His company had undertaken the cartographic work for the SPRI 1974 maps in a most proficient manner, and we were well pleased with the outcome. The folio was going to be a much larger and demanding project, but I suspected the company's experience with and empathy for a further Antarctic undertaking would count significantly. On 26 November 1979 I telephoned David Fryer at his offices in Henley-on-Thames, and we had a wide-ranging conversation in which I sought his ideas and suggestions on realising our vision. He suggested several options. One was to approach a large-circulation publisher of 'popular' atlases that would include 'many pictures', such as *Reader's Digest* or *The Times*. In this way, he commented, we would probably get our scientific maps printed and maybe make some money, but we would lose much control of the project. The second option was to engage a more specialised publisher such as Phillips or Bartholomew, but similar issues of editorial leverage would be evident. Fryer went on to consider likely costs. He suggested that for a short print run of 15 sheets and including printing, binding, and marketing, a selling price might be £90 per volume. We conferred over the process, should we undertake the production entirely ourselves. Fryer and I defined four elements: (i) compilation (this would be undertaken at the SPRI and for which we had funds), (ii) editorial (again a SPRI responsibility), (iii) cartography (would have to be undertaken professionally and might cost £10k–15k), and (iv) printing (this would be by an outside contractor and estimated at between £4k and £5k). It was a very helpful discussion, and based on these initial ideas, two short notes were drafted for publication on the folio concept (see chapter 13, footnote 269); they were couched, unashamedly, in ambitious language. The next priority was the funding.

14.2 Funding

The RES grant from the NERC would finance staff and data analysis. To realise our ambition of producing the folio we would still require additional support for the cartographic drawing, production costs, and printing. For these we now turned to the commercial world and in particular the large petroleum companies for sponsorship. In the late 1970s and early 1980s there was increasing interest by the petroleum industry in exploration of the polar regions for their hydrocarbon potential. The discovery of oil on the North Slope of Alaska had been decisive, as had the exploration of oil deposits in the south-central West Siberian Plain and gas to the north. The 1973–74 oil crisis had been a critical factor for many Western governments. The Organization of the Petroleum Exporting Countries (OPEC) declared an oil embargo against nations that were considered to be supporting Israel during the Yom Kippur War. This caused a 'shock' rise in the price of oil from US$3 per barrel to nearly $12 globally—some 400%, with some countries such as the US experiencing even higher prices.[285] A similar shock occurred in 1979

[285] These prices seem, in the third decade of the twenty-first century, unimaginably small, but it was the magnitude of change that caused the economic problems at the time.

as a result of the revolution in Iran, doubling the price of oil, and was followed shortly thereafter by the Iran–Iraq conflict, which compounded the problems and led to, among other outcomes, the global recession.

The turmoil from these situations gave further impetus to the oil industry to seek ever more remote areas and identify future and more secure reserves. Antarctica was now in their sights, albeit long term.[286] It seemed likely that some of the large multinational petroleum companies might be interested in sponsoring our glaciological *and* geophysical folio of maps, as they would be presenting new data about this region. I made an initial list of those that I thought we could approach: Gulf, Shell, Texaco, British Petroleum, and Atlantic Richfield. I had RTZ written down for their possible mineral interests. Recall from chapter 12 that the Dufek Massif flight lines of John Behrendt were to some extent stimulated by the rising awareness of mineral potential and the negotiations for a regulatory Antarctic convention.

Phillips had not been featured on my list, but by chance I was contacted by Nick Wright, a former graduate student acquaintance in the Department of Earth Sciences in Cambridge whose research had been in the Arctic and was now working for the company. Phillips wanted a briefing on Antarctic geology and geophysics, so Wright had made contact. I was pretty up front and indicated that while I would be pleased to have discussions, maybe Phillips, in return, would be interested in supporting the compilation of a unique geophysical folio of Antarctica. The seed was sown, and shortly thereafter an invitation arrived to visit Bartlesville in Oklahoma. Luckily, I was able to accept and accommodate it in my visit to the US in February 1980.

Bartlesville, at that time, was a strange place. It was a company town; almost everyone working for Phillips or was dependent upon it. A remarkable feature of the city was the splendid 19-storey Price Tower built in 1956 to a design by Frank Lloyd Wright (and bought by Phillips in 1981) set amongst the low-level offices and business facilities of the company scattered around the centre of the town, many of which were linked by a series of busy underground walkways.

I was introduced to Bob Bird of the Future Ventures Section (FVS) and the head of Exploration and Geophysics (R Mellow). The morning consisted of a talk about Antarctic sub-ice geology, and after lunch I had a discussion with Bird and Mellow in which I was able to present our ideas of the folio, as

[286] See, for instance, Cameron, P J (1981) The petroleum potential of Antarctica and its continental margin, *Australian Petroleum Exploration Association Journal* 21 (1): 99–110.

well as other geophysical interests in the Antarctic. I listed what Phillips would get out of sponsoring the folio: early release of material to them, our input to Gondwana reconstructions, acknowledged sponsorship of a prestige publication, and being 'in on the ground floor' of Antarctic activity.

At the close of the day the general conclusion was that the folio was a project they should sponsor from both the FVS and research and development. Phillips decided they could work it into their Industrial Associates Program. The largest amount they were providing under this schedule was $55,500 to Lamont-Doherty Earth Observatory at Columbia University, and approximately $10,000 each to Scripps Institution of Oceanography in La Jolla, California, and Woods Hole Oceanographic Institution in Massachusetts. I was asking for £41,000 over two years, which was considered to be in the right ballpark per year.

There was discussion, as well, about potential Antarctic geophysical fieldwork programmes: first, a study of the Wilkes Basin using a highly portable remote seismic system with data telemetering to satellites; this could be deployed along with ground RES by Twin Otter out of McMurdo and, second, a magnetic and marine multichannel seismic profiling in the Ross Sea. There was further discussion with staff in the FVS regarding plate reconstructions and an inspection of their impressive Landsat image processing facility. The overall result of this visit to Phillips was very favourable, and I awaited their formal agreement to support our work. There would be further visits to Bartlesville.

The SPRI had significant connections with British Petroleum. The company had been among the first to discover oil at Prudhoe Bay on the shores of the Arctic Ocean in the late 1960s and had sought Arctic information and advice from the Institute in those early days. Charles Swithinbank had been invited to travel on the large ice-strengthened oil tanker SS *Manhattan* on its demonstration voyage through the Northwest Passage in 1969. The Institute had maintained connection with Geoffrey Larminie, one of British Petroleum's senior exploration geologists who had worked in Alaska, and was area manager, Alaska. By the early 1980s, Larminie, a pipe-smoking, mustachioed, and highly approachable individual was general manager of the company's Environmental Control Centre.[287] I contacted him in late October 1980 regarding possible involvement in the folio project. Helpful as

[287] Larminie continued to be a good friend to the polar world as a member of the NERC Polar Science Committee and long-term supporter of both the BAS and SPRI.

ever, he advised interest would more likely come from the exploration side of the company and introduced me to John Martin, general manager of exploration for the company's worldwide activities. As a field geologist Martin had also worked in Alaska undertaking exploration that led to the Prudhoe Bay discoveries. With Larminie's assistance I arranged to meet with him at Britannic House, British Petroleum's headquarters in London on Friday, 13 June. The inauspicious reputation of the date was not lost on me. Martin was welcoming and interested. Following a discussion in which I outlined our project, I left materials with him and an estimate of the sort of sponsorship we were looking for. I emphasised that this was a consortium effort, and we were seeking other partners. Shortly after, I received a positive response from him with the offer of an initial allocation of £20,000. We were in business!

14.3 Folio Gains Momentum

Work was proceeding on the principal topics that the folio would comprise, and a core set had been identified:

1. Introduction to Radio-echo Sounding
2. The Surface of the Antarctic Ice Sheet
3. The Bedrock Surface of Antarctica
4. Ice Sheet Thickness and Volume
5. Driving Stresses in the Antarctic Ice Sheet
6. Isostatically Adjusted Bedrock Surface of Antarctica
7. Residual Magnetic Field in West Antarctica
8. Depth to Magnetic Basement in West Antarctica
9. Internal Layering in the Antarctic Ice Sheet
10. Radar Characteristics of the Ross Ice Shelf
11. Ice Sheet Drainage Basins, Balance and Measured Ice Velocities and Subglacial Water
12. Geological Units of Antarctica

To move forward with these dozen sheets, the focus was on an initial tranche, numbers 2 to 9, and individuals at the SPRI were ascribed responsibility: Millar was to concentrate on layering, and Jankowski on West Antarctica and magnetics. Jordan and I would focus on assembling the full continental details of the ice sheet surface, the ice thickness, and the sub-ice bedrock topography. Others would follow in due course: the isostatic map showing the bedrock surface following the removal of the ice overburden and the upward relaxation of the earth's crust, and the map sheet depicting driving stresses.

Much of what we wanted to achieve, and archive, relied on the collation and manipulation of our enormous quantity of data, as well as useable information from other sources—ice thickness, surface and bedrock elevation, surface temperatures, accumulation, total magnetic field, ice velocity, dielectric absorption, and coastline information. The list was long and highly variable in formats. Various database versions had been produced over the years as our needs evolved, data quantity increased rapidly, and software technology became more sophisticated. Cooper described the data as a collection of files, not all in the same format, though broadly speaking the data they contained were much the same. Over time, different workers had followed different rules for saving data—some had retained data only where there was a bedrock return, while others retained all the information. Consolidation into an easily accessed digital database was a major requirement; this was Cooper's responsibility. He set about giving an integrated structure to the collections and successfully developed a directory-driven database using the existing material assembled by his predecessors. It was agreed that at some place in the folio we should explain this data-handling capability in some detail.

14.4 Scales and Map Projections

There were crucial decisions to be taken at an early stage following the discussion with David Fryer. What scale should we select for the maps? We viewed a number of other compilations from around the world. The American Geographical Society folio series of loose-leaf sheets of Antarctica was a useful guide, but most of their maps were at small scale, admittedly a result of the sparse information at that time. The map of subglacial topography, for example, depicted Antarctica at 1:13 600 000. We were also aware that there were surface topographic maps produced of Antarctica at 1:5 million. This latter scale was more appropriate for our volume of data, but we realised a physically large chart would be the result. In the end we settled on a scale of 1:6 million for the main continental maps of the ice sheet surface and sub-ice bedrock. Even so, the outcome would be a paper page with dimensions of 1.35 m × 0.75 m. If they were to be in a folio format, folding would be necessary, and that was going to be an interesting exercise in Antarctic origami! Some of the other parameters could be depicted on maps at a smaller scale, and we agreed on 1:10 million as convenient. The compilation scale, we decided, should be twice that of the final published maps, so the surface and bedrock sheets were prepared at 1:3 million. This necessitated working on at least four large Mylar[288] plots.

Handling these was quite a business but was fortunately made easier by the splendid map room facilities on the top storey of the SPRI. This was a spacious, airy room with considerable light from the north through its large picture windows. Several banks of map presses afforded much space to lay out and work on draft sheets, and technical drawing tables, an optical enlarger, and other cartographic aids were available. Here Sue Jordan and other staff would spend a great deal of time engrossed on the emerging folio.[289]

Another decision confronted us—selecting an appropriate map projection. For the polar regions we considered several, recalling that all projections cannot preserve shape, area, and straight-line direction. The *gnomonic projection* uses radials projected from the point at the centre of the earth. Its main advantage is that all great circles become straight lines, regardless of aspect, so it is very good for navigation purposes. It is not, however, conformal (i.e. does not keep local shapes), and it does not preserve areas. For our purposes the navigational aspects were not relevant, but shape was. The *Lambert azimuthal equal-area projection*, which projects from a sphere to a disc, has the property of maintaining area but distorts shape. We decided finally on the *polar stereographic projection*, which projects meridians as straight lines originating at the pole, with true angles between them. This projection is conformal but distorts scale and area. These distortions can be minimised by choosing carefully the standard parallel (tangent plane), which we took at 71°S. That this projection had been used extensively for polar maps was an advantage and made our transfer of some data from these other compilations easier.

[288] Mylar is chemically matted drafting film suitable for pen-and-pencil drawing and used as a substrate in many drafting and artistic applications providing a durable medium capable of withstanding extensive revisions and changes.

[289] Sue Jordan (personal communication, December 2020) reflected that 'there was always a very pleasant and conducive working atmosphere on this top floor; research students, assistants and full-time staff, visiting scholars and lecturers from around the world, discussions with staff from other organisations'.

14.5 Coastline and Other Details

At an early stage in the preparations, it became apparent that an up-to-date digital coastline of Antarctica was needed; it would be essential for plotting data at a variety of scales and on different projections. It might appear as surprising that there was no definitive map or chart showing the coast of the continent. For most parts of the world the 'coastline' is a fairly easily defined quantity, particularly at a small scale, such as we would be using. In Antarctica, however, nothing is ever very simple, and we were facing two main problems. First, there was lack of precision mapping of coastal margins due to remoteness, limited access by survey vessels due to ice, and large errors in position resulting from limited navigational accuracy by early surveyors. Second was agreeing what constituted 'coastline'. This could be a distinctive rocky shoreline, or it could be an indeterminate ice-bound coast. In the latter case, the ice front may have fluctuated, making the position of the boundary time-variant. We had also to confront defining the inland boundaries of ice shelves, since these are floating. For several large ice shelves such as the Ross, Ronne, and Filchner Ice Shelves we had to identify the grounding lines, where the ice comes afloat. This was of interest scientifically as well as cartographically; it had not been undertaken rigorously before. There was much to be debated, and numerous decisions to be made. In addition, we had made some significant discoveries during the RES campaigns, such as a shift of the inner boundary of the Filchner Ice Shelf, that would be incorporated in the maps.

Charts and maps from worldwide sources at the largest scales were assembled and assessed for their relative and absolute accuracies. This involved examining the age of maps and complexity of the features and using informed common sense. It was very helpful that the SPRI carried a comprehensive collection of polar maps to consult, and the BAS had others we could call upon. Those selected were then digitised, mostly by Jordan, a fairly physically arduous process—tagging all ice cliffs, rock outcrop, ice shelf, and glacier tongue fronts appropriately. Several other important geographical features were also recorded, such as the lateral boundaries of ice streams and ice rises. In regions where the accuracy of mapping was considered to be low and coastal details unreliable, we undertook to entirely remap the coastline using Landsat Multispectral Scanner (MSS) imagery tied to geodetically determined ground control points. The coast between 120°E and 160°E, 105°W and 115°W, and 0° and 49°W was treated in this way. Absolute

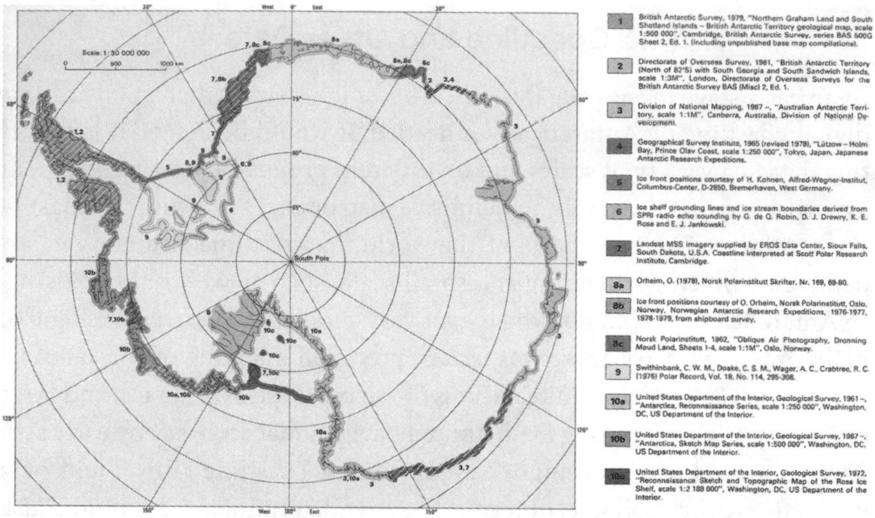

Figure 14.3. Antarctic coastline showing the various sources used in its compilation. (From Sheet iv of the SPRI Folio).

accuracy is dependent on the number of identifiable control points, which are often few. Nevertheless, the relative accuracy and degree of detail from the satellite images made them invaluable. The coastline database in the end comprised over 200,000 points (Figure 14.3)[290]). As the digitisation proceeded it was possible to tag segments according to the type of boundary: ice shelf, ice stream or outlet glacier, ice walls that were not floating but grounded on a rock base, and rock coast. These details later enabled us to calculate the lengths and frequency of these coastline categories.

By midsummer, work was progressing steadily. On 8 August 1980 we had a visit from Vladimir Kotlyakov and Natalia Dreyer from Moscow. Kotlyakov was an old friend of many years. He was a Soviet academician, a well-known glaciologist, and a member of the IAGP committee (Figure. 14.4). He was now head of the Institute of Geography in the Russian Academy of Sciences and editor-in-chief of a project to produce an Atlas of Snow and Ice Resources, supported and funded in part by ICSI[291] and UNESCO. This was to be a major publication of 300 pages and maps, and there would be a

[290] Cooper, APR; Drewry, D J; and Jordan, S (1983) 'Geophysical Database and Antarctic Coastline Compilation', in Drewry, D J (ed.) (1983) *Antarctica: Glaciological and Geophysical Folio*, Cambridge: SPRI.

[291] International Commission for Snow and Ice.

Figure 14.4. Vladimir Kotlyakov in Norway 1980.

special section of 20 pages on Antarctica. Kotlyakov shared his plans and enthusiasm with us as fellow 'cartographers'. He was seeking the latest data and primary materials. It was very interesting to learn of this project, as we respected the Soviet work in producing some excellent atlases and related mapping materials; we were using a number of them to assist in compiling our own folio. We were able to indicate our progress, and we agreed to furnish Kotlyakov with some of our collated data. His institute's timetable was to complete the compilation by 1984 and send the results to the cartographers in 1987–88. Our material would be published well before, so there would be no conflict of interest.

The RES group involved in the folio met on a regular basis. In September we were considering adding a sheet on 'Antarctic surface temperatures and surface accumulation'. Neither of these parameters was crucial to the folio, as they were not derived from our radio-echo soundings, but we were convinced that a 'Glaciological' Folio should present these important components of the Antarctic physical system. We would have to compile the data ourselves from as many sources as we judged reliable. It would be quite a task.

At the end of October, we were discussing the contents of the introductory sheet: Would we illustrate the different RES recording modes (A, Z and ESM)? Would we have illustrations of the radar equipment used, the aircraft, antennas, and other systems? We would need to discuss navigation—the methods and, importantly the uncertainties. We would need to explain how we reduced the data and describe its reliability. There was much to contemplate.

By the next meeting, a month later, progress was pleasing. Cooper was able to report that the heights, thicknesses, and bed elevations had been plotted. Jordan was following up the methods of measuring accumulation. The Ronne-Filchner Ice Shelf material was completed.[292] Work was ongoing on the coastline, as well as on the contours for the Transantarctic Mountains and other mountain regions. Neil McIntyre was assisting Cooper on the sheet depicting the basal shear stress. Jankowski's voluminous tasks on the magnetics were continuing, in addition to his work on the large-scale depiction of the Dufek Massif intrusion. I was working on several of the sheets and the text that would comprise the front end of the folio. These would include a preface and, importantly, listings of participants in the field seasons—acknowledging those who had worked on data reduction, as well as all the individuals and agencies that had been party to the RES work and ultimately to the folio. Recognising the support from the US Navy (VXE-6) aircrews who flew the many RES missions was especially important. It was going to be a long list. Overall progress was good; activity was marching forward.

14.6 Place Names

It was self-evident that we would populate our maps with appropriate place names, identifying coastal features, mountains, glaciers and other ice types, Antarctic research stations, and many other characteristics. We could extract these from the official gazetteers of names that were held in the SPRI library, as well as from officially published maps, and work was set in train to combine these in our drafts. A much more intriguing prospect was naming the new features we had identified on the ice sheet surface and beneath the ice. Several small-scale maps had been produced of the subglacial topography, as it was known from earlier seismic-gravity soundings, and new place names had been ascribed to the main geographical elements, such as the Wilkes Subglacial Basin in East Antarctica, the Byrd Subglacial Basin, and Bentley Trench in West Antarctica. Now we were confronted by many new features, basins, and mountain massifs. It was a rare and intriguing situation—to name continental-scale geographical structures for the first time.[293]

[292] This map would form the basis of a paper in *Nature* with contributions from the BAS and the Alfred Wegener Institute, discussing the morphology and dynamics of the ice shelf (Robin et al. (1983) (footnote 138)).

[293] I had an initial cheeky idea that sprang from the location of several sub-ice features lying within French-claimed Terre Adélie. We should perhaps name them after French wines! Highlands after Bur-

It should be recalled that there are official procedures for the proposal and adoption of geographical place names, and a permanent committee exists in the UK, as in many other countries, for carrying these out. In 1932 a Sub-Committee on Names in the Antarctic was established in the UK, renamed in 1945 as the Antarctic Place-Names Committee, and continues to approve names, under the aegis of the Foreign, Commonwealth and Development Office.[294] To advance the new names to be depicted on the folio sheets I contacted Geoffrey Hattersley-Smith, who was the place-names specialist at the then Foreign and Commonwealth Office dealing with polar regions. I knew Geoffrey quite well, and it may be recalled that it was through his cooperation that the first airborne RES took place in Arctic Canada in 1966, when he was with the Defence Research Board of Canada. Now retired and living in the UK, he was working part-time alongside John Heap, the skillful and highly experienced head of the Polar Regions Section, in Whitehall. I explained to Geoffrey what we were looking for, and he agreed to examine certain of the names we were proposing and, in addition, to make suggestions himself. The list is given in Table 14.1.

It is worth noting that Ice Streams A–E in Marie Byrd Land had been so named unofficially in the 1970 report of the 1969–70 season and had been used widely thereafter by scientists working in this region.[295] Some years late the US decided to rename these features after prominent US researchers who had worked on the boundary between the ice sheet and the Ross Ice Shelf. I thought this a missed opportunity to celebrate some of the early SPRI pioneers who had originally identified these features. There was better luck on the other side of the West Antarctic Ice Sheet, where we discovered a distinctive ice stream flowing into the inner reaches of the Filchner Ice Shelf. We named it the Institute Ice Stream—for the SPRI. In later years a large ice stream entering the Ronne Ice Shelf was designated the Evans Ice Stream in recognition of the work of Stan Evans. In the 1975 map compilation of East Antarctica, we identified the principal surface domes of the ice sheet, labelling them Domes A, B, and C. This shorthand became well entrenched in the Antarctic vernacular. Nevertheless, we proposed, but keeping the initial letter the same, to give them classical names. Thus A

---

gundies, and basins after Bordeaux clarets. Perhaps French wine makers would send us a couple of cases for each one named!

[294] Hattersley-Smith, G (1991) The history of place names in the British Antarctic Territory, *British Antarctic Survey Scientific Reports* No. 113 (parts 1 and 2).

[295] Robin, G de Q ; et al. (1970) (footnote 116).

Table 14.1. New Geographical Place Names in Antarctica Approved by the UK Antarctic Place-Names Committee

Ice Sheet Surface Feature	Sub-Glacial Feature
Institute Ice Stream	Vostok Sub-Glacial Highlands
Hercules Dome	Belgica Sub-Glacial Highlands
Titan Dome	Resolution Sub-Glacial Highlands
Talos Dome	Porpoise Sub-Glacial Highlands
Dome Argus (originally Dome A)	Southern Cross Sub-Glacial Highlands
Dome Circe (originally Dome C) later 'Concorde'	Vincennes Sub-Glacial Basin
Ridge B (Later Boreas Ridge)	Aurora Sub-Glacial Basin
Valkyrjedomen	Adventure Sub-Glacial Trench
Taylor Dome	Astrolabe Sub-Glacial Basin
Ice Stream A (also called 'Mercer')	Webb Sub-Glacial Trench
Ice Stream B (also called 'Whillans' and a branch 'Van der Veen')	Zélée Sub-Glacial Trench
Ice Stream C (also called 'Kamb')	Peacock Sub-Glacial Trench
Ice Stream D) also called 'Bindschadler')	
Ice Stream E (also called 'MacAyeal')	

became Argus, and C became Circe. B had become a ridge, and we retained the name Ridge B on the map but considered it should be Boreas Ridge. The smaller domes of the ice sheet inland of the Transantarctic Mountains had similar classical underpinnings. Between the South Pole and the mountains, Hercules Dome was to celebrate the Hercules C-130 aircraft that had supported the RES programme. Titan Dome was named after the mainframe computer at Cambridge University that had been used for much of the early data reduction and the many calculations. Talos Dome had a more intriguing origin. Talos in antiquity was a bronze giant that protected the island of Crete. Almost every day I would pass a statue of Talos by Michael Ayrton that had been erected in Cambridge close to the Guildhall and the University Computing Centre. I thought, given these other classical names, it would also make a suitable name for the feature in northern Victoria Land.

Names for sub-ice features in East Antarctica were somewhat less exotic and were proposed by Hattersley-Smith after early exploration ships. It should be noted that the Astrolabe Subglacial Basin is where we located the deepest ice in Antarctica at that time: 4776 m. The Webb Subglacial Trench recalled Major Eric Webb, who served with Douglas Mawson on his Australasian Antarctic Expedition in 1911–14 to Commonwealth Bay. I had the great fortune to meet Major Webb when he came to visit the SPRI in the late 1970s. I recall he was a tall, spare man retaining his military bearing and with sparkling engaging eyes. He died in 1984.

14.7 Enter the Cartographers

On 19 December 1980 we held a meeting with David Fryer at his offices in Henley-upon-Thames. From SPRI Sue Jordan and I attended, and David Fryer was accompanied by his cartographic manager, Bob Hawkins. I had already signalled to Fryer that subject to these discussions and an agreement on costs we would be pleased to proceed with them to produce the folio. This was an immensely positive and helpful meeting that clarified several issues and set the pace for our later activities. It is worth recounting some of the detail.

We commenced by reprising the philosophy behind the folio and why we were keen on a loose-leaf format—we wanted sheets that could be extracted, laid-out, and used! And we discussed the general layout. The issue of the coastline was examined, and we agreed to send a Mylar copy of the compilation Jordan had made. We agreed that for some artwork we would prepare detailed drawings from which Fryer could produce films. Colour photographs could be incorporated easily but would be expensive owing to the colour-separation process; black-and-white photos presented no problems. The crunch in our discussions came when Hawkins said they would require mock-ups of all the sheets as soon as possible to develop an appreciation of the balance of the folio and that the first sheet should be ready for work by them at the same time. The pressure was on and clearly going to escalate.

Fryer's agreed to look into producing a suitable binder for the folio and they would investigate printing costs. They told us it might be necessary to print several sheets at once, as this would be more economical—possibly a run of four to five sheets in one batch. We agreed to make four runs of five sheets, with the first set commencing at the end of 1982; these would include several 'introductory' sheets. Then, we turned to costings and more scheduling. Fryer's proposed a sum of £2500 per sheet as an average. This was agreed in principle, excluding any colour photography. We considered the schedule—with the end of 1982 as a suitable deadline, the production rate would be one sheet per month! The dialogue had been very amenable, and Jordan and I returned to Cambridge satisfied that we had achieved an important milestone in our project.

Regular meetings followed the Christmas and New Year break to discuss the many details needed to be settled and the ever-present matter of the timetable! Finalising the coastline was a priority, as well as producing a complete plot of all the RES flight lines, selected for the quality of the navigation, from all seasons. It was agreed to depict the sections of each flight line where ice-thickness data had been obtained. A debate was also developing over aspects of the contouring. It was intended to merge the contouring of the subglacial topography with the bathymetry on and beyond the continental shelf. This would ensure there was no artificial break between the subglacial surface and the seafloor. We decided to use the latest GEBCO Antarctic sheet for the latter but add in new data for Ross and Weddell Seas embayments. Given that the ice shelves are floating, and the RES does not penetrate the water column, we assembled the most recent set of seismic depth determinations and maps from RIGGS for the seafloor beneath the Ross Ice Shelf, courtesy of Bentley's group in Madison, and much older data for the Ronne and Filchner Ice Shelves. We were struck by the remarkable concordance of onshore and offshore contour patterns. To our huge relief, this meant the merging of the two datasets was relatively straightforward.

On 10 March 1981 Bob Hawkins and Mary Spence (cartographic editor) came to the SPRI for discussions. These turned on numerous technical points—hill shading, colours, line weights, the geographical graticule—we agreed that we would depict longitude at 15° intervals and latitude at 5°. I don't think we realised there were so many such building blocks to constructing our dream.

14.8 The Maps Unfold

14.8.1 Ice Sheet Surface

By the end of the month Jordan had pretty much completed the ice sheet surface contouring, and we had spent some time together discussing numerous particulars. The data quantity had exploded. The RES data comprised 101,000 data points covering approximately half the continent. We added to the database several thousand determinations made on oversnow traverses using pressure or barometric altimeters, as well as the very accurate geodetic levelling along oversnow traverse lines completed by Soviet, Japanese, and US scientists. A new era of satellite radar altimetry was also emerging. At SPRI we were beginning an involvement in this exciting scientific opportunity that would last many more years. Seasat data had become available to a latitude of 72°S, and we integrated data provided by Jay Zwally from NASA Goddard Space Flight Center into the map, which assisted greatly in coastal areas of East Antarctica. We also incorporated a number of satellite geoceiver[296] elevation determinations made on the ice sheet surface by various parties. By far and away the most numerous of the additional height determinations for the body of the ice sheet beyond the coverage by the RES data came, surprisingly, from an entirely different and unexpected source—radio altimeters carried by balloons floating above the continent. It is worth a little excursion into this important addition to the database.

An Israeli scientist, Nadav Levanon, at the Department of Geophysics and Planetary Sciences, Tel-Aviv University, was working on height determinations over the Antarctic Ice Sheet from constant-density, free-floating meteorological balloons. Some 411 such balloons were launched in the Southern Hemisphere during the Tropical Wind, Energy Conversion and Reference Level Experiment (TWERLE) during 1975.[297] They floated at an altitude of about 12.5 km, and although not part of the official experiment, many drifted south and 'wandered' across the Antarctic. Each balloon carried three sensors transmitting data every minute: radio altimeter, pressure sensor, and an ambient temperature sensor. Levanon had calculated that the surface heights using the pressure and altimeter readings were accurate to about 60 m and had worked with Bentley in Madison to produce a surface elevation map for Queen Maud Land; they published their findings in *Nature* in 1979.[298] The accuracy was worse by a factor of 2 than ours from RES.

[296] An early form of GPS.

[297] Levanon, N; Julian, P R; and Suomi, V E (1977) Antarctic topography from balloons, *Nature* 268 (5620): 514–16; Levanon, N (1982) Antarctic ice elevation maps from balloon altimetry, *Annals of Glaciology* 3:184–88

[298] Levanon, N; and Bentley, C R (1979) Ice elevation map of Queen Maud Land, Antarctica, from balloon altimetry, *Nature* 278:842–44.

Levanon had made contact with the SPRI, writing to Robin in October 1977, and subsequently visited Cambridge to outline his research, and we were fascinated by his findings. After further discussion he agreed to allow SPRI to use the TWERLE data in the folio sheet on the surface configuration. The somewhat poorer accuracy of the TWERLE data was compensated for by their widespread, continental coverage, particularly their contribution in areas outside our RES grids. We used approximately 5000 data points from the balloons.

The ice sheet surface elevations and the ensuing map were crucial to several other studies, so it was essential to derive accurate values for the database. It will be recalled from chapter 7 that in producing the first RES map following the 1971–72 season we performed a series of calculations to adjust the elevations and produce an optimum plot of data. We applied the same process to the more extensive data that had now been acquired, incorporating an additional number of control points or 'fixes'. Once more, we used a random-walk procedure to minimise and redistribute the errors at the crossing-point nodes of flight lines. The elevation at any one grid node will, after many thousands of 'visits' during computer-generated random walks, converge on a stable solution with errors redistributed over the entire grid. As was performed for the 1974 maps, an independent analysis by least-squares adjustment confirmed the results. This meant that systematic errors of the RES altimetry well away from control points were at worst, 30 m.

All the adjusted elevations within the SPRI database were plotted at a scale of 1:3 million. Jordan and I had now agreed the contour intervals for the main maps: 100 m for the surface, 250 m for the bedrock, and 500 m for the ice thickness. These intervals determined the colour coding applied to the track, which aided the process. Contouring, carefully managed by Sue Jordan, was then undertaken both by hand and using a software routine. Frequently, this combined knowing the strengths and limitations of the physical data, geographical intuition, and common sense.

The results were compared and discussed, and a final version was produced. It was satisfying to see the features of the ice sheet surface appear in increasing detail. For example, while the general configuration in East Antarctica conformed largely with the 1974 map, there were intriguing differences. Dome B, with a 3800 m maximum height disappeared and became Ridge B. At Dome C the 120 m-wide shallow col between the closed 3200 m

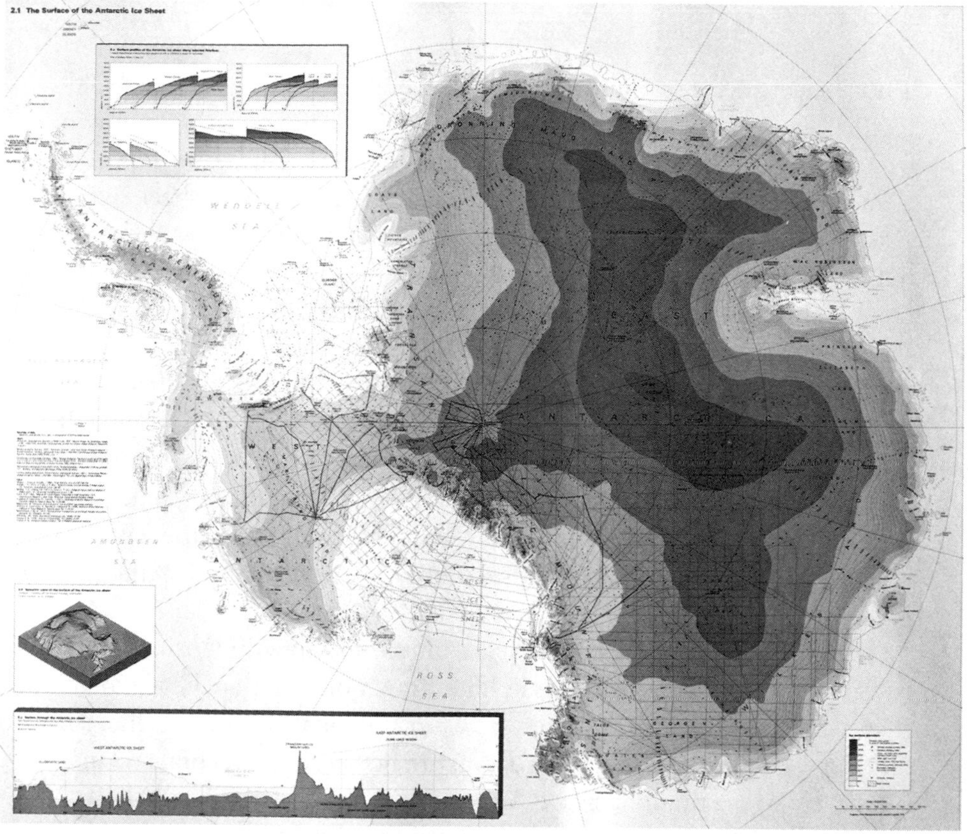

Figure 14.5. Surface of the Antarctic Ice Sheet. Original scale 1:6 million. Contour interval 100 m. Also depicted are several representative profiles of ice sheet flowlines (top left), an isometric 3 D ice surface plot, and at the bottom left a detailed cross section of Antarctica from west to east. (From Sheet 2, SPRI Folio).

contour and the main ridge of the ice sheet also disappeared.[299] Coastal contours for much of Wilkes Land were filled in with reasonable confidence. Inland of northern Victoria Land a minor ice dome was discovered at 2300 m (Talos), as well as another shallow dome between the Queen Maud Mountains and the South Pole (Titan). In addition to the maps, it was decided to generate hypsometric curves for the continent as a whole and also certain geographical regions; these were to be depicted on the folio sheet (Figure 14.5).

[299] More recent satellite-derived elevations and detailed ground surveys have identified a shallow dome some 30 km South of Dome C at 3233 m termed 'Little Dome C'.

14.8.2. Flowlines

The ice sheet surface contours enabled a separate map to be created depicting the general lines along which the ice would be flowing, under gravity, from the interior domes to the continental margins. This is an important element in understanding the large-scale behaviour of the ice sheet and assists in delineating ice drainage basins—the catchments of ice streams and outlet glaciers. Integrating the accumulation of snow/ice within the catchment enables the ice discharge to be calculated and compared with, for instance, measured values at a particular outlet. These data can hint at whether the drainage basin and that part of the ice sheet is in balance or is gaining or losing mass over time, a crucially important finding related directly to the rise or fall of world sea levels and climate change. Many recent studies show net loss and that Antarctica is now a significant contributor to sea-level rise principally from West Antarctica. By contrast, the East Antarctic Ice Sheet is thought to be close to balance or even experiencing marginal gain in mass.[300]

Whilst the outward ice flow is a continuum in three dimensions, it can be represented by carefully chosen lines drawn onto a map. Glaciological theory indicates that ice will flow at 90° (orthogonal) to the regional surface slope. Paul Cooper set up a program to calculate the direction and magnitude of the maximum surface slopes in overlapping boxes of 1° latitude (111 km × 111 km) and to show the derived flow 'vectors' on a map of the surface contours. These could then be used to draw in the flowlines. Figure 14.6 shows the flowline compilation along with the principal ice divides. The lines were constructed to provide an approximate measure of the relative convergence and divergence of the ice flow. At about half the distance up the flow line from the ice edge in each drainage basin a contour line was selected and divided equally along its length. Lines were then drawn up and down the flow from these divisions to produce the detail on the map.[301]

Several distinctive features emerge from scrutiny of this map. A very prominent and almost circular drainage basin discharges through the gigantic Lambert Glacier into the Amery Ice Shelf (1.379 M km^2). Both the Ross and Ronne-Filchner Ice Shelves draw in ice over extensive areas of East

[300] The IMBIE team (2018) Mass balance of the Antarctic Ice Sheet from 1992 to 2017, *Nature* 558:219–22, https://doi.org/10.1038/s41586-018-0179-y.

[301] This map should be compared with a more recent satellite-derived flowline map in Figure 17.2.

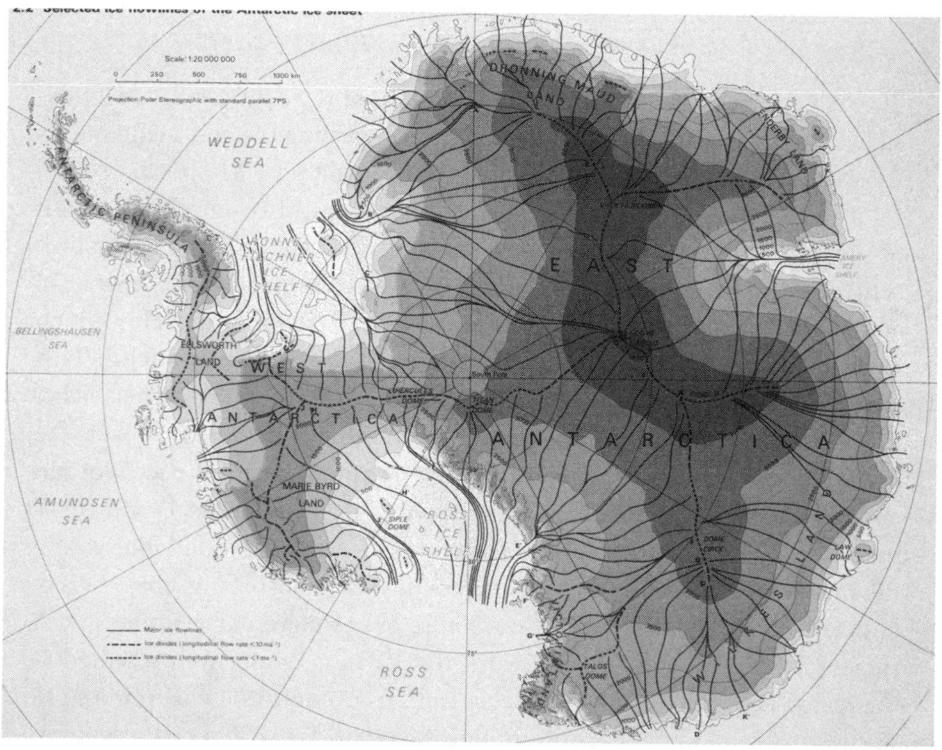

Figure 14.6. Ice sheet surface, ice divides, and selected ice flowlines. (From Sheet 2, SPRI Folio).

as well as West Antarctica. On the Ross Ice Shelf, the ice draining from East Antarctica through the outlet glaciers in the Transantarctic Mountains can be seen to be forced westwards by the much greater discharge of the ice streams in Marie Byrd Land, where the accumulation rate is considerably higher. Between some of the drainage basins there are more stagnant areas of very slow flowing ice; these have been identified as important accumulation zones for meteorites that have fallen into the ice sheet over several hundreds of thousands of years. Here there was yet another intriguing and unexpected story from Antarctica![302]

[302] Whillans, I M; and Cassidy, W A (1982) Catch a falling star: Meteorites and old ice, *Science* 222:55–57; Drewry, D J (1986) 'Entrainment, Transport and Concentration of Meteorites in Polar Ice Sheets', in *Workshop on Antarctic Meteorites* (Annexstad, J; et al. (eds.)), LPI Technical Report 86-01, Houston, Texas: Lunar and Planetary Institute, 37–47.

14.8.3 Compiling Statistics, Writing Papers

The reduction of the RES data and the growth of the database made it possible to extract various interesting statistics regarding the continent and its ice sheet. At the time it was considered these would provide informative insights into and answer some basic questions about this significant region of the world. It has already been shown that the coastline compilation presented in the folio allowed a breakdown of the coastal types. Other, and arguably more fascinating, questions were such as the following: How much ice is there in Antarctica and its various geographical regions, such as East and West Antarctica, the Antarctic Peninsula, and the major ice shelves? What is the average ice thickness, and where is the deepest ice? What is the highest, lowest, and average elevation of the land beneath the ice? Jordan and I with help from Jankowski and two short-term assistants, David Dickins and Gillian Howard, worked on separating out the data and integrating the information to derive these numbers. It was an absorbing, challenging, and revealing exercise, and we were able to compare our estimates with those from earlier studies. This work resulted in a paper we presented at the Antarctic glaciological conference held in Columbus, Ohio, referred to in the next section.[303] The bottom line was that Antarctica contained approximately 30 M km^3 of ice, its area was about 14 M km^2. The average thickness of the grounded ice was 2450 m, and the greatest thickness was 4776 m, located in Terre Adélie in a deep inland trough, and consequently, the deepest or thickest ice in the world.[304] The latter was an interesting statistic that caught the imagination of *The Guinness Book of Records*! Some correspondence ensued with the editors, and they were provided with the appropriate data.

14.8.4 Third International Symposium on Antarctic Glaciology

A significant opportunity to introduce the SPRI RES work on the folio arose with the symposium (TISAG) held at The Ohio State University in Columbus in September 1981 under the paternal hosting of Dr Colin Bull, then

303 Drewry, D J; Jordan, S R; and Jankowski, E (1982) Measured properties of the Antarctic Ice Sheet: Surface configuration, ice thickness and bedrock characteristics, *Annals of Glaciology* 3:83–91.

304 See also footnote 6 for more recent figures. In chapter 17 we shall see that such statistics, along with many features we identified, were either superseded or given much more detail by later RES surveys.

the dean of the School of Mathematics and Sciences. Bull had undertaken geophysical measurements in Greenland in the early 1950s and subsequently, from his base at Victoria University in Wellington, New Zealand, organised a groundbreaking programme of research in the ice-free valleys of southern Victoria Land.

Several SPRI staff attended the conference (Robin, Neal, Millar, and Cooper, and the author), as well as a subsequent workshop on radio glaciology and ice sheet modelling conducted through the auspices of World Data Center A for Glaciology based at the Cooperative Institute for Research in Environmental Sciences at the University of Colorado Boulder.[305] I had written to Bob Hawkins asking that the proof of the ice sheet surface map be sent to the SPRI by 31 August, as Robin was leaving for Columbus shortly thereafter and would be able to hand-carry it to the conference. This would be an outstanding opportunity to display the detail of this compilation and get feedback from our fellow scientists. And so it proved.

David Millar gave a talk on acidity levels in ice sheets based on the notion that some deeper layers, below 1000 m, observed by RES may be caused by acidic ice formed following large volcanic eruptions. He found good agreement with acid profiles measured in ice cores. The value of internal reflecting horizons, although less well resolved than the layers in ice cores, is that they can be traced over wide areas rather than at point sources.[306] Millar also coauthored a paper with Gordon Robin examining the flow of ice at depth around subglacial peaks and expanded the concept of an echo-free zone towards the bed that had been proposed in earlier studies. The lateral flow in mountainous terrain was considered to break up the vertical and lateral continuity of RES layers seen at shallower depths.[307]

Paul Cooper, Neil McIntyre, and Gordon Robin crafted another paper that derived driving stresses in the Antarctic Ice Sheet. These are the forces that determine the flow of the ice sheet and result from the down-slope force of gravity acting on a column of ice of a given thickness. They were able to map the patterns over substantial areas of East and West Antarctica and

[305] (1982) 'Glaciological Data Report GD-13', in *Workshop Proceedings: Radio Glaciology, Ice Sheet Modelling*, University of Colorado Boulder: World Data Center A for Glaciology, 87pp. This meeting was informative; it hinted at the way in which radio glaciology was developing and particularly how data would be stored in the future and archived for wider use.

[306] Millar, DHM (1982) Acidity levels in ice sheets from radio-echo-sounding, *Annals of Glaciology* 3:199–203.

[307] Robin, G de Q; and Millar DHM (1982) Flow of ice sheets in the vicinity of subglacial peaks, *Annals of Glaciology* 3:290–94.

discuss these findings. In areas where the bed is at the pressure melting point, with the possibility of lubrication reducing the friction, driving stresses are low, for instance, in the region of the ice streams of West Antarctica. It also appeared that smaller scales of bedrock topography have only minor effects on the driving stresses, as most of the deformation and flow of the ice takes place above the level of the bed relief.[308] In due course, this study was to make up a separate folio sheet.

Chris Neal attended the meeting and gave a presentation on results from the Ross Ice Shelf. He focused on the small-scale roughness of the ice/water interface (of the order of 50 cm) by examining the fading patterns from the RES returns and ascribing certain characteristics to bottom freezing and melting regimes.[309]

In these and other papers the team were publishing it was possible to discern studies of increased sophistication in the analysis of the radar records. They demonstrated the power of the technique to gain more penetrating insights into the electrical and physico-chemical properties and related larger-scale dynamic behaviour of this large ice sheet and characteristics of the enigmatic continent on which it rests. It seemed the capacity to surprise was limited only by the intellectual veracity of the investigators. From almost every question a possible fresh line of enquiry would arise. It was also the case that many of these findings would be incorporated into sheets of the folio.

14.8.5 Bedrock Surface

The map of the bedrock surface was due to be completed by the first week in October. Sue Jordan and I had pored over it for some considerable time, teasing out the intricacies of the contours. We puzzled over the emerging topography of a land never seen before and unlikely to be viewed by human eye for many thousands of years, by which time the ice sheet may have melted away. Computer contouring provided a first but 'untutored' impression of the landscape. Thereafter we used our geological and geographical intuition to shape the contours, maintaining fidelity with the lines and points of actual data to create a meaningful landscape. It was also important to link

[308] Cooper, APR; McIntyre, N F; and Robin, G de Q (1982) Driving stresses in the Antarctic Ice Sheet, *Annals of Glaciology* 3:59–64.

[309] Neal, C S (1982) Radio-echo determination of basal roughness on the Ross Ice Shelf, *Annals of Glaciology* 3:216–21.

the contours to the details of the terrain exposed above the ice such as along the immense sweep of the Transantarctic Mountains and the imposing Ellsworth Mountains block. We used the excellent USGS 1:250 000 scale Topographic Reconnaissance series maps to aid us in this task. As already described, we likewise had to pay attention to the continental extremity—the coastal bathymetry—and knit together our bedrock contouring with the submarine morphology (Figure 14.7).

14.8.6 Ice Thickness

After the bedrock map, attention turned to the ice thickness and isostatic maps. The former was relatively straightforward and would contain some novel block diagrams conjured up by Mary Spence, and tables to display the statistics we had derived. Their styles and colours added a distinctive and eye-catching element. New data were available on the Ronne-Filchner Ice Shelf, which was the Cinderella of the two largest ice shelves, principally because of its inaccessibility. Behrendt's early seismic soundings and the relatively scant RES lines were the only geophysical information available, but it was considered worthwhile to incorporate an inset map of the ice thicknesses. We had used the RES data between the Ellsworth Mountains and the Pensacola Mountains to redefine the grounding line of the ice shelf (Figure 14.8).

The Ross Ice Shelf sheet never materialised, and its details in much simplified form are given in this and the other continental sheets.

14.8.7 Isostatic Bedrock

The map depicting the bedrock following removal of all the ice and allowing for the surface to rise fully was going to be an interesting exercise and important for a genuine comparison of the sub-ice landscape and inferred geology with other continents. It is well known from early studies in Scandinavia and later in North America that when the ice sheets that covered these continental areas retreated at the end of the last Ice Age two major changes took place. First was a rapid rise in sea level from the melted ice, of the order of about 120 m. The second and much slower effect, over many thousands of years, was the uplift of the land, termed *isostatic adjustment*. It is important to reflect that the contours of the subglacial surface in Antarctica represent a rather ephemeral topography that has resulted from the

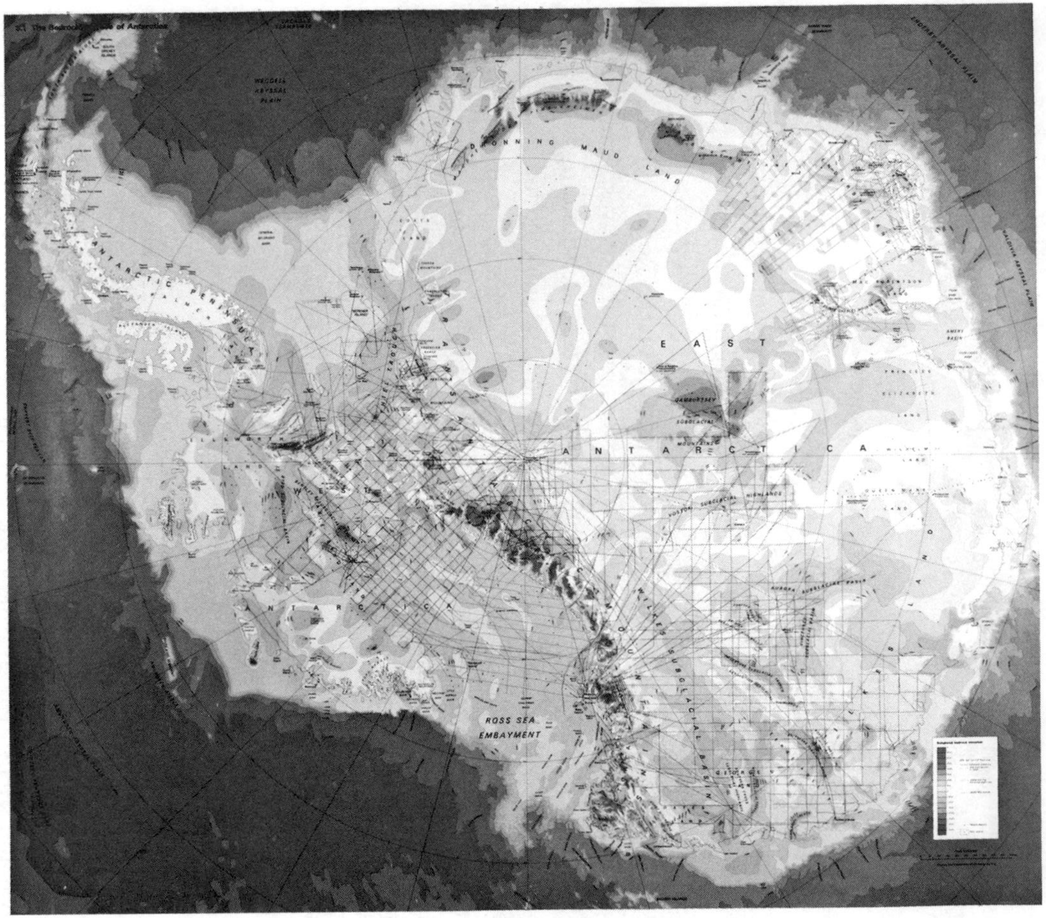

Figure 14.7. The bedrock surface of Antarctica. Original scale 1:6 million. Contour interval 250 m. Note the continuous contouring from the land onto the continental shelf and then into the deep ocean. (From Sheet 3, SPRI Folio).

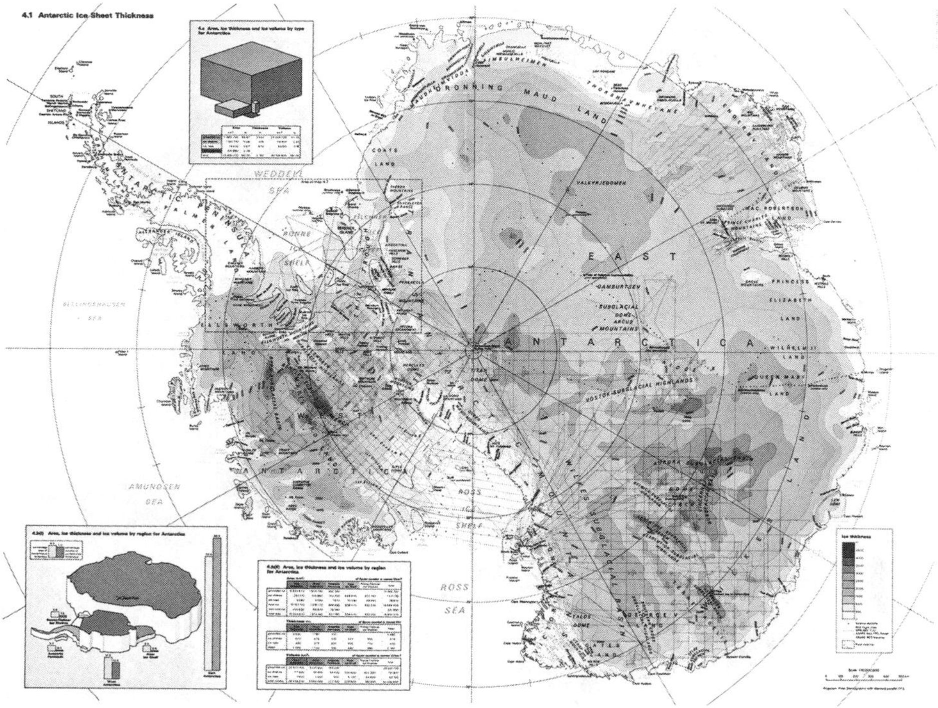

Figure 14.8. Ice thickness with information on areas and ice volumes. Contour interval for the continental ice sheet 500 m, for the Ronne-Filchner Ice Shelf 100 m. Original scale 1:10 million. (From Sheet 4, SPRI Folio).

depression and warping by the weight of the overlying ice, some 2.7×10^{13} tonnes. By removing the ice overburden, it would be possible to restore the Antarctic to its pre-glacial configuration, excluding any erosion that may have taken place. From geophysical studies it is known that the depression from ice loading is roughly proportional to the ratio of the densities of ice and deeper mantle rocks (i.e. about 1:3). The spatial pattern of this loading or warping is made more complex by the rigidity or stiffness of the earth's crust, which dictates, according to various geophysical studies, that the load at any point will be transmitted by an amount that decreases exponentially to zero at a radius of about 350 km (the *deformation function*). To make this correction, we considered Antarctica a matrix of blocks free to move up or down independently, and Jankowski and Cooper undertook a set of computations. As the ice is removed an individual block will move up but will be affected by neighbouring blocks according to the deformation function. It was assumed, at the beginning, that the crust is currently in approximate isostatic equilibrium, as suggested by regional gravity anomalies. After the

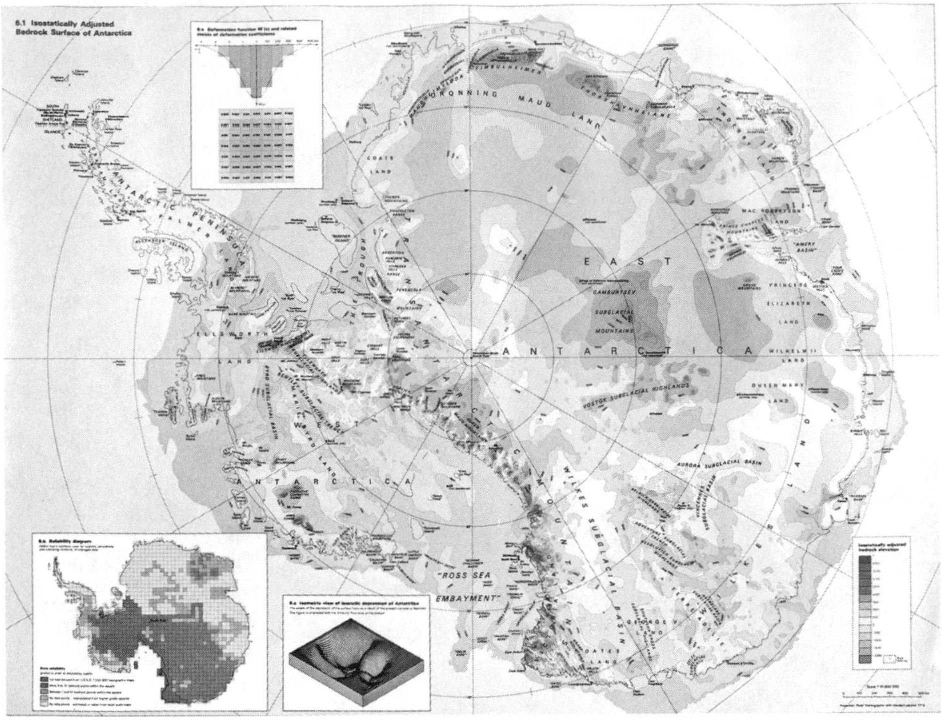

Figure 14.9. The isostatic rebound map of Antarctica. Contour interval 500 m. Original scale 1:10 million. (From Sheet 6, SPRI Folio).

calculations were completed, the new elevations were plotted and contoured to realise a new 'adjusted' rock surface (Figure 14.9). A smaller map showed the contours of uplift reaching a maximum of 950 m in central East Antarctica and 500 m in West Antarctica. A small 3-D diagram was incorporated on the folio sheet depicting the extent of the depression. It was interesting to observe that much of East Antarctica now lay above the adjusted sea level, whilst West Antarctica retained its deep seaway between the Ross Sea and Ellsworth Land—the expression of the West Antarctic Rift.

14.8.8 Magnetics Sheets

Ed Jankowski had completed and submitted his thesis in June 1981[310] but not before he had spent some time contributing to the folio, compiling two sheets on the results of the airborne magnetic measurements. The survey

[310] Jankowski, E J (1981) 'Geophysical Studies of West Antarctica, PhD thesis', University of Cambridge, 293pp.

over two seasons in West Antarctica was the first to amalgamate total magnetic field observations with ice-thickness measurements from RES. This combination provided a powerful tool for earth scientists to explore direct comparisons between bedrock topography and residual magnetic anomalies, as well as between elevation of the magnetic basement and the ice/rock interface, and enabled quantitative estimates of the magnetisation of the bedrock.

With our emphasis on presenting measured properties, we decided to display the data as profiles along the flight lines. This involved calculating the residual magnetic field from the total field recorded during flights (compensating for the effects of the aircraft, undertaking some smoothing to eliminate high-frequency noise, and removing the earth's regional magnetic field). The first map[311] set out the residual field, but the large separation of flight lines prevented any meaningful contouring, although some broad regional trends could be discerned. A separate map was included of the flight lines and residual anomalies over the Dufek Massif igneous intrusion showing its overall extent (some 50,000 km^2) and the high-frequency anomalies along the closely spaced flight lines over the exposed rock areas of the massif and the adjacent Forrestal Range. The greatest anomaly was 3600 nT (see Figure 14.1).

The second of Jankowski's map sheets was on the depth to the magnetic basement (Figure 14.10).[312] Using the residual magnetic data, he calculated the maximum depths to the magnetic basement[313] incrementally along all the profiles and also used the data to estimate and contour the thickness of the non-magnetic material that may be overlying the magnetic basement—which is presumed to be sedimentary strata. In some cases, such as the head of the Filchner Ice Shelf, the bedrock was interpreted to be underlain by several kilometres of sediments. Some 300 separate determinations or depth solutions were listed on the sheet.

Jankowski and I wrote a paper summarising much of the work in West Antarctica. His results, combining the radio-echo sounding with the airborne magnetics, were significant in providing fresh insights into the perplexing sub-ice geology of this region. We were increasingly certain this was a major rift zone, with evidence of sub-ice volcanism, between the East Antarctic

[311] Jankowski, E J (1983) 'Residual Magnetic Field in West Antarctica', Sheet 7 in Drewry, D J (ed.) (footnote 1).

[312] Jankowski, E J (1983) 'Depth to Magnetic Basement in West Antarctica', Folio Sheet 8 in Drewry, D J (ed.) (footnote 1).

[313] The technique used is termed *Werner deconvolution*.

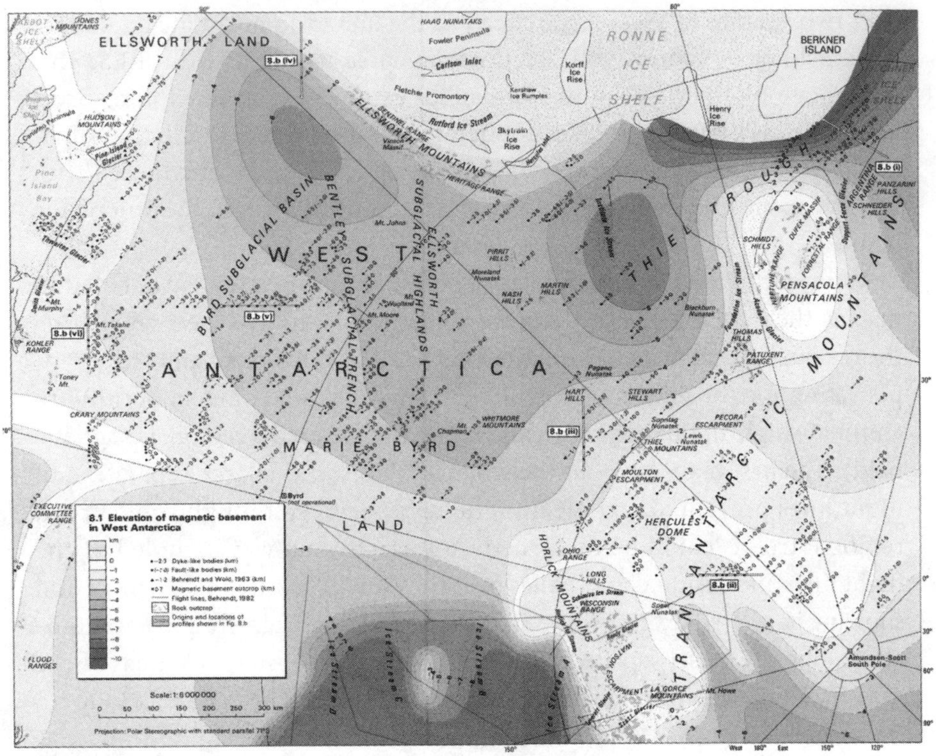

Figure 14.10. Height/elevation of the magnetic basement in West Antarctica. Original scale 1: 6 million. The difference between this surface and the present-day sub-ice topography infers a depth of non-magnetic material (most likely sedimentary strata). This quantity is shown on an adjacent map on the same folio sheet. (From Sheet 8, SPRI Folio).

craton, the edge of which was the impressive escarpments of the Transantarctic Mountains to the east, and the partly exposed volcanic provinces of northern Marie Byrd Land. We submitted the paper to *Nature* as one of its longer 'articles', and it was published in early 1981.[314]

14.8.9 Internal Layering

The last but certainly not least important facet of RES portrayed in the folio was the internal reflecting horizons or stratification within the ice sheet. This phenomenon, it will be recalled, was discovered in the earliest ground-based

[314] Jankowski, E J; and Drewry D J (1981) The structure of West Antarctica from geophysical studies, *Nature* 291 (5810): 17–21.

SPRI soundings in northwest Greenland, and layers were widely observed from the airborne surveys in Antarctica. Preben Gudmandsen reported recording similar layers in the Greenland Ice Sheet.[315] Reflections, with substantial horizontal continuity, were intriguing features that formed the focus of a number of studies (and students!) at the SPRI. What was their origin? Were they an artifact of the radio-echo system? Were they more or less visible at certain frequencies and pulse lengths? How deep could they be seen? What was their lateral extent? What do they tell us about ice flow and past surface and basal conditions?

The first presumption was that the internal reflections on the RES records are caused by changes in the dielectric properties of the ice.[316] Second, the position and strength of the observed reflections can be affected by the choice of radio frequency and hence wavelength. Layers appear to be well detected at 60 MHz. At higher frequencies greater scattering and absorption may mask the returns from them. Their definition can also be altered by choice of pulse length. At higher pulse lengths (e.g. 1000 ns) there appears to be a merging or integration of the individual layers from those seen at say 250 ns and 60 ns (Figure 14.11[317]).

In a study by Julian Paren and Gordon Robin[318] the fall-off in reflection coefficients in the upper part of the ice sheet (500 m to possibly as deep as 1000 m) was surmised to be due to changes in ice density, and the pattern correlated well with the observed density profile from core holes. Below these upper levels there are virtually no density variations, and the reflections here have to be attributed to some other mechanism affecting the electric properties of the ice. Figure 14.12[319] shows the plot of reflection coefficient with ice depth in the vicinity of Dome C that was presented in the folio; it provides a very clear distinction between these two parts of the ice sheet.

Layer reflections at depth are thought to result principally from variations in the chemical composition of the ice due to the presence of aerosols and dust that have been transferred to Antarctica by global atmospheric circulation and precipitated onto the ice sheet surface via snowfall and

[315] Gudmandsen, P (1975) Layer echoes in polar ice sheets, *Journal of Glaciology* 15 (73): 95–101.

[316] That is, electrical permittivity, often described by the term *loss tangent*.

[317] Millar, DHM (1981) 'Radio-echo Layering in Polar Ice Sheets', PhD dissertation, University of Cambridge, 177pp.

[318] Paren, J G; and Robin, G de Q (1975) Internal reflections in polar ice sheets. *Journal of Glaciology* 14 (7): 251–59.

[319] Millar, DHM (1983) Folio Sheet 9, 'Internal Layering in the Antarctic Ice Sheet' (footnote 1).

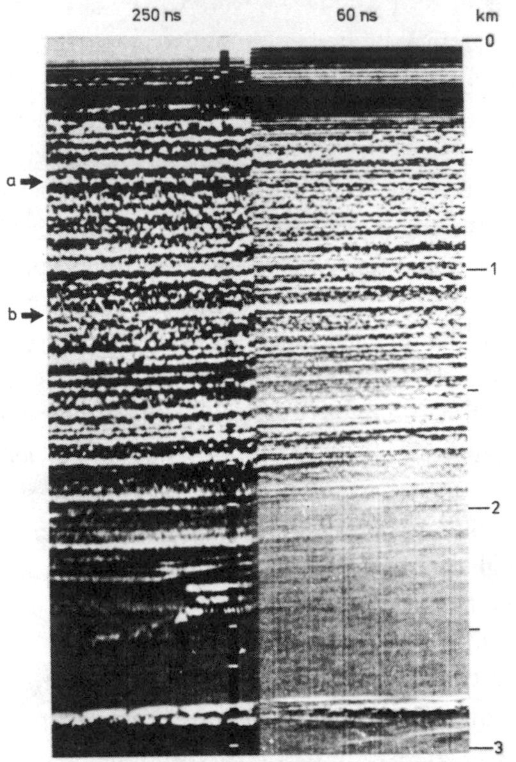

Figure 14.11. Example of the change in position and some integration of layers with a change in pulse length; 250 ns is equivalent to about 21 m, and 60 ns to about 5 m. Arrow "a" indicates two layers shown as one in the longer pulse, and "b", a case where there are no apparent changes. (Courtesy David Millar).

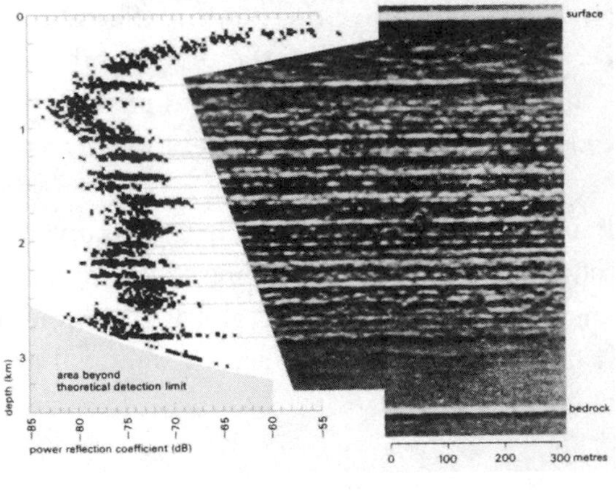

Figure 14.12. Power reflection coefficients with depth (left) in the vicinity of Dome C, East Antarctica, correlated with layers seen on the RES image. A strong trend in the upper 500–600 m is due to the influence of increasing density. Below that height variations result from changing acidity levels. (Courtesy David Millar).

subsequently buried. In particular, acidic impurities, the products of volcanic activity, have been viewed as the most likely causes, being deposited a year or two after a major eruption. There is a very strong correlation between RES layers and measured acidity levels in ice cores from Greenland and Antarctica. It is also considered that in areas of strong shear-flow at depth the ice may produce pronounced crystal alignments (fabrics) that can generate layer-like reflections.

It was also apparent that as the ice flows from the centre of the ice sheet the lower layers are affected by the underlying topography. The reflecting horizons bend over hills and mountainous terrain and sink down where there are valleys or troughs. The effect of the bed topography gradually diminishes (or is damped out) towards the ice surface (see Figure 14.13[320]). This deformation is able to provide considerable information about the dynamics of ice flow. Another striking feature, referred to earlier, was the absence of layers towards the base of the ice sheet even though the RES system could detect echoes at or below this level. This basal layer-free zone (BLFZ) is mostly seen in mountainous terrain and results from the tilting, bending, and break-up of the continuity of layers as the ice flows over and around irregular topography. This was an important conclusion, as the phenomenon could limit the chronology of ice cores drilled to bedrock, where the lowermost sections might not increase uniformly in age with depth.

A further fascinating feature of the layers is that since they were formed at the surface at the same time, they can be used as horizons of equal age (isochrones). Their variation over long distances can thus assist in interpreting the long-term behaviour of the ice sheet and in deducing changes that may result from climate shifts. This concept was taken forward at an early stage at the SPRI and has been the object of investigations by researchers over the last two decades and one of the primary objectives of the SCAR AntArchitecture Action Group. Internal layers can be used as input to ice sheet models as a record of the ice sheets' past behaviour.[321] Some studies, in addition, have used layers to deduce accumulation rates in the past.[322]

[320] Millar, DHM (1983) Folio Sheet 9, 'Internal Layering in the Antarctic Ice Sheet' (footnote 1).

[321] https://www.scar.org/science/antarchitecture/home; Ashmore, D W; Bingham, R G; Ross, N; Siegert, M J; Jordan, T A; and Mair, DWF (2019) Englacial architecture and age-depth constraints across the West Antarctic Ice Sheet, *Geophysical Research Letters* 47:e2019GL086663, https://doi.org/10.1029/2019GL086663.

[322] Schroeder, D M; et al. (2020) Five decades of radioglaciology, *Annals of Glaciology* 61 (81): 1–13, https://doi.org/ 10.1017/aog.2020.11.

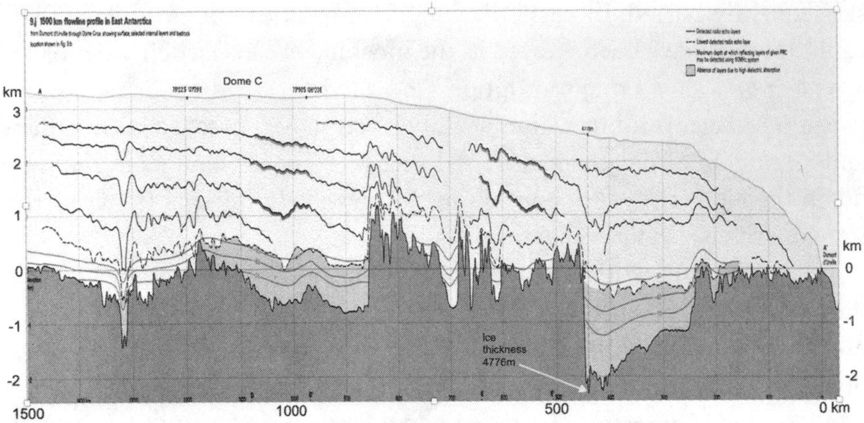

Figure 14.13. Profile through the East Antarctic Ice Sheet from Dumont D'Urville (right) through Dome C towards Vostok Station showing long-distance continuity of selected layers from RES. These are the soundings along the line of the French Traverse to Dome C described in sections 6.1, 7.6, and 8.8.1. Vertical displacement of layers due to flow over irregular topography is also shown. The deep trough inland of the coast was the location of the deepest ice measured during the NSF-SPRI-TUD programme (4776 m). (From Sheet 9, SPRI Folio).

If there is sufficient information of the rate of snow accumulation and surface velocity, it is possible, knowing the depths, to calculate the age of the RES layers if it is also assumed the ice sheet has remained stable over relatively long periods of time. In northwest Greenland one of the early studies of this sort suggested an age of the first observable layer as AD 957 ± 100 years. Millar showed this range to be correlated with the eruption of Eldgja in Iceland (AD 934 ± 2 years).[323] Ian Whillans, during a short sabbatical visit to the SPRI, calculated the ages of prominent layers in the vicinity of Byrd Station in West Antarctica at 1,500, 5,500, and 35,000 years ago. In recent studies, linkages have been made between layers and deep ice cores where depth and age have been adduced, such as at Dome C, Vostok, Dome A (Argus), and Dome 'Fuji' (Valkyrjedomen), as well as at the EPICA site in Dronning Maud Land. For example, investigations of internal reflecting horizons in central East Antarctica by an international group have traced 26 horizons across the Dome C region, 19 of which could be dated from the EPICA core at Dome C. These cover the last four

[323] Millar, DHM (1982) (footnote 306).

glacial cycles, with the youngest dated at 10 ± 0.25 ka and the oldest dated at 366 ± 5.78 ka.[324] Such studies may assist the identification of future ice core drill sites, particularly the exciting prospect of accessing the continent's 'oldest' ice. Layer studies will continue to intrigue and offer a valuable insight into ice sheet behaviour past and present.

[324] Cavitte, MGP; Young, D A; Mulvaney, R; Ritz, C; Greenbaum, J; Ng, G; Kempf, S D; Quartini, E; Muldoon, G R; Paden, J; Frezzotti, M; Roberts, J; Tozer, C; Schroeder, D; and Blankenship, D D (2020) Ice-penetrating radar internal stratigraphy over Dome C and the wider East Antarctic Plateau, US Antarctic Program Data Center, https://doi.org/10.15784/601411; Cavitte, MGP; et al. (2020) A detailed radiostratigraphic data set for the central East Antarctic Plateau spanning the last half million years, Preprint, *Earth System Science Data Open Source Discussions*, https://doi.org/10.5194/essd-2020-393.

15

The Last Push

It was apparent that to complete the folio, as well as to extend several new lines of research, the RES programme needed more money. Following the return of the team from TISAG in Ohio in late September 1981, attention focused on writing new proposals, particularly to the NERC, for additional resources. Thankfully, this bid for finance was successful, and an extension to the RES grant was confirmed; it enabled positions to be funded and the Antarctic RES work to be continued to the end of October 1985. In particular, it would cover the salaries of Neal, Jordan, and Cooper. Two other substantial research grant proposals were also submitted at that time that would consolidate future expansion into a full-scale glaciological programme in Svalbard in collaboration with Norsk Polarinstitutt and allow the SPRI to emerge as a significant contributor to the satellite study of polar ice sheets as part of the accelerating interest in earth remote sensing by the European Space Agency.

15.1 The Folio Completed

In late October/early November we were working to a timetable of completing the first eight sheets in February/March 1982 with the remainder by the end of that year. The Surface and Bedrock, Ice Thickness, and Isostatic Bedrock sheets were already with the cartographers, and the layouts for the other sheets on magnetics and layers were approaching readiness. Jankowski would be leaving to take up a post as geophysicist with British Petroleum at the end of the year, so it was imperative to have his input before departure, although we continued to work on the magnetics sheet into early 1982. Colour proofs arrived at SPRI in small batches, with the first tranche in March. It was a regret that the map of the Ross Ice Shelf did not materialise. It was hoped that a joint map combining Bentley's RIGGS surveys might eventuate following correspondence. Nevertheless, the RIGGS data and analyses were collected into a series of papers and published in 1990 by the American Geophysical Union.

Richard Fifield, the editor of *New Scientist*, had become interested in the SPRI work and commissioned an article that would showcase results of the folio. It would be an excellent way to give the folio publicity and to explain our research to a wider audience. I worked on drafting this article, which was published in late July.[325] It incorporated a number of figures depicted in the sheets of the folio which we supplied to Fifield in proof format, alongside photographs and other illustrations. The article was titled 'Antarctica Unveiled', and the introductory strapline devised by *New Scientist* read: 'What lies beneath the ice sheets of the southern polar continent? This autumn sees publication of a unique folio of maps that are the fruits of a decade's geophysical research and international cooperation'. The team was delighted with this article and the interest that it generated. It was a disappointment, therefore, that the timetable for the publication was about to receive a serious blow.

Something of a slowdown in the cartographic work at Fryer's had been evident for a little while. Financial difficulties caused by the 1982 recession had resulted in David Fryer seeking a buyer for his company. Fortunately, in the wings was an existing customer and publisher based in Lebanon. GEOprojects officially took over the whole of Fryer's business in May 1982, retaining Fryer as general manager. At the SPRI we were unaware of these negotiations, but the impact on completing our work was noticeable—we had not received a proof since the summer. Discussions with Bob Hawkins continued; he had also been retained as the cartographic manager in the Henley Office. The Henley end of the new company had been guaranteed seven years of work, but they now had an urgent 112-page atlas to complete by a deadline of 16 November. Needless to say, the new owners insisted that their projects take precedence over that of previous Fryer clients. It was understood that following this atlas our Antarctic folio would be back into production. These developments were highly regrettable. Although there was a verbal indication that the folio would proceed, we knew nothing of the new owners and their plans. In particular, I was concerned that should we complete the first batch of eight sheets satisfactorily, the follow-on with the last tranche might not be so straightforward. It was a worrying time given the already high intellectual and monetary investment. Nevertheless,

[325] Drewry, D J (1982) Antarctic unveiled, *New Scientist* 5 (1315): 246–51.

our intention was to move on, but the inevitable prospect was for the project to be six months or more behind schedule. These uncertainties continued through the autumn and drifted into the winter. Work on the two magnetics sheets we had prepared did not get started by GEOprojects until early December. I was then in touch on a weekly basis to follow progress and maintain a degree of pressure.

In January 1983 Hawkins informed me they were now dealing with the type-patching of place names on the sheets and had three highly experienced cartographers (Valerie Newbury, Hilary Hopgood, and Anne Kent). This appeared to be progress. The fabrication of the binder was now underway. A company called Rexine had been contracted to manufacture a sturdy ring binder which would have a 'leather-cloth' outer cover with title, logo, date, and publisher on the front and be 57 cm × 40.5 cm (22.5 in. × 16 in.) (Figure 15.5). Five hundred had been ordered. The binder containing the folio sheets would then be placed inside a cardboard container, ordered from a company called Tillotsons Corrugated Cases (now D S Smith), in Burwell, Cambridgeshire (and more usually producing packaging for crisps/crackers, biscuits, and soaps, but now Antarctica!).

Matters seemed to be moving ahead, but because of the slippage we were having to juggle a number of projects simultaneously—the next phase of our Svalbard venture had gone live, and we were in the thick of planning a substantial airborne campaign for May. Furthermore, we had formally initiated the Satellite Altimeter Group (SAG) and were becoming increasingly enmeshed with colleagues in other institutes and with ESA. In March we were expecting but did not receive further folio proofs. At the end of the month, I wrote a detailed letter to Fryer setting out our continuing concerns and the need to complete our project. The schedule we had planned the previous year had slipped by months, and we still had no prospect of a final printing or publication date. I explained that as time went by, our credibility in producing and delivering the folio was diminishing. We had, of course, given commitments to organisations to provide these maps for their planning purposes. Our flyer and advance publicity had borne fruit, and many libraries around the world, as well as individuals, had signed up to purchase copies. These were added pressures to deliver. I stated that we needed a realistic schedule and hoped we could aim for a date in late May. This, I indicated would tie in with my attendance at an important Antarctic geoscience meeting in the US in June, and I was hoping to take two copies with me. I also raised the matter of the second tranche of five maps. I set out two possible scenarios. The first was to reduce the number of sheets to the status of preface pages. This, I suggested, would allow us to complete the essence of the folio in the first printing. We might, I said, have to forgo the later sheets, although this would be very unsatisfactory. The second alternative, which I described as more appealing, was to go to another company to work on the final sheets. There would, however, be issues of stylistic compatibility and continuity. I mentioned to Fryer that in discussion with Hawkins it was suggested that Mary Spence be seconded to another company on a temporary basis to oversee the work.

May turned into June, and the delays continued; it was highly frustrating. Final proofs were due as I left Cambridge for the two conferences—on Antarctic geoscience, and ice and climate interactions—in the US. But another serious setback had arisen. We had hoped to go straight to printing the sheets but discovered there was a problem with paper! We needed high-quality, high-stability paper and had hoped to purchase the required quantity, in advance, the previous year but we were unable to do so owing to cash flow. When we were in position to obtain the paper in the earlier part of the year (1983), we discovered no stocks were available and had to request that a special batch be manufactured. This put us back a further three months!

The printing, folding, and assembling of the sheets into the folio was to be undertaken by Cook, Hammond and Kell. The printing was subcontracted to George Philip & Son Ltd, a long-established company. In August, Sue Jordan and I accompanied Bob Hawkins to the printing works in West London to see the first sheets run off. Sue and I had never witnessed this process, and our emotions were running high, and the atmosphere was extremely tense! The printing hall was immense, and on the floor stood several colossal German-manufactured four-colour printing presses and all the associated stacks of cut paper. Printing plates had been produced from the cartographic films and mounted into one of the machines. We were told that once the machine started, it would send through many scores of sheets, at high speed, which operators called the run-up. When it did, feelings were palpable, expectations high, responsibility hugely worrisome. The large sheets were disgorged with great rapidity into a vast hopper at one end. We took off various copies to inspect them—to check alignment and the correct colour balance—so these could be adjusted for the print run.

We returned to Cambridge, both of us elated—as the sheets looked superb—and exhausted from the nervous energy we had expended; this was

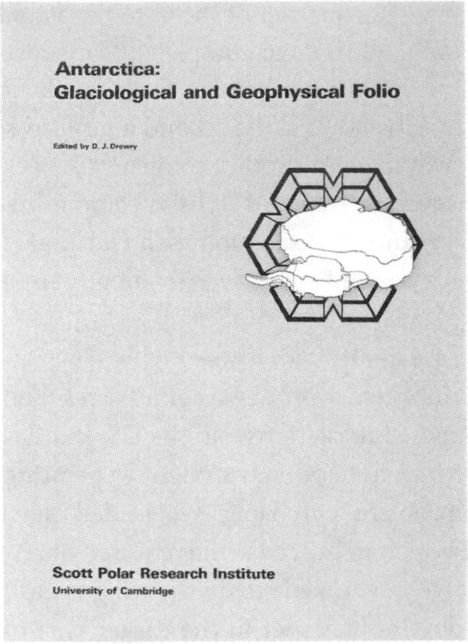

Figure 15.1. The front of the folio binder. (The height is 57 cm.). (Courtesy D Fryer).

after all the culmination of years of work and of many hands. Not all sheets were printed the day we visited, but by early September the task had been completed. That was not the end of the process, however. All the sheets had to be folded. Those with maps at 1:10 million scale were machine-folded twice, forming three panels. The largest at 1:6 million scale, had to have a section cut out of the top left-hand corner and were creased down in both directions. The sheets were to be hand-folded at Cook, Hammond and Kell, as well as having the left-hand leading edge, which stood proud of the folding, hole-punched for later insertion into the binder. But fate had one last card up its sleeve. When we received the sheets, we discovered that the folding was sub-standard ("appalling workmanship" I wrote in my day book following a meeting at the SPRI on 26 October with GEOprojects and Cook, Hammond and Kell). We also spotted some defects in printing, and the first two sheets had to be reprinted to achieve the correct colour balance.

These deficiencies were remedied, and the folio was finally assembled and could be distributed to those who had already ordered copies and to partners and other organisations involved in Antarctic scientific work. From my perspective and those of my colleagues in SPRI who had toiled valiantly for many years on the project it was a truly outstanding outcome (Figure 15.1).

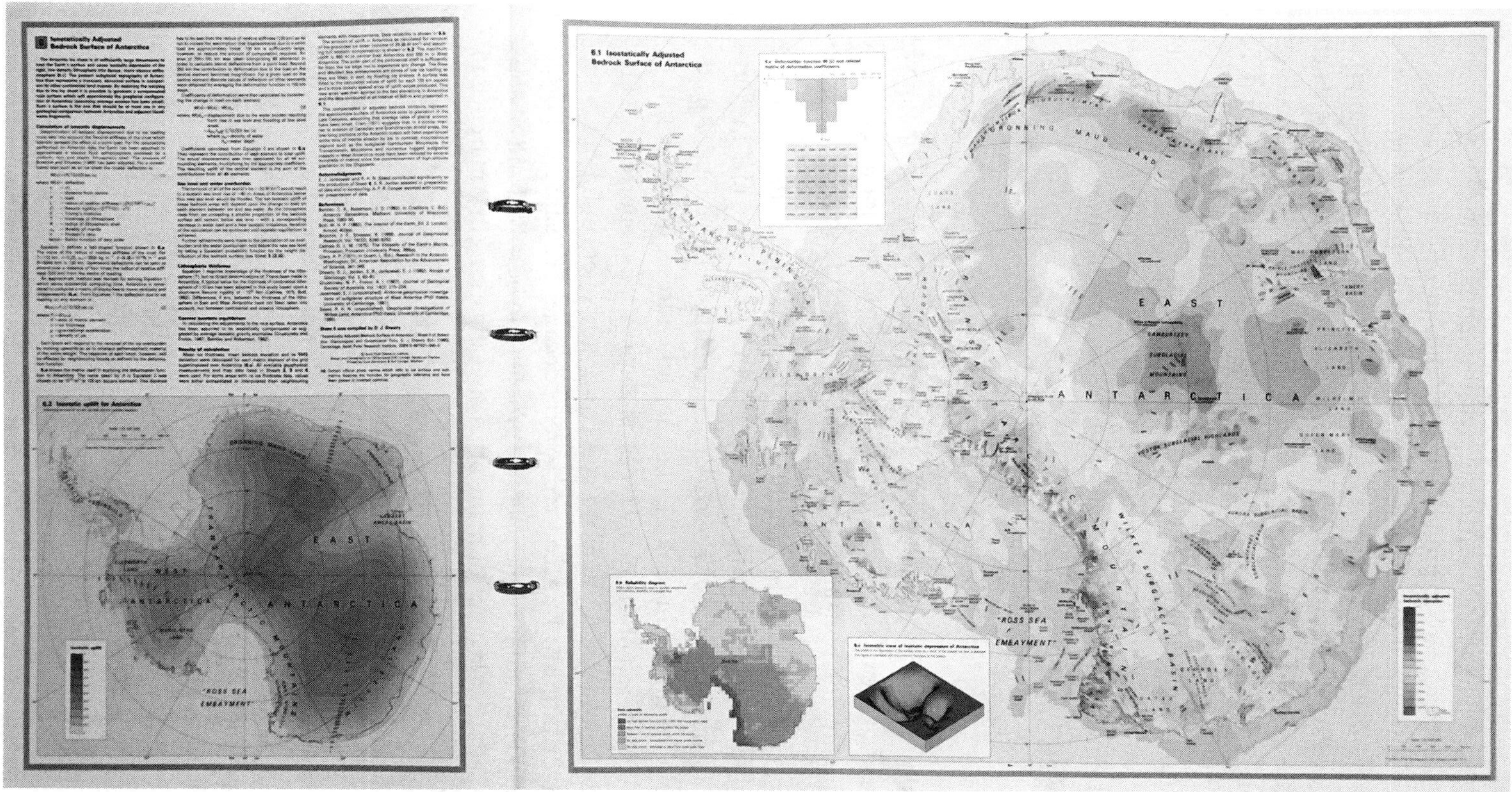

Figure 15.2. Folio showing a typical sheet in the loose-leaf binder. This sheet, which has been unfolded, depicts the isostatically compensated bedrock surface at scale 1:10 million. Explanatory text with references is on the upper half of the left-hand side on all sheets.

The finished folio was an aesthetically pleasing compilation of maps and diagrams, the cartographic production was high in originality and of the highest quality, and it contained an abundance of scientific information, so much of it entirely new. It revealed, in astounding detail, the geographical features of the remotest and least explored of our planet's continents and presented scientific studies to aid in understanding the important role this region plays within the integrated environmental systems of our world. For Gordon Robin, who along with Stan Evans had been the progenitors of RES and the Antarctic programme, it was a fitting triumph and a most suitable culmination of Robin's time as director of the SPRI; he had taken early retirement in September of 1982 but had emeritus status and was still working in the Institute.

15.2 The Folio Reviewed

Thankfully, the folio was well received, and there were reviews of it in various journals. One of the most detailed was by Professor John Sutton FRS. He commented on several aspects, as well as reflecting that

> [t]here can be no doubt that this Folio reports one of the century's major advances in knowledge of the Earth. [It] . . . provides in summary form the results of a remarkable scientific achievement. An admirable feature of this Folio is the care taken to set out the limits of accuracy of each of the many types of observations reported. . . . The application of remote sensing and computer science did not come about fortuitously. This required several decades of sustained effort, a deep understanding of the scientific problems to be solved and an ability to see the relevance of many other investigations in Antarctica to the study of the ice sheet.[326]

Claude Lorius (Laboratoire de Glaciologie et Géophysique de l'Environnement Centre National de la Recherche Scientifique, Grenoble, France) in his review for the *Polar Record* wrote, 'We must therefore congratulate the SPRI and the editor for presenting this up-to-date compilation of physical information about Antarctica in an integrated and attractive form. This folio will certainly be useful for many research topics concerning the still

[326] Sutton, J (1984) *Geographical Journal* 150 (3): 409–10.

poorly-known ice-sheet and continent, and will stand as a major contribution in the field for many years'.[327]

In 1984 the folio was presented with the prestigious John Bartholomew and Sons Award for Thematic Mapping of the British Cartographic Society. The award is given for originality and excellence in the field of thematic cartography with emphasis on effective communication of the intended theme or themes.

[327] Lorius, C (1984) *Polar Record* 22 (138): 330–31.

16

The Aftermath

The Antarctic phase of the RES programme at the SPRI had come to its denouement with the publication of the folio. The team was thinning out, and individuals were migrating into new and challenging, often adjunct, areas of science, such as satellite remote sensing and a major glaciological programme in Svalbard, and eventually other Arctic archipelagos. Nevertheless, for several years following, there would be papers, articles, and reports emanating from research conducted on the data that had been collected. Indeed, new insights into the collections would emphasise the contemporary value of this programme. Thirty or more years later, the SPRI RES data were being reprocessed and image-enhanced by a group at Stanford University in California led by Dustin Schroeder in collaboration with Julian Dowdeswell at SPRI (Figure 16.1).This involved carefully preparing the many thousands of metres of the 35mm RES films, which were then scanned at high resolution at the SPRI using equipment from Lasergraphics, a company based in Irvine, California, with a scan rate of up to 30 frames per second.[328] The end result was some 80 terabytes of data.[329] Figure 16.2 shows the remarkable improvement in the resolution and definition of the radar returns following this processing. Given the enormous investment of energy and resources in acquiring these RES data, to which this book attests, it has been immensely pleasing to see they are available for future scrutiny and research. They continue to yield valuable insights into Antarctica and are enabling comparison with more recently collected radar profiles un-

[328] Than, Ker (2017) Vintage film provides Stanford scientists new insights about Antarctica, *Stanford News*, December 15.

[329] Dustin Schroeder and the SPRI (2018) Stanford-Cambridge Radar Film Digitization Project, http://purl.stanford.edu/vk620zs1672; Schroeder, D M; Dowdeswell, J A; Siegert, M J; Bingham, R G; Chu, W; MacKie, E J; Siegfried, M R; Vega, K I; Emmons, J R; and Winstein, K (2019) Multi-decadal observations of the Antarctic Ice Sheet from restored analog radar records, *PNAS* 116 (38): 18867–73, https://doi.org/10.1073/pnas.1821646116.

Figure 16.1. Preparations for digital scanning of SPRI RES 35mm film records by Dustin Schroeder and Jessica Daniel (Stanford University) at the SPRI. (Courtesy Dustin Schroeder).

derpinning the investigation of noticeable changes to the ice sheet as a result, most likely, of global climate change.[330]

With Chris Neal's departure and following the two trial seasons of the impulse radar on the Aletsch Glacier and Agassiz Ice Cap, the momentum was lost and we did not pursue that technical line of research further. Ed Jankowski had joined British Petroleum, and we took the analysis of the airborne magnetics data no further. David Millar left for an associate editorship at the science journal *Nature*, curtailing the investigation of layering and associated dielectric studies. Neil McIntyre bridged the transition between the airborne RES studies and satellite investigations and assisted with field work in Svalbard. His studies combined satellite altimetry, Landsat imagery, and RES. Surface topographic roughness of the ice sheet was shown

[330] Schroeder, D M; and SPRI (2019) (footnote 329).

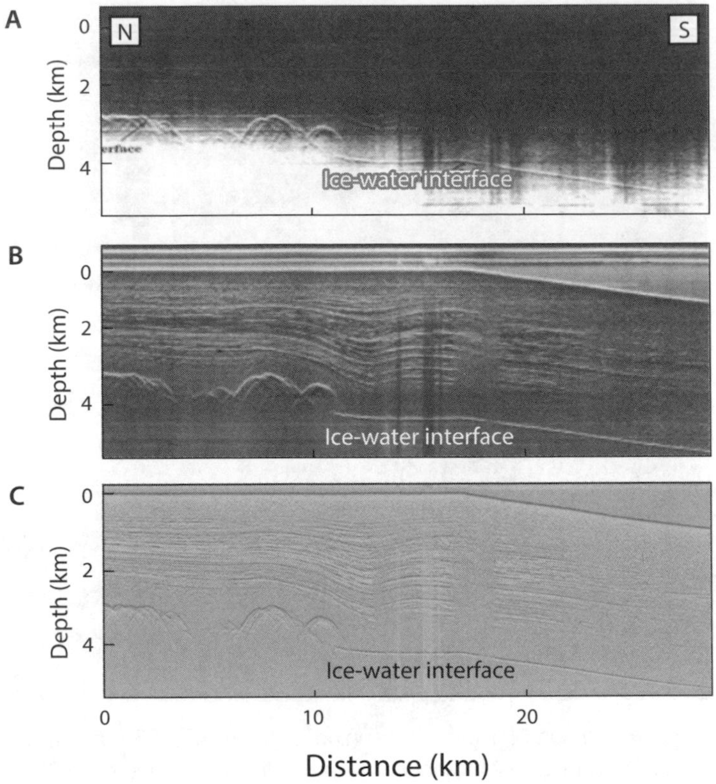

Figure 16.2. Increased resolution of the SPRI RES records from the Stanford scanning. Top: 1974 original print record of part of subglacial Lake Vostok. Middle: intermediate-resolution scan. Bottom: high-resolution direct scan. (Courtesy Dustin Schroeder).

to increase towards the thin ice of coastal regions, where temperature and velocity variations have significant effects. Temperature within the ice was also shown to be important in the response of the ice surface to bedrock irregularities, as was the presence of water layers at the bed. McIntyre extended his investigations to the transition to high-velocity flow in outlet glaciers such as those cutting through the Transantarctic Mountains and in coastal zones of both West and East Antarctica. He showed this flow is initiated abruptly in response to bedrock steps such as sub-glacial fjord heads.

Paul Cooper and Michael Gorman, with their typical energy and enthusiasm, moved into the Svalbard Glacier Geophysics team and also contributed to the work of the satellite altimeter group. They participated in the first season of airborne radio-echo sounding in Nordaustlandet, Svalbard, and in subsequent ground-based studies. Sue Jordan worked for a further period assisting on various Antarctic-related matters before her contract ended, and she moved to the mapping group (MAGIC) at the BAS. The investigation of sub-ice lakes which had risen to great prominence with the discovery of Lake Vostok was to continue with Gordon Robin (in retirement), Julian Dowdeswell, and Michael Gorman, along with Julian's student and later postdoc Martin Siegert. Siegert's work on sub-ice lakes and the use of RES to study glaciological processes more generally has been a major contribution.

16.1 Svalbard

The RES capability in the SPRI was now focused on the north. In a report to NERC in December 1980, relating the progress of the Antarctic research grant, I wrote:

> This (Svalbard) project would be seen to move on from our current radio-echo sounding studies (Grant 2291), principally in Antarctica. By 1982 synthesis of our Antarctic material in a major folio will be nearing completion, experimental testing of a new 1 GHz echo sounder completed and the results of our 1980 Spitsbergen reconnaissance fully worked-up. A project in Svalbard will thus provide a new thrust, maintain our expertise and momentum in remote sensing (radar) techniques and keep our team together while allowing expansion into glaciological modelling and ice/climate relationships in the Arctic which we believe is of growing scientific and practical importance.

The 1980 reconnaissance had demonstrated that the SPRI MkIV 60 MHz equipment was suited to the shallower though warmer ice in Spitsbergen. Cooperation with Orheim and Liestøl at the Norsk Polarinstitutt continued to be highly productive. In discussion with Orheim in 1981 the notion was considered of extending the successful reconnaissance RES activity in Spitsbergen to the investigation of the large ice masses on the northeastern island of the Svalbard archipelago (Nordaustlandet), which included the Austfonna ice cap—the largest in the European Arctic, at 8,000 km^2. It had been known for some time that parts of this ice cap 'surged', perhaps on a periodic basis. There were strong analogies with what was considered at that time might occur to the vast ice sheet in West Antarctica under climate warming. The first cooperative campaign was planned for the spring of 1983 with an airborne RES survey over the whole ice cap. There had been very little work on this island (actually comprising two ice masses, the larger Austfonna and the smaller Vestfonna), principally owing to inaccessibility. It had been subject to glaciological investigations by a Swedish campaign led by Professor Valter Schytt in 1957 and 1958 as part of the country's national contribution to the IGY.

In January 1982 I flew to Oslo for meetings regarding the project and then took the sleeper train to Stockholm. After I arrived late owing to 'frozen brakes' (even in Sweden such malfunctions happen), the meeting with Schytt was very congenial. I had known him for several years as a highly respected scientist in the polar glaciological world. He had wintered with Robin and Swithinbank in Antarctica in 1949–52 on the Norwegian-British-Swedish Expedition (NBSE) and had been responsible for the first deep ice core drilled in Antarctica (to 100 m). He was now a senior professor at Stockholm University. Schytt provided copious data and advice from his expeditions, as well as a delightful dinner with his wife, Anna Nora, and an invigorating sauna. I returned to Cambridge with renewed enthusiasm for this project. An application was written and submitted to the NERC for funding, as well as for PhD studentships, and was approved in November 1982.[331]

A new and significant development was digital recording. The system was designed and fabricated by Michael Gorman, with software for the microprocessor written and developed by Paul Cooper. It was clear that the RES group were still making important innovations, and this was one of the first fully digital systems to be designed and constructed.[332] The oscilloscopes, with their bulky cassettes of 35mm film, were gone, and now the receiver output was digitised by a fast sample-and-hold circuit.[333] Furthermore, the high density and wide dynamic range of the radar records made possible investigations that had previously not been attempted on a large scale, such as the power reflection coefficient of the bed. The digitised ice-thickness information was recorded on magnetic tape and interleaved with navigational information (see below). The antenna comprised two half-wave dipoles, one

[331] NERC Grant application: "Glaciological measurements and modelling of surging ice caps in Svalbard. June 1981. The award was £61,500. In 1986 an extension of £56,900 was applied for to NERC.

[332] Various attempts to undertake digital recording have been made and are reviewed briefly by C R Bentley in Bogorodsky, V V; et al. 1985 (chapter 5, n. 43).

[333] The radar waveform was digitized at a frequency of 8 s^{-1} and divided into 256 points, representing range resolution of 100 ns. Returned pulse strength information was also contained in the digital record.

Figure 16.3. Part of the SPRI-NP team in Longyearbyen during RES operations. Left to right: Ed Murton (pilot), Cooper, Gorman, Orheim, and Williams (air crew).

to be mounted under each of the wings of the aircraft, with the wing acting as a reflector, giving a forward gain of about 8 dB. One antenna was used for transmission, the other for receiving.

The 1983 field campaign involved the remaining members of the Antarctic RES team: Cooper, Gorman, McIntyre, and Drewry, and a new research student, Julian Dowdeswell.[334] From the NP the participants were the two Olavs—Liestøl and Orheim—and two experienced surveyors, Knut Svensen and Trond Eiken (Figure 16.3). To undertake the airborne RES and to deploy field parties on the ice cap surface the NP had hired a ski-equipped De Havilland Twin Otter from the British Antarctic Survey. Garry Studd and Ed Murton (pilots)[335] and 'Taff' Williams (flight support) proved invaluable,

[334] Dowdeswell had graduated from Cambridge in geography. He had taken the final-year course Glacial Geologic Processes taught by me in the Tripos and had organized a successful expedition to Iceland. After Cambridge, he went to the Institute of Arctic and Alpine Research in Boulder, Colorado, to undertake work for a masters' degree with John Andrews. He returned to Cambridge and the SPRI for his PhD. In a highly successful career, he continued the Svalbard work at the SPRI, moved to Aberystwyth, Wales, where he established a vibrant glaciological group and likewise at Bristol University before returning to Cambridge to a professorship and as director of the SPRI from 2002 to 2022.

[335] Ed Murton was a gifted and adaptable pilot. On leaving the BAS he worked for British Airways, later piloting the Concorde—a far cry from bush flying in Antarctica. I had the unexpected privilege of flying with him on a trip to New York on the Concorde, and he invited me into the very cramped flight deck for a memorable landing at John F Kennedy International Airport in New York City.

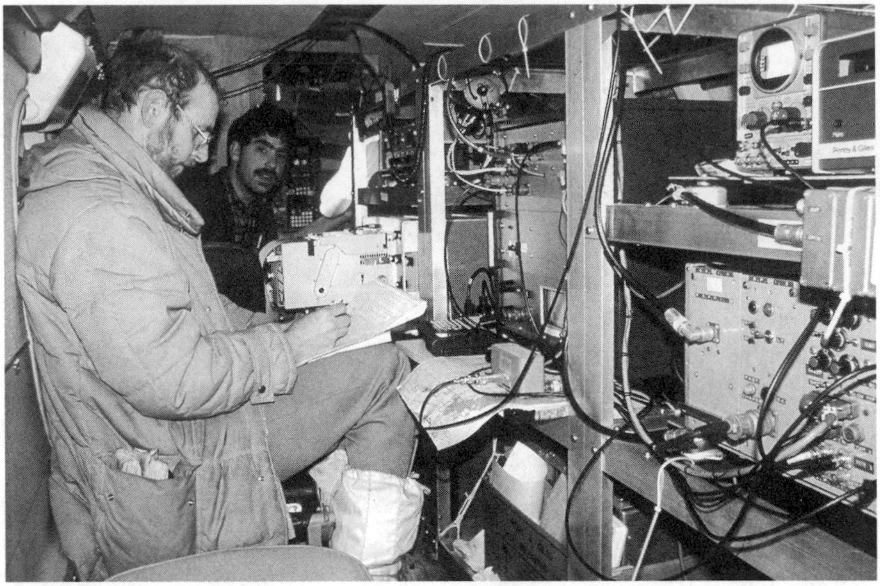

Figure 16.4. SPRI RES equipment installed in BAS Twin Otter. Michael Gorman (left) and Paul Cooper (right) in the cramped operating area.

highly experienced, and helpful companions. The aircraft was fitted out with the RES equipment prior to deployment to Longyearbyen in Spitsbergen (Figure 16.4). The navigation for the ice-thickness survey was to be assisted by a Motorola mini-ranger system: four radio transponders were positioned at previously selected locations on the ice cap and interrogated by the distance measuring system installed in the aircraft. This gave high-precision navigation of the order of 5 m. A further development was to have the navigation and ice-thickness digital data on the same record and in fully computer-compatible form, thus making the reduction of the data post operations speedier (Figure 16.5).[336]

Eight missions were flown out of Longyearbyen, totalling 32 hours on-station; the completed flight lines are shown in Figure 16.6. In addition, a number of glaciers were sounded in Spitsbergen, supplementing the 40 that were surveyed in 1980. There was also a flight on 26 April to sound the small ice cap on Kvitøya (White Island) on the eastern extremity of the archipel-

[336] Gorman, M R; and Cooper, APR (1987) A digital radio echo-sounding and navigation recording system, *Annals of Glaciology* 9:81–84.

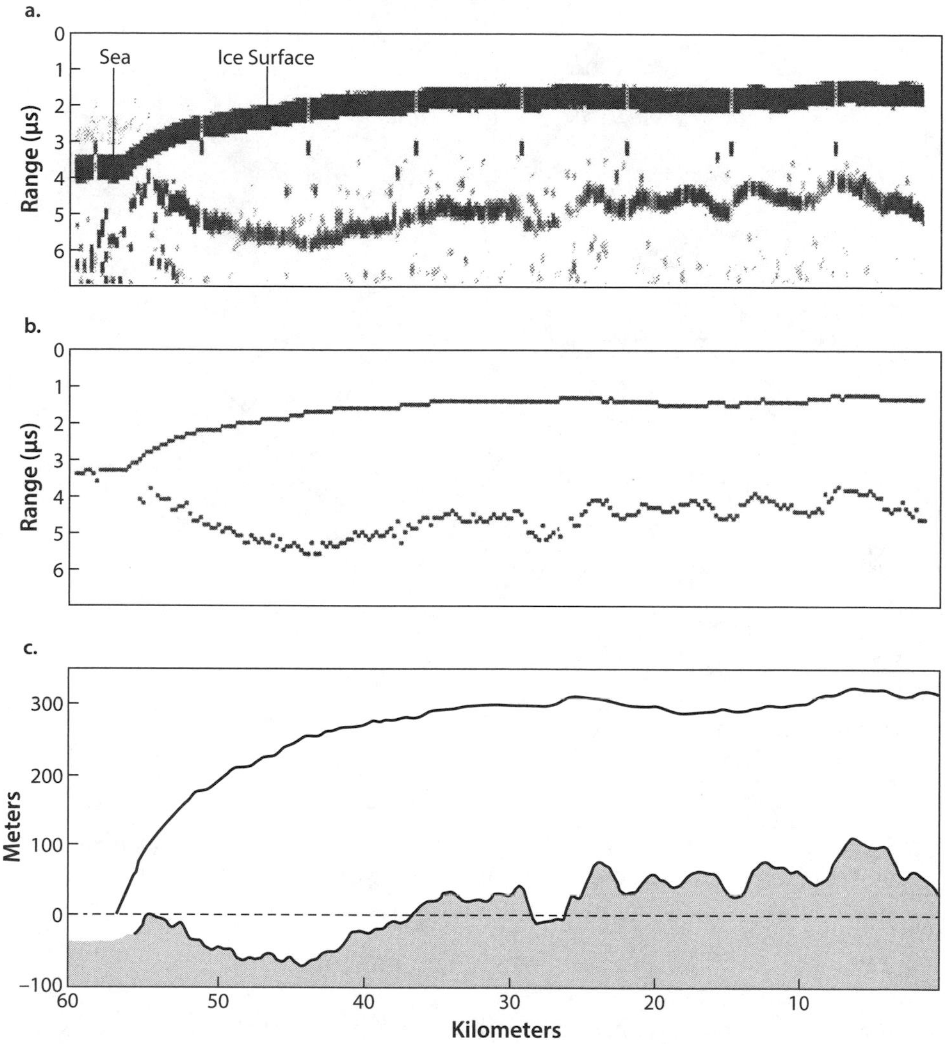

Figure 16.5. SPRI MkIV 60M Hz digital record through the margin of Austfonna, Nordaustlandet, Svalbard. Figure 16.6 shows the location of this RES record. Profile a: the 'raw' record. Steps in the profile are at 15 m intervals. The tones show the relative echo strength. Note clutter echoes from internal reflectors and surrounding rock surfaces. b. Output from the bottom/surface tracking algorithm. c. The interpreted ice and bedrock surfaces.

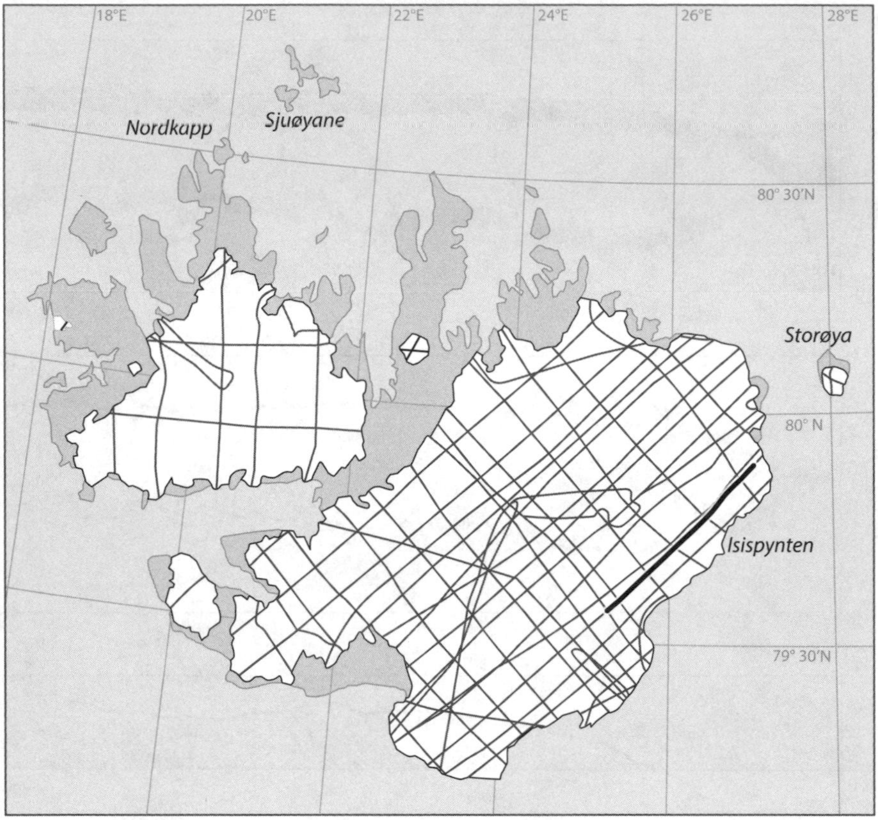

Figure 16.6. Flight lines over the two ice caps of Nordaustlandet. Vestfonna is the smaller to the northwest. The southernmost extension of Austfonna is the surging margin of Bråsvellbreen. Kvitøya lies 100 km due east of Storøya. The profile in Figure 16.5 is marked.

ago.[337] The island is known for the discovery in 1930 of the remains of S A Andree and his companions, A Strindberg and K Fraenkel, after the crash of their ill-fated balloon attempt on the North Pole in 1897.

Ours was an interesting flight, since we were on the very edge of Soviet airspace. It should be recalled that in the early 1980s the Cold War was still a reality, and the Arctic Ocean was the principal theatre for any possible confrontation using ICBMs launched from nuclear submarines and land bases. The air navigation charts we were using in the Twin Otter displayed

[337] Bamber J L; and Dowdeswell, J A (2017) Remote-sensing studies of Kvitøyjøkulen, an ice cap on Kvitøya, North-East Svalbard, *Journal of Glaciology*36 (122): 75–81.

very clear warnings regarding the nearby Soviet airspace, with indications that the Norwegian-Soviet border should be kept at a significant distance. It was 26 April when, having completed some additional lines on Nordaustlandet, we flew from Storøya east towards Kvitøya. The weather was relatively calm, but there was low stratus, which we kept beneath. Shafts of sunlight penetrated the cloud layer into the gloom below as we traversed a dull, leaden sea with some covering of sea ice. Ahead of us loomed the drear outline of the low ice cap on the island. We commenced our flight lines—longitudinal and then a series of orthogonal cross-lines; it would take what seemed two very long hours. At the extreme eastern end of the island, we had to fly out well beyond the coast close to the Norwegian-Soviet median line before turning back to make our next run. I recall there was silence in the aircraft, as we all knew we were being tracked by Soviet radars as we got closer to Ostrov Viktoriya (Victoria Island), just on the other side of the boundary. The oppressive weather enhanced my feeling that we were hanging on the very edge of the Western world, beyond which was a certain darkness. But then the aircraft turned, and we headed back towards the island. It was a memorable experience.

In addition to the RES flights, we deployed ground-based parties to undertake a small programme of glaciological measurements (Figure 16.7). The two groups measured shallow-ice temperatures, accumulation, and ice surface velocities. Two holes were drilled to approximately 10 m, one near the summit of Austfonna, at 760 m asl and the second on the newly identified Isdomen at 374 m asl.

The campaign was extremely effective, and we obtained very good results using the SPRI equipment. An overview was published in the *Polar Record* in 1985.[338] Bed echoes were recorded over 75% of Austfonna and 50% of Vestfonna. In the latter area, signals were attenuated considerably by scattering from ice lenses and other internal inhomogeneities which we considered were due to surface melting and refreezing conditions prevalent on a low-altitude ice cap. The same problem was encountered on Kvitøya. The ice on Austfonna was found to be about 400 m thick—500 m within the central area and more than 600 m in isolated basins. The surging Bråsvellbreen lobe was 250 m thick and grounded on the continental shelf. On Vestfonna the ice was about 300 m thick. The season formed a significant part

[338] Drewry, D J; and Liestøl, O (1985) Glaciological investigations of surging ice caps in Nordaustlandet, Svalbard, 1983, *Polar Record* 22 (139): 357–78.

Figure 16.7. Julian Dowdeswell on Austfonna during 1983 field operations on this isolated ice cap. (Courtesy J A Dowdeswell).

of Dowdeswell's thesis and led to a further two intensive seasons of surface glaciological work on Austfonna, in 1985 and 1986.

An important part of our proposal to study Austfonna in particular was to model its behaviour and thus come to understand better why it surged periodically. We were keen to use a finite element (FE) approach, as this had been demonstrated as an effective modelling technique for glaciers. One of the most suitable FE programs had been developed by Roger Hooke at the University of Minnesota. I wrote to him seeking permission to use his program, which we could develop further for our purposes. He most generously and quickly provided a copy.

The FE method is used a great deal in engineering studies, so I talked to a good friend in the Mechanical Engineering Department in Cambridge, Rod Smith, enquiring whether he would be interested in collaborating. He responded very positively with the offer to have one of his research students, Wilf Nixon, work on this project as part of his activities, alongside Paul Cooper, Julian Dowdeswell, and a new research student at the SPRI, Len Watts. This was an excellent development. Smith was keen on mountains and the Arctic. He and I and had both participated in expeditions, although on separate occasions, to the Stauning Alps in East Greenland, so he did not need much encouragement to come onboard! The collaboration led to an interesting initial study of Finsterwalderbreen.[339] Watts' thesis was on finite element simulations of ice mass flow, and besides the Finsterwalderbreen example, he worked on several other case studies including Austfonna, using the flowline that had been measured by the SPRI-NP programme in 1983 (see footnote 206).

The RES operation in Nordaustlandet was the last major airborne campaign by the SPRI for some time to come. In the early 1990s the SPRI MkIV was used once more in a cooperative venture with the Russian Academy of Sciences to sound small ice caps in Franz Josef Land, and a new 100 MHz system was developed for a further collaborative sounding project in Severnaya Zemlya in 1997.[340] In 2002 the same 100 MHz system was deployed by Dowdeswell and Gorman in collaboration with Canadian scientists to sound several of the ice caps on Devon and Ellesmere Islands, marking the final operation using SPRI RES equipment.[341] The discussion of these later investigations lies beyond the scope of the present book. The Svalbard work continued with the ground-based glaciological studies on Austfonna and elsewhere in Spitsbergen that produced satisfying and new research and engaged fresh research students, notably Jonathan Bamber, Jefferson Simões, and Len Watts.

16.2 Satellite Altimeter Group

The extension of the SPRI RES work to study ice surfaces using radar altimeters from polar or near-polar orbiting satellites resulted in the formation of a new research group, the Satellite Altimeter Group (SAG). It had been important to diversify SPRI's glaciological research interests as the Antarctic RES programme wound down and the folio was completed. The initial link-up with the Rutherford Appleton Laboratory (RAL) and the interest of their scientists, particularly David Croom and John Powell, had been a helpful start. It was agreed to prepare a joint proposal related to

[339] Nixon, W A; et al. (1985) Applications and limitations of finite element modelling to glaciers: A case study, *Journal of Geophysical Research* 90:11303–11.

[340] Dowdeswell, J A;, Gorman, M R; Glazovsky, A F; and Macheret, Y Y (1996) Airborne radio-echo sounding of glaciers on Franz Josef Land in 1994, *Materialy Glyatsiologicheskikh Issledovaniy Khronika* 80:248–54; Dowdeswell, J A; et al. (2002) Form and flow of the Academy of Sciences Ice Cap, Severnaya Zemlya, Russian High Arctic, *Journal of Geophysical Research* (*Solid Earth*) 107, issue B4, https://doi.org/10.1029/2000JB000129.

[341] Dowdeswell, J A; et al. (2004) (footnote 182).

Figure 16.8. The ERS-1 satellite. (Courtesy ESA).

experiments in support of the forthcoming ERS-1 altimetry project. ERS-1 was Europe's first remote-sensing satellite, launched in 1991 and operated until March 2000, exceeding its planned lifetime (Figure 16.8). Amongst its payload was a nadir-pointing radar altimeter operating in the Ku-band (13.8 GHz), principally for precise measurements of ocean height. Such an instrument could also provide detailed profiling of the great ice sheets and sea ice, as had been attempted using the GEOS-C and Seasat satellites. The radar tracking systems and understanding the nature of the returned pulses from ice sheets, however, were still in their infancy. More research was needed, and SPRI's experience with airborne radars fitted the bill.

A proposal to further this work at SPRI was to be submitted to both NERC and SERC (Science and Engineering Research Council), as these two groups had formed a joint climate committee. Vernon Squire, who had recently completed his PhD in the SPRI Sea Ice group, was going to be a co-principal investigator with me. We assembled an original and interesting research proposal to the NERC side of the joint committee for funding.[342]

[342] NERC Grant Application, 'Climate-related remote sensing of polar ice' June 1981.

We had endless queries, including a visit to SPRI by a subset of the committee who went over the minutiae of our funding application. We did not receive an adjudication until April 1982! This was pretty disgraceful treatment and quite excessive scrutiny. We eventually received a positive response, but NERC cut the award by almost half. Squire was funded partly on this grant, as was a new member of the team, Eva Novotny, an astrophysicist who came from Liverpool University to join the group in late 1982.

As the ERS-1 satellite project moved forward we perceived that a series of pre-flight studies and simulations of the performance of the radar altimeter over a variety of ice-covered surfaces would deliver a very substantial contribution. With our NERC grant to support such work we entered collaborations with the Mullard Space Science Laboratory (MSSL), part of University College, London, alongside RAL. I was content that the lead for the applications to ESA for these several studies should be our colleague Chris Rapley at MSSL. The Svalbard activity was creating substantial new research, work was ongoing to complete the folio, and a raft of papers were awaiting completion. Moreover, I had been invited by ESA to join the ERS-1 Radar Altimeter Team, representing the cryosphere community, which made significant demands on time and effort. Over the next four years the collaboration with MSSL and other partners worked remarkably well, and several joint contracts were secured that enabled us to maintain momentum in our altimeter group and undertake worthwhile research.[343]

There was no doubt that the future 'mapping' of the surface of the great ice sheets (although not the bed, for which RES is still the major tool) was going to be undertaken from polar-orbiting satellites with a variety of sensors—in particular, altimeters—and with high precision. We were keen and glad to have entered this arena at a relatively early stage; new students were joining the SAG (Neil McIntyre was already heavily involved in using both RES and satellite data, and Mark Drinkwater was engaged in a fully altimeter-based research topic for his thesis). Radar depth sounding would continue to find new applications; the more exotic and exciting have been the exploration of the icy surface of extraterrestrial planetary bodies.[344]

[343] An example of the several studies that were completed is the first of the reports to ESA: *A Study of Satellite Radar Altimeter Operation over Ice-Covered Surfaces* (May 1983) ESA Contract No. 5182/82/F/CG(SC), 224pp.

[344] Schroeder, D M; et al. (2020) Five decades of radioglaciology, *Annals of Glaciology* 61 (81): 1–13. This provides a résumé of progress in this field.

17

Reflections

In the decades following the completion of the Antarctic folio and the Svalbard campaigns, the development of RES and its deployment to investigate burgeoning questions regarding the great ice sheets passed to other research groups around the world, and advances in technology and interpretation have flourished and proceeded apace. New and sophisticated radars have been constructed, tested, and flown in Antarctica and Greenland and deployed on ground-based projects. Digital recording and sophisticated processing of radar scans have enabled faster, more quantitative investigations and easier correlation with other information such as ice sheet mass balance and velocity measurements, as well as ice core studies and satellite observations. With the introduction of GPS, navigation—the Achilles heel of the early RES missions—has improved by orders of magnitude. And extensive airborne missions flown remotely by drone are becoming a reality. The detail and quality of the RES data have been astonishing. Their analysis has led glaciologists to new and exciting constructs adding depth and fresh insight into the architecture, behaviour, and environmental couplings of ice sheets, and have enabled powerful modelling of future scenarios. Notwithstanding the temptation to elaborate and discuss these innovative and stimulating developments, it is not the purpose of this book to recount them in any detail; several comprehensive reviews in recent years have provided useful summaries of these advances.[345] Rather, I have chosen first to mention some of the research groups that have taken these studies forward and thereafter describe a limited number of examples of the progress in investigat-

[345] Plewes, L A; and Hubbard, B (2001) A review of the use of radio-echo sounding in glaciology, *Progress in Physical Geography* 25 (2): 203–6; Bingham, RVG; and Siegert, M (2007) Radio-echo sounding over polar ice masses, *Journal of Environmental & Engineering Geophysics*, https://doi.org/10.2113/JEEG12.1.47; Allan, C (2008) A brief history of radio-echo sounding *IEEE Earthzine*, 26 September; Schroeder, D M; et al. (2020) (footnote 344); Popov, S (2020) Fifty-five years of Russian radio-echo sounding investigations in Antarctica, *Annals of Glaciology* 61 (81): 14–24.

ing and representing aspects of Antarctic glaciology since the publication of the folio. Such studies are increasingly being directed at investigating the vital role of the ice sheet in global climate change (such as influencing world sea levels) and monitoring its physical changes, which are of the utmost importance. In earlier chapters I have referred to the contributions of past and present RES and ancillary glaciological studies of Antarctica in tackling this subject, one of the greatest issues of our time.

17.1 Some New Brooms

In the UK, the locus of RES technical developments and Antarctic fieldwork moved to the BAS, where the group established by Swithinbank was led by Chris Doake and later under the effective leadership of Hugh Corr. They were continuing with both electronic improvements and extensive surveys on the Antarctic Peninsula and the Ronne and Filchner Ice Shelves and were progressing into areas of West and East Antarctica. A low-frequency broadband impulse system was designed for ground-based operations (Deep Looking Radio-echo Sounder—DELORES), and an airborne radar (PASIN) first deployed in 2004–5 has proved an exceptionally capable instrument. A further development was phase-coherent radars such as the Autonomous phase-sensitive Radio-echo Sounder (ApRES) under the guidance of Keith Nicholls at BAS and in cooperation with scientists at University College, London. Originally designed for determining the basal melting beneath floating ice shelves, it has proved extremely successful and versatile, and is being deployed by many research groups worldwide.[346] A collaboration led by geophysicist Fausto Ferraccioli between the BAS and scientists from Italy resulted in highly detailed airborne geophysical surveys in East Antarctica combining variously RES, magnetometry, and gravity measurements.[347]

[346] King, E C (2020) The precision of radar-derived subglacial bed topography: A case study from Pine Island Glacier, Antarctica, *Annals of Glaciology* 61:154–61, https://doi.org/10.1017/aog.2020.33; Nicholls, K W; et al. (2015) A ground-based radar for measuring vertical strain rates and time-varying basal melt rates in ice sheets and shelves, *Journal of Glaciology* 61 No. 230, 2015 https://doi: 10.3189/2015JoG15J073. Lok, L B; Brennan, P V; Ash, M; and Nicholls, K W (2015). Autonomous phase-sensitive radio-echo sounder for monitoring and imaging Antarctic ice shelves, 2015 8th International Workshop on Advanced Ground Penetrating Radar (IWAGPR), 2015, pp. 1-4, https://doi: 10.1109/IWAGPR.2015.7292636.

[347] Ferraccioli, F; et al. (2009) Aeromagnetic exploration over the East Antarctic Ice Sheet: A new view of the Wilkes Subglacial Basin. *Tectonophysics* 478 (1–2): 62–77.

Several other countries have developed their own programmes as RES has become the standard geophysical technique for depth sounding the ice. Italian geophysicists have surveyed the region around Dome C and at Lake Vostok.[348] In the US, Bentley's group continued for some time, but Ken Jezek moved to The Ohio State University, and Don Blankenship to the Jackson School of Geosciences at the University of Texas, Austin (Austin Institute for Geophysics), which thereafter became one of the primary centres for US RES work conducting intensive multi-sensor campaigns in West Antarctica.[349] Blankenship also worked in collaboration with Martin Siegert (who had moved to Bristol University) and Julian Dowdeswell at the SPRI on Arctic ice masses. At Columbia University's Lamont-Doherty Earth Observatory in New York, Robin Bell and Michael Studinger became highly engaged in the geophysical exploration of the East Antarctic Ice Sheet and of Lake Vostok (see the AGAP project, section 5.10). A group of active radio scientists and technicians at the University of Kansas, initially under R K Moore and later under the leadership of Prasad Gogineni, developed new and sophisticated radar systems in the Center for Remote Sensing of Ice Sheets (CReSIS) (for example, a Multi-Channel Coherent Radar Depth Sounder—MCoRDS).[350] Scientists in Russia have continued to progress in their RES research. Yura Macheret, a highly skilled and approachable scientist, visited the SPRI in the early 1980s and with whom we discussed Svalbard soundings, published a valuable book in 2006, although in Russian. This summarised research undertaken in the Polar Urals, Caucasus, Altai, Tien-Shan, Kamchatka, and Svalbard, as well as on several of the Russian Arctic islands.[351] International conferences in RES such as those organised by the International Glaciological Society have proved fertile occasions to bring together this growing community, present new results, and stimulate new and younger scientists from around the world.[352]

[348] Tabacco, I E; et al. (2003) Airborne radar survey above Vostok region, East Central Antarctica: Ice thickness Lake Vostok geometry, *Journal of Glaciology* 48 (160): 62–69.

[349] Peters, M E; Blankenship, D D; and Morse, D L (2005) Analysis techniques for coherent airborne radar sounding: Application to West Antarctic ice streams, *Journal of Geophysical Research* (*Solid Earth*) 110, no. B6, https://doi.org/10.1029/2004JB003222.

[350] Gogineni, S; Chuah, T; Allen, C; Jezek, K; and Moore, R (1998) An improved coherent radar depth sounder, *Journal of Glaciology* 44 (148): 659–69.

[351] Macharet, Y Y (2006) *Radio-Echo Sounding of Glaciers* (in Russian), Moscow: Scientific World, 302pp.

[352] International Symposium on Radioglaciology and Its Applications, Madrid, Spain, 9–13 June 2008; Five Decades of Radioglaciology, Stanford, California, 8–12 July 2019.

17.2 The Surface Configuration of the Ice Sheet

With uncanny prescience Gordon Robin wrote a paper, presented at a meeting in Ottawa, Canada, in 1965, titled 'Mapping the Antarctic Ice Sheet by Satellite Altimetry', in which he enunciated all the elements for future successful satellite programmes.[353] He had the vision of acquiring synoptic data at continental scale without the vicissitudes of weather and conventional logistics. In the introduction he wrote: 'It is proposed that the Antarctic ice sheet should be surveyed by a radio altimeter . . . carried in a satellite in a polar obit.' Today, satellite altimetry has transformed the mapping and study of the surface of the great ice sheets in Antarctica and Greenland. A generation or more after Robin, SPRI doctoral students Jonathan Bamber and Mark Drinkwater became significant international players in this important field. After completing his PhD on RES of Svalbard glaciers, Bamber (at Bristol University) has led a vigorous and effective international group working on the surface mapping, mass balance, and flow of ice sheets, employing, principally, Earth Observation techniques.[354] Using the data from the ERS-1 altimeter, Bamber produced a digital elevation model for the ice sheet surface and compared this with a digitized version of the SPRI folio map within the region covered by ERS-1 (to 81.5°S). He found that only 37% of the area covered by both agreed to better than 50 m, and there were significant deviations in coastal regions, where both the satellite and the SPRI data were sparse.[355] Progress had been made. The SCAR Antarctic Digital Database, managed by the BAS,[356] now provides the latest surface topography of the continent and is depicted in Figure 17.1. The database is accompanied by the Reference Elevation Model of Antarctica (REMA).[357]

Mark Drinkwater's thesis at SPRI on radar altimetry studies of polar ice positioned him for a role with the Earth Sciences Division of NASA's Jet Propulsion Laboratory (JPL), where he was a senior research scientist

[353] Robin, G de Q (1966) Mapping the Antarctic Ice Sheet by satellite altimetry, *Canadian Journal of Earth Sciences* 3 (6): 893–901.

[354] In 2015 Bamber was elected president of the European Geosciences Union and in 2019 was elected a Fellow of the American Geophysical Union.

[355] Bamber, J L (1994) A digital elevation model of the Antarctic Ice Sheet derived from ERS-1 altimeter data and comparison with terrestrial measurements, *Annals of Glaciology* 20:48–54.

[356] Scientific Committee on Antarctic Research (SCAR) (May 2020) *Antarctic Digital Database*, Version 7.2, www.add.scar.org; Paul Cooper, who worked extensively on the SPRI Folio, later moved to the BAS and played a significant role in constructing and operating ADD.

[357] Howat, I M; Porter, C; Smith, B E; Noh, M.-J; and Morin, P (2019) The reference elevation model of Antarctica, *The Cryosphere* 13:665–74, https://doi.org/10.5194/tc-13-665-2019.

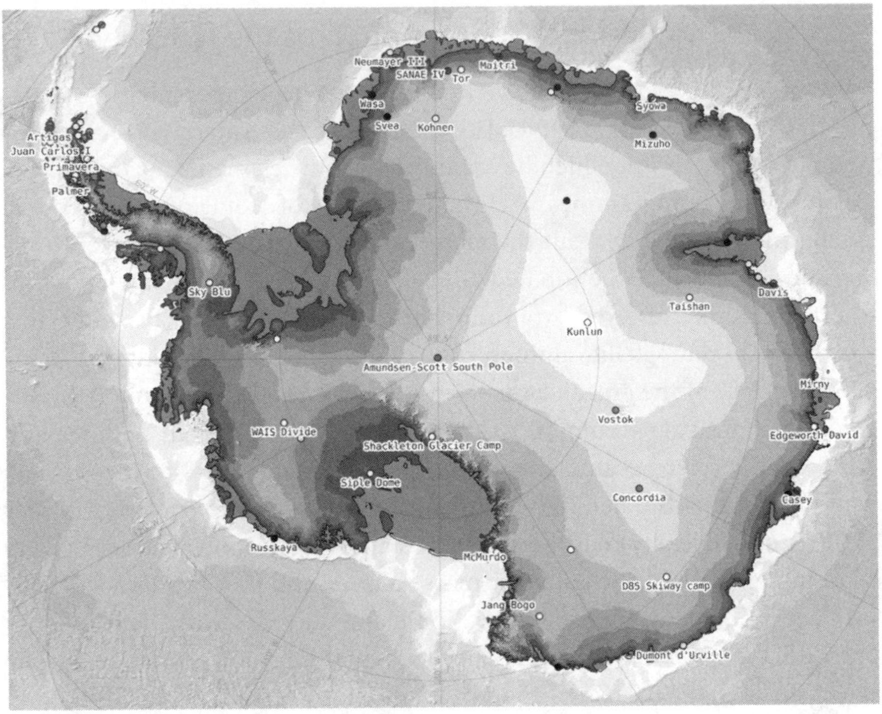

Figure 17.1. Contours of the surface of the Antarctic Ice Sheet. (Courtesy BAS, UKRI).

before moving to the European Space Agency as head of the Ocean and Ice Unit and then to the influential role of head of the Mission Science Division involved in developing ice-observing satellites such as CryoSat-1/2, SMOS, and the Swarm minisatellites in the Earth Explorer series of missions, as well as in interpreting these data over climate-sensitive regions of Antarctica.[358]

ERS-1 and successor international satellite missions—such as, ICESat, Cryosat, GRACE—have supported the modelling of a range of continental-scale glaciological processes—ice flow and ice mass losses and gains, the defining of ice drainage basins, and pinpointing ice shelf and ice stream grounding lines, as well as the movement of ice fronts.[359] For example, using

358 For example, Hogg, A E; et al. (2018) Mapping ice sheet grounding lines with CryoSat-2, *Advances in Space Research* 62 (6): 1191–202.

359 Liu, Y; et al. (2015) Ocean-driven thinning enhances iceberg calving and retreat of Antarctic ice shelves *PNAS* 112 (11): 3263–68.

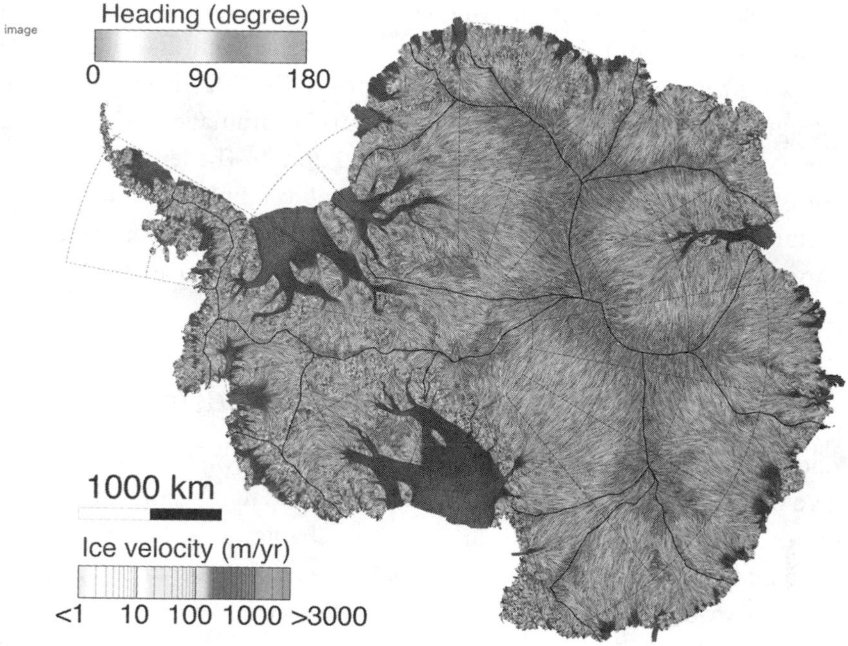

Figure 17.2. Ice velocities and flow pattern over the Antarctic Ice Sheet using SAR interferometric phase data (interior) and speckle-tracking in fast-moving zones. The velocity is logarithmic. The principal ice divides, and hence major drainage basins are also designated. (From Mouginot et al. (2019); see footnote 360). Courtesy the authors and American Geophysical Union).

sequential images and synthetic aperture radar (SAR) interferometric phase data over a 25-year period, Mouginot and colleagues were able to map ice sheet velocities with a precision of 20 cm per year and in direction of 5° over 70%–80% of Antarctica (Figure 17.2[360]).[361] Much of the data emanating from these studies is available at centres such as NASA's Distributed Active Archive Center at the National Snow and Ice Data Center.[362]

As climate change has mercilessly affected the polar regions the value of satellite surveillance has been amply demonstrated and is providing warnings of shifts and alterations of global significance such as ice loss and sea-level rise. One of the first studies to show real, significant ice sheet change

[360] Mouginot, J; Rignot, E; and Schaech, B (2019) Continent-wide, interferometric SAR phase, mapping of Antarctic ice velocity, *Geophysical Research Letters* 46 (16): 9710–18.

[361] Mouginot, J; et al. (2019) (footnote 360)

[362] https://nsidc.org/daac.

from satellites was by Duncan Wingham and colleagues that led directly to the large-scale multidisciplinary programme in the Amundsen Sea sector. Following this, for example, over the period 1979–2017 Eric Rignot and colleagues calculated the ice discharge from 176 drainage basins with an estimated contribution to sea level of 14 ± 2.0 mm.[363] The lengthening time series of observations collected along repeat orbits combined with the high resolution of imaging sensors are enabling datasets to be compared and changes in dynamics and structure to be monitored in unprecedented detail.

17.3 The Subglacial Bedrock Landscape

Following the cessation of the NSF-SPRI-TUD radio-echo programme there was a pause in the acquisition of new ice-thickness data from Antarctica at continental scale. As reported earlier in this chapter, new groups were established in the US and other countries focused on regional data acquisition. Of particular note were the NASA missions undertaken by IceBridge between 2009 and 2019 in Antarctica and Greenland.[364] This comprehensive programme was, to some extent, the successor to the SPRI RES campaign. The aircraft, which included DC-8 and P3-B models, supported a wide range of sensors and not simply new radars for ice-depth sounding.[365] Laser altimeters were the primary instruments, as IceBridge was conceived as a means of 'bridging' the time gap in polar observations between those conducted using NASA's ICESat (2003–2010) and ICESat2, launched in 2018.[366] This ensured there was a continuous collection of data regarding changes in land and sea ice at a time of increasing importance for the study of climate change impacts. While several of these missions focused on West

[363] Wingham, D J; et al. (1998) Antarctic elevation change from 1992 to 1996, *Science* 282 (5388): 456–58, https://doi.org/10.1126/science.282.5388.456); Rignot, E; Mouginot, J; Scheuchl, B; van den Broeked, M; van Wessemd, M J; and Morlighema, M (2019) Four decades of Antarctic Ice Sheet mass balance from 1979–2017, *PNAS* 116 (4): 1095–103.

[364] For Greenland NASA established the Program for Arctic Regional Climate Assessment (PACA) in 1993, which provides an annual venue for scientists to discuss results and develop new projects for research on and understanding the changes occurring to the Greenland Ice Sheet.

[365] The instruments were constructed by the Center for Remote sensing of Ice Sheets (CReSIS) at the University of Kansas and comprised a multichannel coherent radar depth sounder operating over a wide range of frequencies and receiver systems for noise reduction. Another suite of ice-depth radars was flown by the University of Texas Institute for Geophysics (UTIG).

[366] Other instruments included magnetometers and gravimeters, digital systems, and surface temperature sensors.

Antarctica, it was recognised that the East Antarctic Ice Sheet was being much less well investigated. With its considerably larger ice volume and possessing many outlet glaciers with the potential to discharge large quantities of ice into the ocean, it was the focus of another project—ICECAP (Investigating the Cryospheric Evolution of the Central Antarctic Plate) involving teams from US, UK, France, and Australia to study its dynamics and evolution.

In the late 1990s the acquisition of RES data was growing apace, and the glaciological community recognised the need to coordinate and rationalise the ice-thickness material. An international consortium, whose origins lay in a European initiative, was established to combine these results and produce a new topographic map and model of the bed of the Antarctic Ice Sheet. In 1990 the European Science Foundation (ESF) and the European Commission (Directorate-General XII) together established an advisory panel for enhancing collaboration in marine and polar science. This was the European Committee on Ocean and Polar Sciences (ECOPS), tasked with crafting a European long-term strategy and with fostering major European projects. The author was a member of ECOPS. In polar science, ice sheet–climate interactions were seen as a key topic, and two major proposals were implemented—EPICA (the European Project for Ice Coring in Antarctica, which fostered the deep drilling at Dome C and later in Dronning Maud Land) and EISMINT (the European Ice Sheet Modelling Initiative). The latter, which the author initially convened, concentrated on achieving a better understanding of the dynamics of large ice masses and the relationship of their behaviour to climate forcing. An essential prerequisite for modelling was an up-to-date depiction of the surface, thickness, and bed. Under the joint sponsorship of EISMINT and the SCAR, an international database of ice-thickness measurements over Antarctica was established, which resulted in a new topographic model of the bed, including the seafloor of the surrounding continental shelf, that would assimilate and update the existing SPRI bedrock dataset. The project was called Bedmap and had an initial working group of 21 scientists from eight countries and was led enthusiastically by David Vaughan from the BAS. After several years of progress, a new model and map were produced in 2000/2001.[367] Data are added continuously, and the latest version (Bedmap2) comprises

[367] Lythe, M B; Vaughan, D G; and the BEDMAP Consortium (2000) BEDMAP—bed topography of the Antarctic, 1:10,000,000 scale map, BAS (Misc.) 9, Cambridge: British Antarctic Survey.

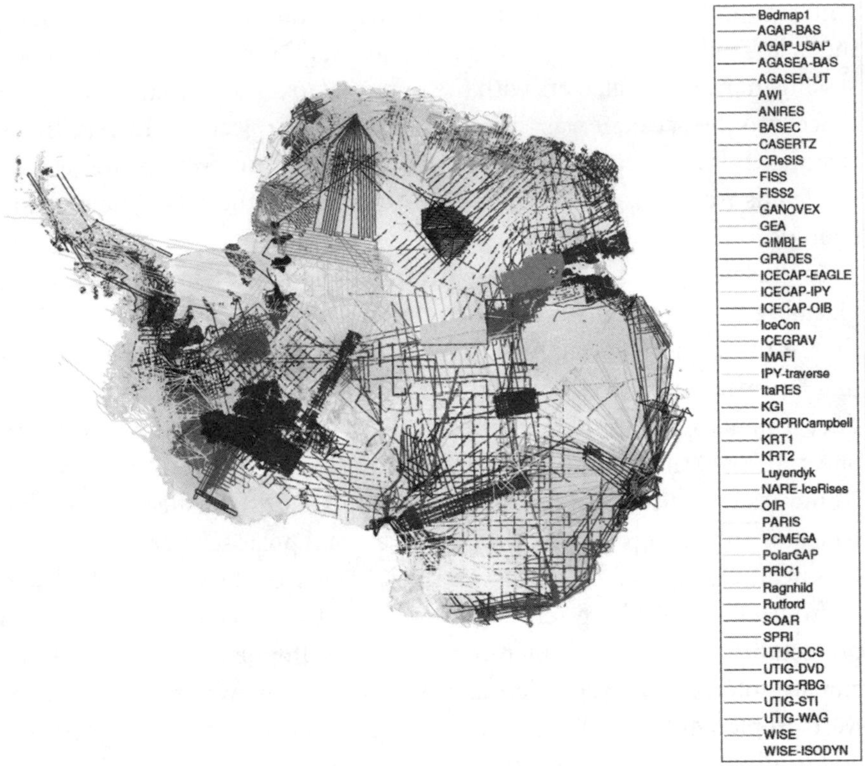

Figure 17.3. Radio-echo sounding flight lines used in the Bedmap compilation to 2019. (Courtesy BAS, UKRI).

some 25 million measurements, over two orders of magnitude more than were used in the first compilation.[368] A further development has been the use of mass conservation, streamline diffusion, and other 'smart' methods to derive bed topography and a bathymetry map of Antarctica in a project called BedMachine. Figure 17.3 illustrates the extensive coverage of Antarctica by RES, including the SPRI work of 35–40 years ago (where the navigation was of sufficient accuracy). A version of BedMap2 is reproduced in Figure 17.4.

To illustrate how our appreciation of the bedrock topography of Antarctica has developed and to illustrate the extraordinary detail that is now

[368] Fretwell, P; et al. (2013) Bedmap2: Improved ice bed, surface and thickness datasets for Antarctica, *The Cryosphere* 7:375–93, https://doi.org/10.5194/tc-7-375-2013; Pope, A (2017) Antarctica Bedmap2 [dataset], University of Edinburgh, https://doi.org/10.7488/ds/1916.

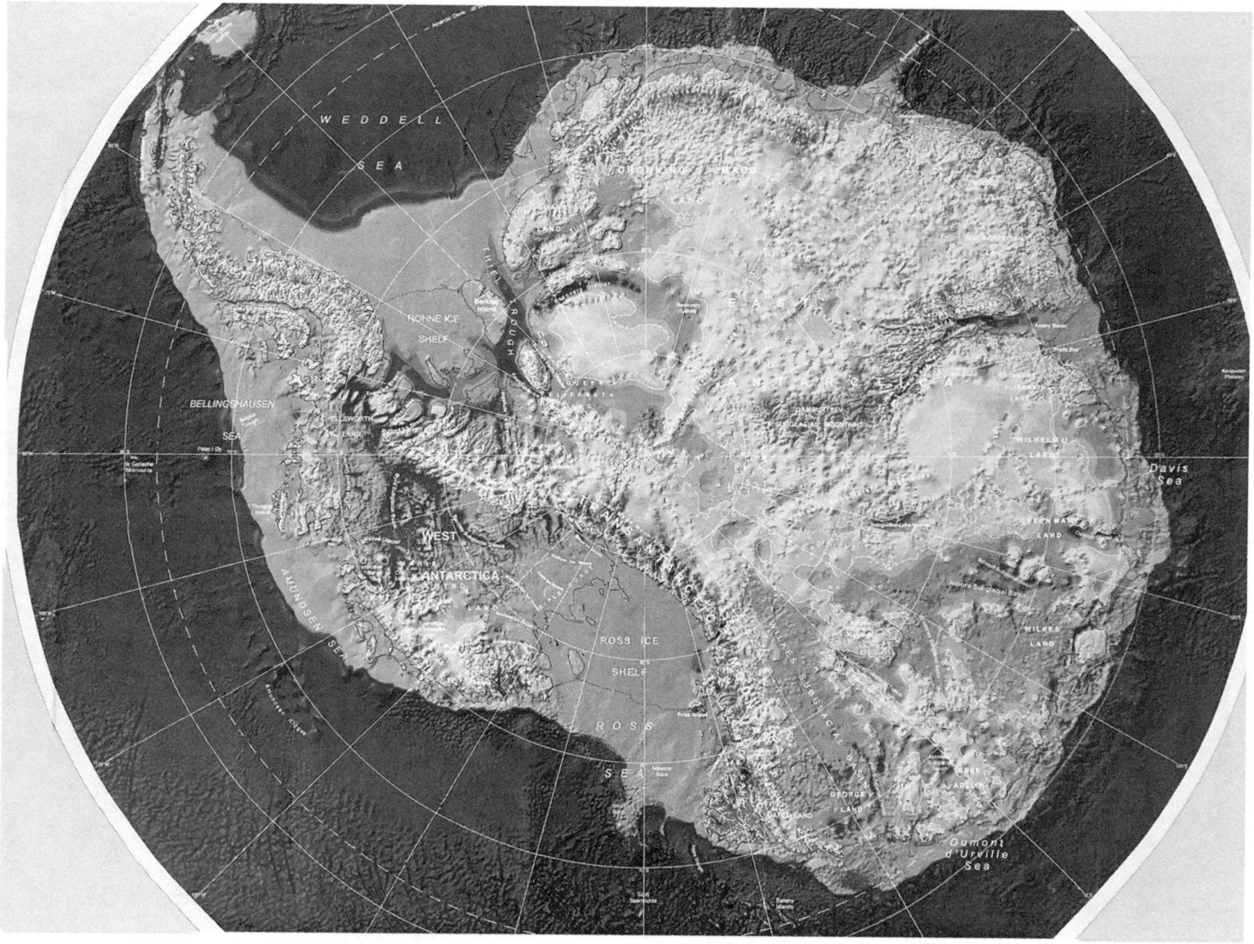

Figure 17.4. Bedmap2 depicting subglacial relief of Antarctica. (Courtesy BAS, UKRI).

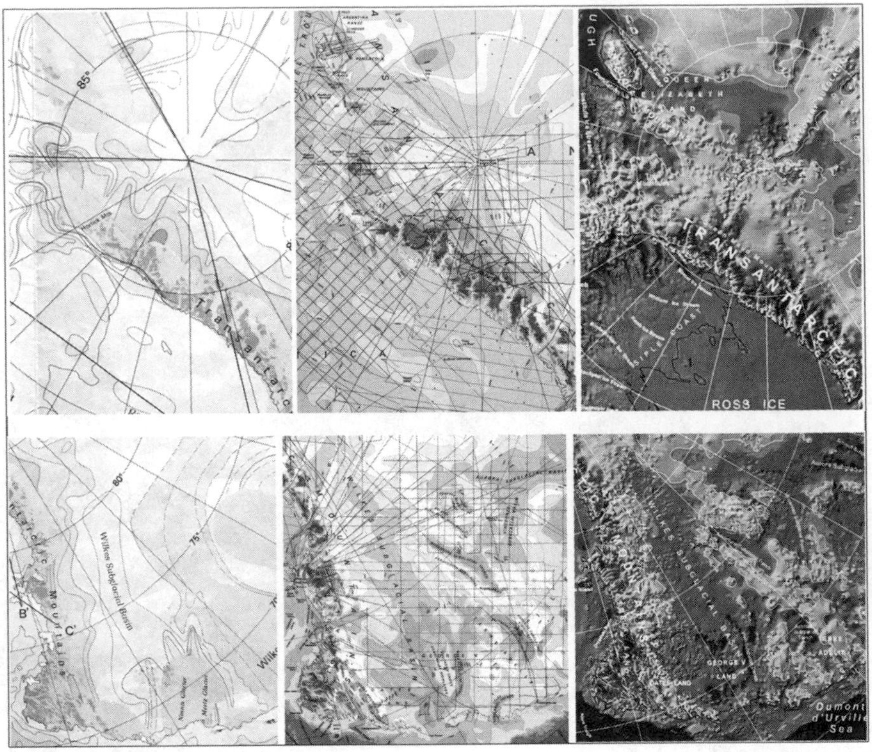

Figure 17.5. Two comparisons of Antarctica's bedrock relief. Left: Bentley (1972); see footnote 368. Centre: SPRI Folio (1983) (Courtesy SPRI). Right: Bedmap2 (Courtesy BAS and Fretwell et al. (2013); see footnote 368).

available, Figure 17.5 shows how selected regions have been depicted in the compilations by Bentley (1964), the SPRI Folio (1983), and Bedmap2.[369]

These contemporary RES initiatives have progressively been directed at investigating the topography around the margins of Antarctica, where ice streams constitute the primary discharge of ice to the ocean. While the focus has been on Thwaites Glacier with its multinational and large-scale field programmes, it has become evident that the continent is ringed by many other outlets whose stability and possible disintegration are in question; deep

[369] Bentley, C R (1972) 'Subglacial Topography', in Heezen, B C; et al. (1972) 'Morphology of the Earth in the Antarctic and Subantarctic', Antarctic Map Folio Series No.16, American Geographical Society.

troughs have been discovered in several locations.[370] The inland configuration of ice streams is critical as to whether they are stable features or, with increased melting due to oceanic and atmospheric warming, may retreat rapidly towards the interior, increasing the release of ice into the ocean and thus contributing to the rise of global sea level.

17.4 Water beneath the Ice

One of the noteworthy and beguiling discoveries of the SPRI programme was the detection of large bodies of water beneath the ice sheet—lakes. Such findings stirred the imagination not only of scientists but of the general public about what these lakes might reveal and contain, locked away in such an extreme environment for hundreds of thousands of years, or even longer. The lakes, and smaller water bodies that have been termed 'ponds', are likely however, to represent a small fraction of the overall area of basal ice that is thawing at the pressure melting point. The widespread presence of thin layers of water (millimetres to a few centimetres in thickness) is directly relevant to the dynamics of the ice sheet and is an important boundary condition to be incorporated into numerical models of ice sheet behaviour.[371]

By the late 1990s, as recounted in earlier chapters, there was growing interest in surveying the largest of the discovered lakes beneath the Russian Vostok Station. Italian, US, and Russian research groups in particular commenced detailed geophysical exploration of subglacial Lake Vostok in 1999. Sergey Popov summarised the information from closely spaced radio-echo sounding flights, and his surface topographic compilation is shown in Figure 17.6, indicating the extraordinary detail now available.[372]

The SCAR took a direct interest in Lake Vostok and established a group of specialists to coordinate the various research projects that emerged. To balance and promote the numerous priorities of the investigating parties with strongly competing interests this coordination evolved into a research

[370] Morlighem, M; Rignot, E; Binder, T; Blankenship, D; Drews, R; Eagles, G; et al. 2020. Deep glacial troughs and stabilizing ridges unveiled beneath the margins of the Antarctic ice sheet, Nature Geoscience 13:132–37.

[371] Oswald, GKA, personal communication, February 2021; Oswald et al. (2018) Radar evidence of ponded subglacial water in Greenland, *Journal of Glaciology* 64 (247): 711–29.

[372] Popov, S (2020) (footnote 345).

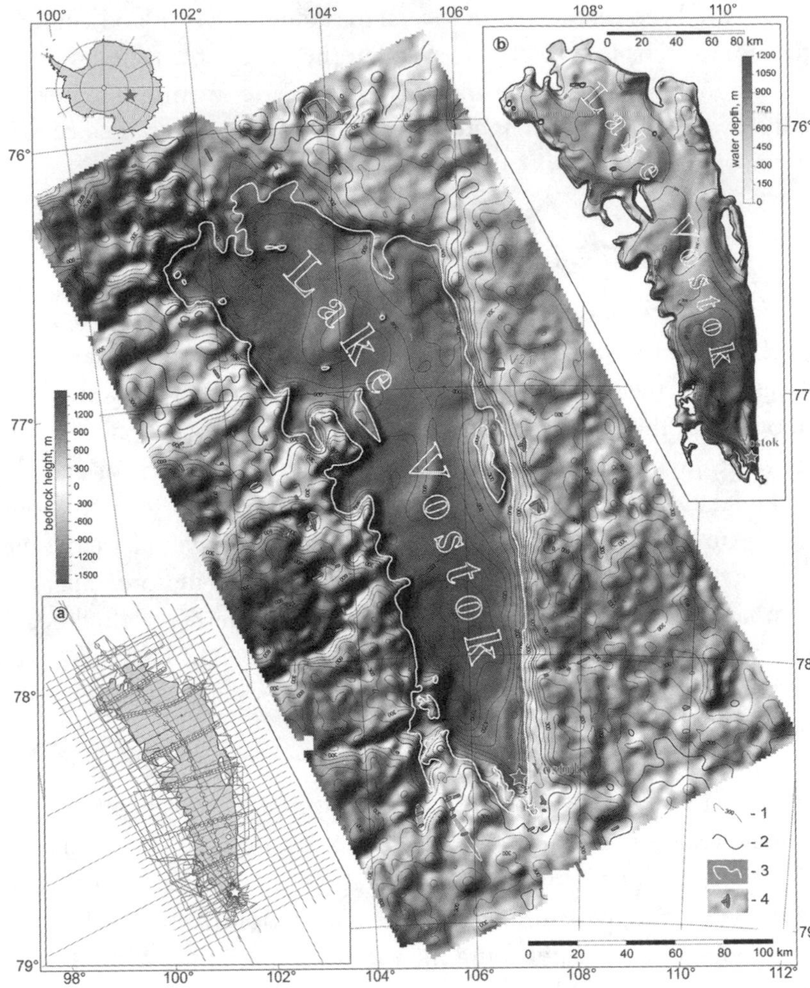

Figure 17.6. Sub-ice bedrock basin containing Lake Vostok from RES and seismic sounding. Inset bottom left shows the sounding lines. Inset top right shows the estimated water depth. (From Popov (2020); see footnote 345). (Courtesy International Glaciological Society).

programme: Subglacial Antarctic Lake Environments (SALE) to cover its broad range of physical, chemical, and biological aspects. A particularly crucial issue was that access to lakes had to avoid any contamination to ensure, amongst other discoveries, that any possible life in the lake would not be compromised.

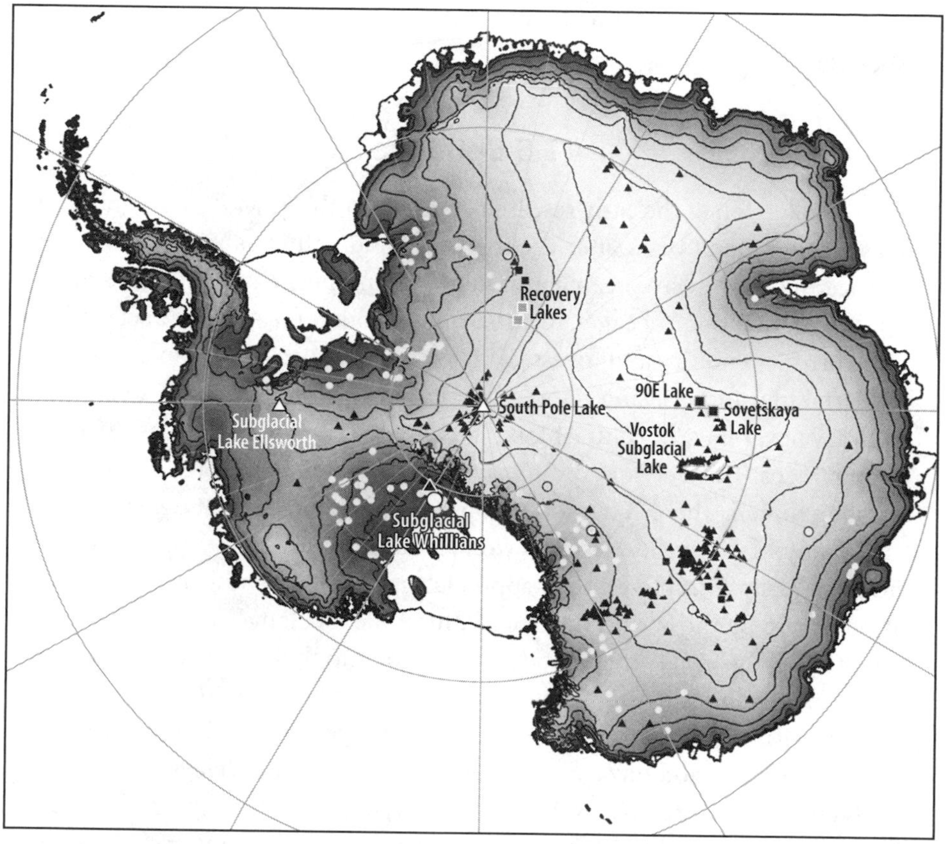

Figure 17.7. Location of subglacial lakes in Antarctica as determined by a variety of methods—RES, seismic sounding, gravity mapping, surface height changes). (Adapted from Siegert et al. (2018); see footnote 374).

The investigation of sub-ice lakes presents enormous logistical, scientific, and engineering challenges, as experienced at subglacial Lake Ellsworth.[373] It is clear that further exploration will identify additional sub-ice lakes; there are more than 400 at the time of writing this book. A map showing a latest inventory is depicted in Figure 17.7.[374] Future studies and investigations will reveal their hidden stories and likely provide remarkable

[373] https://web.archive.org/web/20121231133642http://www.antarctica.ac.uk/press/press_releases/press_release.php?id=2014.

[374] Siegert, M J (2018) A 60-year international history of Antarctic subglacial lake exploration, *Geological Society, London, Special Publications* 461 (1): 7–21, https://doi.org/10.1144/SP461.5.

insights into their characteristics and the processes at the base of the ice sheet.[375]

17.5 Epilogue

This recollection of the early research ventures of the SPRI RES programme was written many years after the incidents and activities mentioned. The pioneering efforts have been continued and, in many cases, superseded by the next generations of eager, talented, and resourceful Antarctic scientists. The thrust of enquiry and investigation moves ever onward. Some of those undertaking studies using remote-sensing data, from satellites in particular, may never have been to Antarctica, living vicarious polar lives through their screens. This does not diminish in any way their contributions and scientific acumen. There is no doubt in my mind, however, that experiencing the manifold dimensions of this wonderous continent firsthand yields an added and significant aesthetic appreciation to the scientific endeavour. It is often said by those who have confronted the challenges and been captivated by its raw beauty that Antarctica has got into their bloodstream. Many of us understand that is an irreversible transfusion.

My SPRI colleagues and I who served on those first airborne missions faced adventure and hazard, endured frustrations and delays, and toiled with analyses and reports welcome the burgeoning interest in Antarctica. A large number of studies by scientists from many nations are revealing more secrets of this remote continent and the land hidden beneath its mammoth ice sheet to which, with good luck and opportunity, we contributed a soupçon of early knowledge. They have led to a recognition of the critical role played by its icy burden in the life-support systems of planet Earth. It was possible even several decades ago to discern clear signs of its vulnerability to a changing external environment. The pace of that change has quickened. Active research on this most important continent gives hope that we will better understand this vital component of our precious world.

[375] Zotikov, I A (2006) (footnote164); Siegert M J; Jamieson, SSR; and White, D A (eds.) (2018) *Exploration of Subsurface Antarctica: Uncovering Past Changes and Modern Processes*, London: Geological Society of London, Special Publications 461:1–6, https://doi.org/10.1144/SP461.15.

Appendix 1

DISPLAY AND RECORDING OF RES DATA

This appendix provides a brief overview of the type of records produced by SPRI RES systems to aid interpretation of the several images in the text. Figure A.1 is a simple representation of the transmission and return of radar signals to, through, and from the ice base when the system is operating from an aircraft.

The transmitted pulse is directed downwards from aerials attached to the underside of the aircraft. The inset shows the returned signal strength as a function of time. The first pulse is the transmission, the strong second spike is the return from the ice sheet surface, and the last, weaker return is from the base.

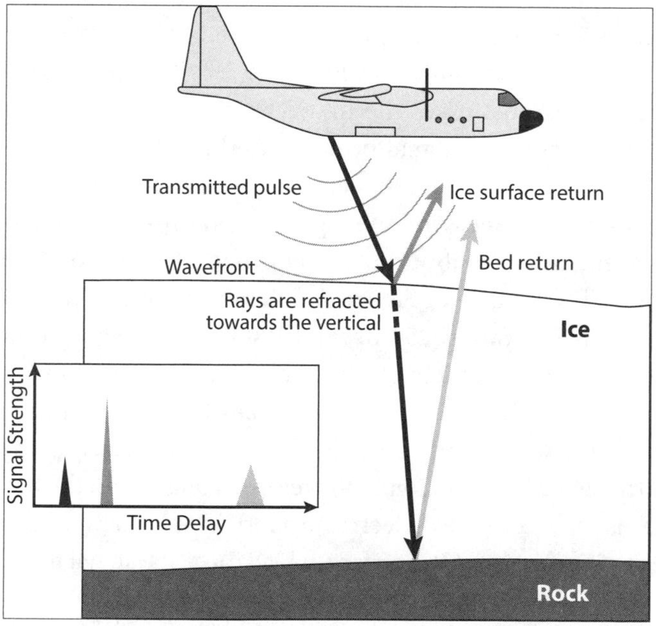

Figure A.1. Basic geometry of radio-echo sounding from an aircraft.

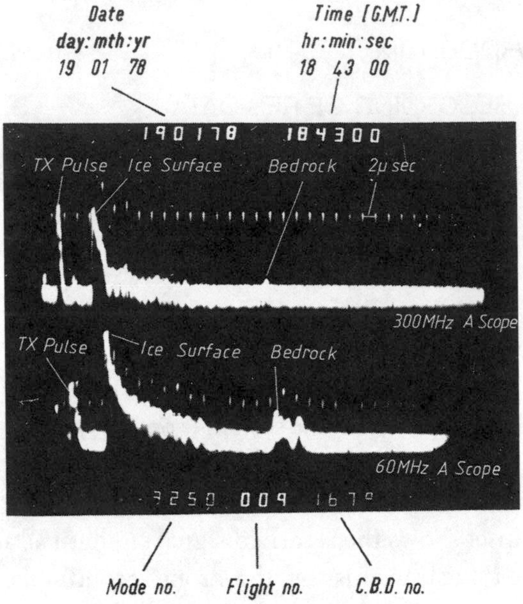

Figure A.2. Key features of an oscilloscope A-scope display for 300 MHz and 60 MHz TUD RES systems as used in the 1977–78 and 1978–79 Antarctic seasons (TX=transmitter pulse, and CBD=a counter number (coded binary decimal)).

Prior to digital recording, the returning signals were displayed on an oscilloscope. An oscilloscope can be set up to present the signals in two modes, called *A-scope* and *Z-scope*.

In the A-scope mode, shown in Figure A.2[376], TX represents the transmitter trigger. In most cases the first return thereafter is the reflection from the ice surface. The signal now passes through the ice, and any significant changes within the ice that alter the dielectric properties can result in a reflection. The decay of the signal strength shows various small-amplitude returns, probably caused by different strata or layers within the ice sheet. The images in Figure A.3 depict three typical pulse-length settings at 60 MHz: 60 ns, 250 ns, and 1000 ns, corresponding approximately to 10, 40, and 170 m in ice, respectively. The selection is important when investigating the details of certain features such as internal layers.

Eventually the wave front strikes the bottom of the ice, and the energy is reflected. The shape and duration of the returned pulse from the bed or from a water layer is related to the dielectric properties of either and the degree of roughness or irregularity of the surface, which cause the signal to be scattered.

[376] This figure and Figure A.4 are reproduced from 'NSF-SPRI-TUD Radio Echo Sounding Programme 1977–78, Guide to Film Interpretation' (compiled by E Jankowski), SPRI, 1978.

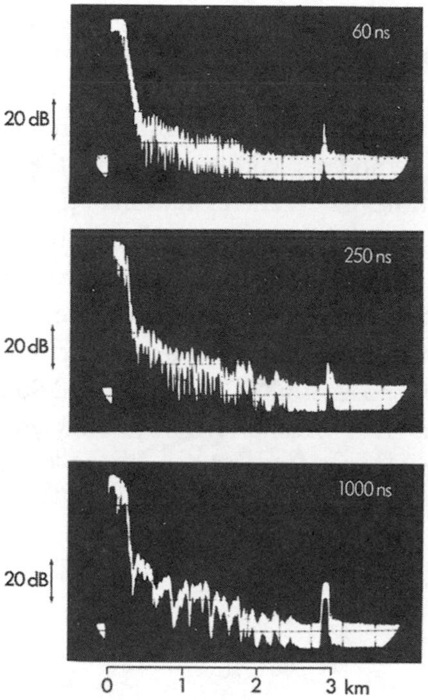

Figure A.3. RES A-scope record, at the South Pole, showing the effect of different pulse lengths (TUD 60 MHz). Vertical scale is the returned signal strength; horizontal axis is the range in kilometres. Note the much finer resolution at the shorter pulse lengths. The return at ~3 km depth is from the bedrock. (Courtesy David Millar).

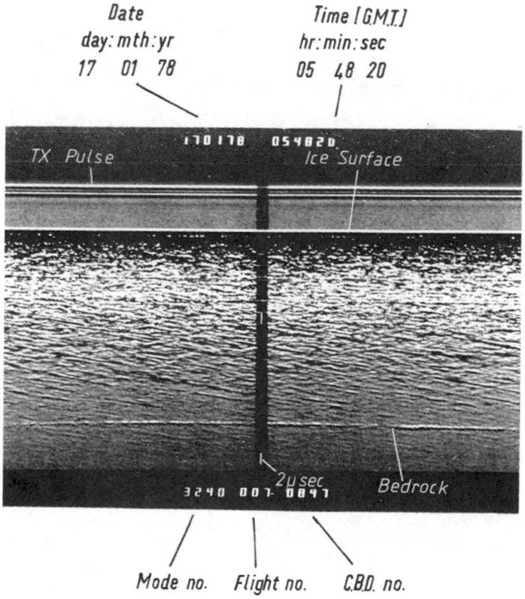

Figure A.4. Key features of an oscilloscope Z-scope display for the 60 MHz TUD RES system as used in the 1977–78 and 1978–79 Antarctic seasons. The display features for 300MHz are virtually identical.

The A-scope was helpful for monitoring but did not produce a visual profile of the ice thickness. In the Z-mode (Figure A.4) photographic film is transported across an intensity-modulated oscilloscope screen with the time base switched off. Weak echoes are integrated, and the continuity of returns is clearly displayed; this was the best method of recording data until the advent of fully digital systems. The brightness, size, and definition of the spot on the oscilloscope were important, as was the sensitivity of the film. The speed of transport of the film was significant, as it controlled the rate of integration of the echoes—an important factor in detecting very weak returns. Special cameras with large cassettes to hold many tens of metres of 35mm film were acquired for these purposes from D Shackman and Sons, a company in the UK. In later seasons (1977–78 and 1978–79) a Honeywell 1856A fibre-optic recording oscillograph was also used.

Appendix 2

SPRI RESEARCH STUDENTS IN THE RADIO-ECHO SOUNDING PROGRAMME

Table A2–1. SPRI Research Students in the Radio Echo Sounding Programme (1962–1987)

Name	Dates	Thesis Title (PhD unless otherwise stated)
Walford, Michael E R	1962–1965	Radio Echo Sounding of Polar Ice Masses
Bailey, Jeremy Thomas	Killed in an Antarctic crevasse accident, 12 October 1965	
Paren, Julian Gerald	1965–1970	Dielectric Properties of Ice
Ewen-Smith, Beverley Michael	1968–1971	Radio Echo Studies of Glaciers
Harrison, Christopher Hanson	1968–1972	Radio Propagation Effects in Glaciers
Drewry, David John	1969–1973	Sub-ice Relief and Geology of East Antarctica
Davis, John Leslie	1971–1973	The Problem of Depth Sounding Temperate Glaciers (MSc)
Oswald, Gordon Kenneth Andrew	1971–1975	Radio Echo Studies of Polar Glacier Beds
Neal, Christopher Scott	1972–1977	Radio Echo Studies of the Ross Ice Shelf
Hargreaves, Neil Duncan	1974–1977	Radio Echo Studies of the Dielectric Properties of Ice Sheets
Rose, Keith Everard	1973–1978	Radio Echo Sounding Studies of Marie Byrd Land, Antarctica
Steed, Richard Hugh Nicholas	1977–1980	Geophysical Investigations of Wilkes Land, Antarctica
Jankowski, Edward Jan	1977–1981	Airborne Geophysical Investigations of Sub-Glacial Structure of West Antarctica
Millar, David Hugh McFarlane	1978–1981	Radio-Echo Layering in Polar Ice Sheets

(Continued)

Table A2–1. Continued.

Name	Dates	Thesis Title (PhD unless otherwise stated)
McIntyre, Neil Forbes	1980–1983	The Topography and Flow of the Antarctic Ice Sheet
Dowdeswell, Julian Andrew	1981–1984	Remote Sensing Studies of Svalbard Glaciers
Bamber, Jonathan Louis	1983–1987	Radio Echo Studies of Svalbard Glaciers
Watts, Leonard Gary	1984–1987	Finite Simulations of Ice Mass Flow
Drinkwater, Mark Roland	1984–1987	Radar Altimetric Studies of Polar Ice

Abbreviations

AARI	Arctic and Antarctic Research Institute, St. Petersburg (formerly Leningrad)
ADD	Antarctic Data Directory
AGAP	Antarctica's Gamburtsev Province Project
AGS	American Geographical Society
ANARE	Australian National Antarctic Research Expeditions
APL	Applied Physics Laboratory, Johns Hopkins University
ARDS	Airborne Research Data System
BAS	British Antarctic Survey, Cambridge, UK
Bedmap/2	Suite of gridded products describing surface elevation, ice thickness, and the sea floor and subglacial bed elevation of the Antarctic south of 60°S.
BLFZ/BEFZ	Basal layer-free zone (sometimes referred to as Basal echo-free-zone)
BP	Formerly British Petroleum
C-121(J)	Lockheed Super Constellation aircraft (J refers to the model)
C-130(F, R)	Lockheed Hercules aircraft (F and R refer to models)
C-141	Lockheed StarLifter aircraft
CHCH	Christchurch, New Zealand
CHS	Canadian Hydrographic Service
COMNAVSUPPFORANTARCTICA	Command Naval Support Force for Antarctica
CRAMRA	Convention for Regulation of Antarctic Mineral Resource Activities
CReSIS	Center for Remote Sensing of Ice Sheets, University of Kansas, USA
CRREL	Cold Regions Research and Engineering Laboratory, Hanover, New Hampshire, USA

CTD	Conductivity-temperature-depth measurement device in oceanography
DEW Line	Distant Early Warning Line (a system of radar stations across the Arctic regions, including Canada, Greenland, Alaska, and Iceland)
DOS	Directorate of Overseas Surveys (UK)
DRB	Defence Research Board (Canada)
DVDP	Dry Valley Drilling Project
DYE-3	DEW Line station located on the ice sheet in southern Greenland
ECOPS	European Committee for Ocean and Polar Science
EISMINT	European Ice Sheet Modelling Initiative
ELRDL	US Army Electronics Research and Development Laboratory (later USAEL)
EPF	Expéditions Polaires Françaises
EPICA	European Project for Ice Coring in Antarctica
ERS-1	European Remote Sensing satellite no. 1
ESA	European Space Agency
ESF	European Science Foundation
ESM	Echo strength measurement
ESTAR	Trade name for Kodak's polyester-base film product
FE	Finite element
FIDS	Falkland Islands Dependencies Survey
FSS	Flying-spot scanning
GEBCO	General Bathymetric Chart of the Oceans
GPRC	Geophysical and Polar Research Center, University of Wisconsin–Madison, USA
GRACE	Gravity Recovery and Climate Experiment (satellite mission 2002–2017 that provided data on mass loss of ice sheets in Greenland and Antarctica)
IAGP	International Antarctic Glaciological Project
ICBM	Intercontinental ballistic missile
ICSI	International Commission for Snow and Ice
IDIOT	Ice Depth Instrument, Operator Transported (SPRI 300 MHz system)

IFR	In-flight rations (US military)
IGY	International Geophysical Year
INS	Inertial navigation system
JATO	Jet-assisted take-off
LGGE	Laboratoire de glaciologie et géophysique de l'environnement, Grenoble, France
LIMA	Landsat Image Mosaic of Antarctica
MAC/MAW	Military Airlift Command/ Military Airlift Wing (US)
MAGIC	Mapping and Geographic Information Centre
MSSL	Mullard Space Science Laboratory, University College, London
NAC	National Airways Corporation (former New Zealand domestic airline, merged with Air New Zealand in 1978)
NASA	National Aeronautics and Space Administration (US)
NBSE	Norwegian-British-Swedish Expedition
NERC	Natural Environment Research Council (UK)
NP	Norsk Polarinstitutt
NSF	National Science Foundation (US)
NWC	Naval Weapons Center (US)
NZARP	New Zealand Antarctic Research Programme
OAP	Office of Antarctic Programs, US National Science Foundation
OAT	Outside air temperature
ONR	Office of Naval Research (US)
OPP/DPP	Office of Polar Programs/Division of Polar Programs (US)
PCSP	Polar Continental Shelf Project
PoI	Pole of Relative Inaccessibility
QMLT	Queen Maud Land Traverse
RAL	Rutherford Appleton Laboratory, UK
RAR	Rapid-access recording photographic film
RES	Radio-echo sounding
RGS	Royal Geographical Society, London

RIGGS	Ross Ice Shelf Geophysical and Glaciological Survey (part of RISP)
RIS	Ross Ice Shelf
RISP	Ross Ice Shelf Project
SAE	Soviet Antarctic Expeditions
SAG	Satellite Altimeter Group, SPRI
SALE	Subglacial Antarctic Lake Environments
SANAE	South African National Antarctic Expeditions
SCAR	Scientific Committee on Antarctic Research
SCR-718	An aircraft radio altimeter
SFIM	Société de Fabrication d'Instruments de Mesure (French company that designed early models of aircraft flight recorders)
SIPRE	US Army Snow, Ice and Permafrost Research Establishment (forerunner of CRREL)
SLAR	Side-looking airborne radar
SNSK	Store Norske Spitsbergen Kulkompani
SPRI	Scott Polar Research Institute, University of Cambridge
TACAN	Tactical air navigation system, used by military aircraft; it provides the user with bearing and distance to a ground or ship-borne station
TUD	Technical University of Denmark, Lyngby
UHF, VHF, HF	Ultra-high-, very-high-, and high-frequency radio bands, respectively
UKRI	United Kingdom Research and Innovation (non-departmental public body bringing together seven disciplinary research councils including NERC)
USAEL	US Army Electronics Laboratory
USARP/USAP	United States Antarctic (Research) Program. The "Research" was dropped in 1985
USGS	United States Geological Survey
USMC	United States Marine Corps
VXE-6	Antarctic Development Squadron Six (US Navy air test and evaluation squadron)
WISP	West Antarctic Sheet Project

Glossary

A-scope display	An oscilloscope mode that displays a waveform of physical phenomena that can be converted to a voltage such as vibrations, temperature, or electrical properties (e.g. current or power). A voltage waveform shows time on the horizontal axis and voltage on the vertical axis (see Appendix 1)
Ablation	All processes by which snow and ice are lost from a glacier, snow cover, or floating ice by melting, evaporation, sublimation, wind erosion, and avalanches.
Absorption (dielectric)	Frequency-dependent loss in transmitted radio energy in ice due to defects in the crystal lattice and the presence of any impurities.
Attenuation	The reduction of the amplitude of a signal, electric current, or other oscillation.
Avionics	All the electronic equipment installed in an aircraft, for example, displays and systems controlling flight, navigation, communications, flight recorders, fuel, and engines.
Balance velocity	For a given cross-sectional area, the mean velocity required to discharge the snow/ice accumulated upstream in a given time assuming steady state.
Bandwidth	The reciprocal of the pulse width in radar systems.
Basal shear stress	The force at the base of an ice mass that balances the down-slope weight of the ice above (can also be considered a frictional term).
Blue ice	Solid glacier ice that has been exposed at the surface by wind and sublimation. See also Sublimation.
Clutter echoes	Unwanted echoes in radar systems that can cause serious performance issues and obscure critical returns.

Cold-based and warm-based glaciers	General terms that describe a glacier that is either frozen to its bed (cold) or whose base is at the pressure melting point (warm).
Critical angle	Greatest angle at which a ray travelling in one transparent medium can strike the boundary between that medium and a second of lower refractive index without being totally reflected within the first medium.
Dead reckoning	A method of calculating the current position of an aircraft using a previously determined position and then successively estimating speed, heading, and course over elapsed time.
Deconvolution	See Migration.
Depth hoar	Large snow crystals at the base of snow layers that form when uprising water vapor deposits or de-sublimates onto existing snow crystals. Depth hoar crystals are bound poorly to each other and form a weak layer in a snowpack.
Dielectric constant	The ratio of the electric permeability of a material to the electric permeability of free space (i.e. vacuum).
Dipole antenna	Simplest and most widely used class of antennas consisting of a horizontal metal rod with a connecting wire at its centre.
Doppler system	Aircraft system for determining speed from measuring the Doppler effect of echoes from three or four beams directed downwards from the aircraft.
Driving stress	Mean stresses within an ice mass that are in balance with variable basal shear stresses and longitudinal stress gradients. Calculated in similar manner to basal shear stress. See also Basal shear stress.
Dyke	A discordant intrusive sheet which cuts across older rocks. See also Sill.
Elastic wave	Motion in a medium in which, when particles are displaced, a force proportional to the displacement acts on the particles to restore them to their original position. Sound waves as used in seismic exploration are an example of elastic waves.
Fading	Variation of the attenuation of a radio signal as a result of irregularities in a rough reflecting surface

	that are of the same order as or greater than the radio wavelength. Fading can be of the order of 10 dB.
Firn	Old snow that has been transformed into a denser material. Snow grains, to some extent, are joined together, but the snow is still porous (density: 400–800 kg m^{-3}).
Flowlines	Expression (usually at the surface) of the three-dimensional path traced by particles of ice within a moving ice mass.
Geothermal heat flux	Flow of heat energy resulting from radioactive decay within the interior of Earth.
Glacier surge	Catastrophic advance of an ice mass in which velocities increase suddenly by up to an order of magnitude. The ice surface may become intensely crevassed. Frontal advance may vary from a few tens of metres to several tens of kilometres.
Glaciomarine sediments (also Glacimarine)	A general term describing inorganic and organic material deposited in a marine setting by a combination of glacier- and marine-related processes.
Gondwana	The southern continents of Antarctica, India, South America, Australia, and Africa (including Madagascar) that formed a separate super-continent following the breakup of Pangaea. See also Pangaea.
Ground effect	Positive influence on the lifting characteristics of the horizontal surfaces of an aircraft wing when it is close to the ground. It increases air pressure on the lower wing surface ('ram' or 'cushion' effect), thereby improving the aircraft lift-to-drag ratio.
Grounding line	Zone where an ice sheet, ice stream, or glacier flowing into the sea comes afloat.
Ice cap	A dome-shaped glacier, usually covering a highland area; considerably smaller than an ice sheet.
Ice front	Vertical cliff forming the seaward face of an ice shelf or floating glacier (2–50 m in height).
Ice rise	A usually dome-shaped mass of ice resting on rock and surrounded by an ice shelf, or partly by

	an ice shelf and partly by sea; no rock is exposed (can reach more than 100 km in size).
Ice sheet	Ice mass of considerable extent and thickness (>50,000 km^2).
Ice sheet modelling	Experiments to make quantitative predictions about the behaviour of ice sheets. They can assist in constraining future responses of ice masses in relation to external forcing.
Ice shelf	Floating plate of ice, usually with a level or gently undulating surface, extending from an ice sheet over the adjacent sea and nourished principally by snowfall.
Ice stream	Part of an ice sheet in which the ice flows more rapidly. Margins are often marked by change in surface slope and the presence of bands of shear crevasses.
Ice tongue	The extension of an individual glacier or ice stream into the open sea, from a few hundred metres to several tens of kilometres in length.
Ice fall	Zone of heavily crevassed ice typically flowing over a rock step.
Impedance	A measure of the opposition that a circuit presents to an electric current when a voltage is applied.
Inertial navigation	Automated dead-reckoning navigation system using accelerometers to sense aircraft motion and gyroscopes for rotation to calculate position and velocity.
Ionosphere	Layer of Earth's atmosphere that is ionized by solar and cosmic radiation comprising electrons and electrically charged atoms and molecules stretching from a height of about 50 km to more than 1,000 km
Isochron	Lines or horizons connecting points of equal age.
Isostatic adjustment	Sinking or rising of large parts of Earth's crust with respect to the denser asthenosphere; caused by a heavy weight placed on or removed from Earth's surface, such as a large continental ice sheet.

Isotope analysis — Method of determining patterns of climatic change over long periods using the ratio of the stable isotopes of oxygen (^{18}O:^{16}O) and hydrogen (^{2}H:^{1}H) as a proxy indicator of temperature.

Jamesway — A portable and easy-to-assemble hut with wooden ribs and an insulated fabric covering, designed for Arctic weather conditions. A version of the Quonset hut created by James Manufacturing Company (Fort Atkinson, Wisconsin).

Magnetic compensation — Procedure comprising a series of aircraft manoeuvres used to calculate and remove any effects of the aircraft and its magnetic signature on magnetic field measurements.

Magnetic susceptibility — The degree to which a material (e.g. rocks) can be magnetized in an external magnetic field.

Migration — A process for reconstructing the space-time positions of echoes returned from a surface.

Moraine — Sediment or rock debris eroded and transported by an ice mass (many forms on glaciers include basal, lateral, medial, and end moraines); usually poorly sorted, often with massive structure, and containing striated and angular clasts. Also known as *till.*

Mylar — Chemically matted drafting film suitable for pen-and-pencil drawing, drafting, and artistic applications, capable of withstanding extensive revisions and changes.

Nunatak — A rocky, isolated peak protruding through an ice mass such as a glacier or ice sheet.

Outlet glacier — A valley glacier that drains an inland portion of an ice sheet or ice cap and flows through a gap in peripheral mountains

Ozone — (O_3) A highly reactive gas in the atmosphere created by the reaction of ultraviolet light with diatomic oxygen molecules (O_2) that splits them into individual oxygen atoms (O) that then combine with other O_2 molecules. The ozone layer, between 15 and 35 km above Earth's surface (the stratosphere), contains a high concentration of O_3 in relation to other parts of the atmosphere and reduces the amount of harmful ultraviolet radiation reaching Earth.

P and S waves	In seismology, P (or primary) waves are compressional elastic waves that are longitudinal in nature, so they travel faster than other waves through the earth and the first to arrive at a seismograph station or exploration geophone. S waves are shear (or secondary) elastic waves that are transverse in nature, meaning that the particles of an S wave oscillate perpendicular to the direction of wave propagation.
Pangaea	A super-continent that existed during the Late Palaeozoic and Early Mesozoic eras. It assembled from earlier continental units approximately 330 million years ago, and it began to break apart about 200 million years ago. It comprised the large northern continent called Laurasia and the southern continent called Gondwana. See also Gondwana.
Periglacial	Applying to terrain affected by proximity to glacial conditions, characterized by frost shattering and other cold-climate processes with or without the presence of permafrost.
Permittivity	(also known as *dielectric constant*) A measure of the amount of electric potential energy, in the form of induced polarisation, stored in a given volume of material under the action of an electric field.
Polar diagram	A plot of the radiation patterns of an antenna indicating the magnitude of the power radiated from the dipole in any given direction.
Polar grid	Navigation system based on the use of a grid, typically oriented parallel to a specified meridian of longitude, overlaid on the appropriate polar stereographic projection navigational chart.
Polar path compass	Alternative to conventional magnetic compass and directional gyro in higher latitudes, where magnetic deviations are problematic. The compass is mounted on a platform kept horizontal by a gyroscope. The directive element must be non-pendulous.
Polarisation	The orientation of the electric field of a radio wave. Polarised waves have a fixed, constant

	orientation and create a path shaped like a plane as they travel through space.
Pressure melting point	The temperature at which ice begins to melt under a given pressure. Ice melts at 0°C under ordinary atmospheric pressure at sea level. At higher pressures, the melting point is lower, and water can remain liquid at temperatures below its ordinary freezing point.
Refraction	Bending of a radio wave as it passes from one material to another with a different density (e.g. from air to ice).
Rodina	A supercontinent predating Pangaea, assembled from 1.3 Ga BP to 750 Ma BP, when it broke apart.
Sastrugi	Sharp, irregular ridges formed on a snow surface by wind erosion and deposition. Ridges are often parallel to the direction of the prevailing wind.
Scattering	Loss of radio signal power in ice due to the presence of air bubbles, ice lenses, pockets of water, solid materials such as sand and gravel, cracks, and fractures.
Sill	A tabular sheet intruded, concordantly, between older layers of sedimentary rock. See also Dyke.
Static and pitot pressure	Pitot-static systems in aircraft comprise sensors that detect the ambient air pressure affected (pitot pressure) and unaffected (static pressure) by the forward motion of the aircraft. These pressures are used on their own or in combination to provide indications of various flight parameters.
Steady state	In glaciology the condition of an ice mass in which its dynamic and thermodynamic processes function in balance over a specified period without significant perturbation. A theoretical concept, never encountered in the real world.
Stinger	Tail-mounted boom on an aircraft housing a magnetometer.
Stratosphere	The second layer of Earth's atmosphere. It extends from the top of the troposphere (at about 8 km height in polar regions) to about 50 km (31 mi) above the ground.

Sublimation	The transition of a substance directly from the solid to the gas state without passing through the liquid state.
System sensitivity (SNR)	The ability of a radar sounder in detecting echoes, defined as the ratio of the transmitted power to the minimum detectable received power at the antenna, measured in dB.
Tidewater glacier	A glacier extending down to sea level and entering the open sea or, more commonly, a bay or fjord.
Tillite	An ancient till or moraine. It is typically highly consolidated. See also Moraine.
Troposphere	Lowest layer of Earth's atmosphere and where nearly all weather takes place. It contains 75% of the atmosphere's mass and 99% of the total mass of water vapour and aerosols. The boundary is the tropopause, at a height of ~8 km in polar regions.
Z-scope	A display function on an intensity-modulated oscilloscope in which the time base is switched off.

Index

Page numbers in *italics* refer to figures and illustrations.